LAND USE SCENARIOS

Environmental Consequences of Development

Integrative Studies in
Water Management and Land Development

Series Editor
Robert L. France

Published Titles

Boreal Shield Watersheds: Lake Trout Ecosystems in a Changing Environment
Edited by J.M. Gunn, R.J. Steedman, and R.A. Ryder

The Economics of Groundwater Remediation and Protection
Paul E. Hardisty and Ece Özdemiroğlu

Forests at the Wildland–Urban Interface: Conservation and Management
Edited by Susan W. Vince, Mary L. Duryea, Edward A. Macie, and L. Annie Hermansen

Handbook of Regenerative Landscape Design
Robert L. France

Handbook of Water Sensitive Planning and Design
Edited by Robert L. France

Land Use Scenarios: Environmental Consequences of Development
Allan W. Shearer, David A. Mouat, Scott D. Bassett, Michael W. Binford, Craig W. Johnson, Justin A. Saarinen, Alan W. Gertler, and Jülide Kahyaoğlu-Koračin

Porous Pavements
Bruce K. Ferguson

Restoration of Boreal and Temperate Forests
Edited by John A. Stanturf and Palle Madsen

Wetland and Water Resource Modeling and Assessment: A Watershed Perspective
Edited by Wei Ji

LAND USE SCENARIOS

Environmental Consequences of Development

Allan W. Shearer

David A. Mouat

Scott D. Bassett

Michael W. Binford

Craig W. Johnson

Justin A. Saarinen

Alan W. Gertler

Jülide Kahyaoğlu-Koračin

CRC Press
Taylor & Francis Group
Boca Raton London New York

CRC Press is an imprint of the
Taylor & Francis Group, an **informa** business

CRC Press
Taylor & Francis Group
6000 Broken Sound Parkway NW, Suite 300
Boca Raton, FL 33487-2742

First issued in paperback 2017

© 2009 by Taylor & Francis Group, LLC
CRC Press is an imprint of Taylor & Francis Group, an Informa business

No claim to original U.S. Government works

ISBN-13: 978-1-4200-9254-7 (hbk)
ISBN-13: 978-1-138-11230-8 (pbk)

Visit the Taylor & Francis Web site at
http://www.taylorandfrancis.com

and the CRC Press Web site at
http://www.crcpress.com

Contents

List of Maps

Color versions of the maps in this book are available on-line at www.landusescenarios.dri.edu.

Sprawlscapes and Timelines

Robert L. France, Series Editor

"Sprawl" as it has become known—the spread of development from urban centers out into the countryside—is coming to be realized as being one of the most serious threats to the integrity of natural systems.[1] Whereas it is possible to comprehend the enormity of urbanizing sprawl from the air,[2] it is best to appreciate its reciprocal, urban densification, with one's eyes and body whilst on the ground. For example, it was only recently while standing with my father on the High Street in the center of downtown Sheffield, England and looking out past the old buildings to views of the rolling hills that I could finally grasp the stories he had long told of his boyhood in that industrial city wherein he and friends would simply hop on a streetcar and in only a few minutes time be out hiking or climbing in the wonderful and wild countryside of the Peak District. Sprawl's absence was reinforced again when, a few months later, I was on one of the medieval pilgrimage routes through southern France toward Santiago de Compostela and where, after what was only a brief amount of time, I suddenly found myself outside the dense city of Montpelier and walking along a windswept plateau empty of any dwellings.

There is a need to develop place-based relationships between environmental policy and implementation, particularly in relation to trade-offs that can often exist among ecological, social, and economic aspirations, and in a form that not only supports but actually encourages objective discussion and decisions about what the future might look like on the ground.[3] The present book, *Land Use Scenarios: Environmental Consequences of Development* by Allan W. Shearer, David A. Mouat, Scott D. Bassett, Michael W. Binford, Craig W. Johnson, Justin A. Saarinen, Alan W. Gertler, and Jülide Kahyaoğlu-Koračin, clearly

demonstrates how the approach of alternative futures scenario modeling enables: (a) predicting impacts of land use alterations on ecological processes, (b) integrating human dimensions into effective planning, and (c) developing an understanding of the uncertainty of impacts and associated risks of various development scenarios. And through its comprehensive discussion of the potential effects of development upon hydrology, air quality, environmental regulations, stakeholder sociology, and public education in addition to those of housing density upon landscape ecology, vegetation patterns, and ensuing biodiversity, the present book is a worthy addition to the multi- and inter-disciplinarity shown by previous volumes in this CRC series, Integrative Studies in Water Management and Land Development.

One of the most pressing issues of land use planning concerns the question of where and how to place more people on the land; i.e., should new developments be smaller and of a higher incidence or be fewer and larger in order to reduce environmental impacts? Will negative impacts be exacerbated or lessened if development is concentrated into particular areas or should it be allowed to be dispersed evenly over the landscape? How do marginal changes contribute to cumulative effects? etc. By creating a series of predictive judgments about what could happen in response to various development scenarios, Shearer et al. use GIS analysis not as an end in itself, an all too common mistake in many environmental assessments,[4] but rather as a means to generate spatially explicit possibilities in which to enable objective decision making by informed stakeholders.

Shearer et al.'s results from the particular case study they examine confirm one long-held heuristic but seldom empirically enunciated belief, namely that low-density development generates the most negative environmental impacts, as well as raising two alarming and hitherto seldom recognized implications of development: (a) indirect negative effects on biodiversity as due for example to elevated stormwater runoff, wildfire suppression, and habitat fragmentation may be just as or even more severe than effects caused by direct habitat loss resulting from site development, and (b) setting aside protected areas without careful consideration to the type and extent of proximal development may ultimately be insufficient to prevent species loss.

As Neils Bohr humorously, but aptly, quipped, "prediction is difficult, especially when it is about the future."[5] The present book does an invaluable service to land use planners faced with this difficulty in terms of spreading development by helping to enunciate agents of change and in illuminating quantitative approaches to predicting the environmental effects thereof. Master empiricist, Bohr, would be happy.

~Harvard University

NOTES:

[1] e.g., Robert L. France, "Preface: Anecdotes of Sprawl," in Robert L. France, ed., *Introduction to Watershed Development: Understanding and Managing the Impacts of Sprawl* (Lanham, Maryland: Rowman & Littlefield, 2006).

[2] Alan Berger, *Drossscape: Wasting Land in Urban America* (New York: Princeton Architectural Press, 2006).

[3] Robert L. France, "Engaging Time: Frameworks for Modeling Scenarios of Alternative Development Futures," in Robert L. France, ed., *Introduction to Watershed Development: Understanding and Managing the Impacts of Sprawl* (Lanham, Maryland: Rowman & Littlefield, 2006).

[4] Robert L. France, "Janus Planning: Using Computer Tools to Look Backward and Forward Simultaneously," in Robert L. France, ed., *Handbook of Water Sensitive Planning and Design* (Boca Raton: CRC Press, 2002); J. Schloss, P. Zandbergen, M. Bavinger, and Robert L France, "Multiple Objectives in Watershed Management through Use of GIS Analysis," in Robert L. France, ed., *Handbook of Water Sensitive Planning and Design* (Boca Raton: CRC Press, 2002).

[5] Carl Steinitz quoted in Robert L. France, "Engaging Time: Frameworks for Modeling Scenarios of Alternative Development Futures," in Robert L. France, *Introduction to Watershed Development: Understanding and Managing the Impacts of Sprawl* (Lanham, Maryland: Rowman & Littlefield, 2006).

Preface

Questions about how to shape and steward landscapes are among the most important asked by individuals and communities. They are also among the most difficult to answer. In part, there are general conceptual challenges that stem from trying to represent, understand, and evaluate coupled socio-ecological systems. But more specifically, there are geographic issues that play out across space and over time. Changes to the landscape can alter local natural processes—for example, by increasing or limiting nutrient flows, by creating barriers between habitats, and by speeding or slowing hydrologic regimes. Each change, seen in the isolation and within a limited spatial context, will produce some recognizable advantages (and perhaps some disadvantages) for the stakeholders. But, natural processes do not yield at the governmental boundaries that typically define administrative areas or ownership boundaries. Therefore, purposeful—and good intentioned—actions that alter natural processes can produce unintended impacts to neighbors. As a result, land managers attempting to maintain or improve the parcels for which they are responsible must consider not only how they might maintain or improve the parcels for which they are responsible, but also anticipate how neighboring owners and managers might change their respective properties.

The central premise of this book follows from the observation that the success of local decisions will be determined, to a significant degree, by factors that are difficult to control or to forecast. And therefore, there is the need to undertake what can be considered "geographic vulnerability analysis" through the development of scenarios about how, where, and when future patterns of land use might develop. In particular, there is the need to examine "critical uncertainties"—those aspects of the future that are most difficult to predict, but that may have the most profound impact on pending decisions.

This kind of approach is exemplified through a study of the region of Marine Corps Base Camp Pendleton and Marine Corps Air Station Miramar, California. Located between San Diego and Los Angeles, the area has recognized ecological importance and continues to attract development. In this context, the largely un-built, but extensively used, military lands provide terrain for fulfilling the stated mission (combat training operations) and for habitat to a large number of species, including several that are listed by the federal government as in danger of extinction. However, ongoing regional growth, which marginally and cumulatively changes natural processes (e.g., by destroying or degrading natural habitat, by accelerating storm runoff, etc.) is resulting in increased land management pressures. Future growth may further exacerbate the situation, and it is prudent to understand how patterns of new development may affect management options. Obviously, concerns about ecosystem management for individual properties within the region are not limited to the military land managers. The Cleveland National Forest and The Nature Conservancy's Santa Rosa Plateau Ecological Reserve are just two of the properties that are also vulnerable to regional development. In publishing the research, the investigators hope to provide information regarding issues of urban development and possible environmental consequences to stakeholders and jurisdictions whose actions may influence the future of the region. More broadly, we hope to aid planners, managers, and stakeholders in other areas to engage spatial contingencies toward the goal of more resilient landscapes.

The research was supported by the United States Marine Corps through a contract issued to the Desert

Research Institute. The study was conducted by a team of investigators from the Desert Research Institute, Harvard University, the University of Florida, and Utah State University. The information contained herein is believed to be reliable, but the investigators and their institutions do not warrant its completeness or accuracy. Opinions and estimates constitute judgments of the research team.

Throughout the course of the study of the region of MCB Camp Pendleton and MCAS Miramar, California, many local stakeholders provided thoughtful observations and insightful comments. The researchers would especially like to thank: Mary Jane Abare, Lupe Armas, Susan Baldwin, Bill Barry, Jim Bartel, MajGen William G. Bowden, Dave Boyer, Slader Buck, Larry Carlson, Sue Carnevale, Pat Christman, Paul Cote, Stephen Ervin, Steven Evanko, Karen Evans, Janet Fairbanks, LtGen Edward Hanlon, Jr., Col David Johns, Jay Kerry, Robbie Knight, Richard Kramer, Wayne Lee, LtCol Thomas Lhuillier, Col David Linnebur, Mike McLaughlin, Larry McKenney, Scott Morrison, Stan Norquist, Jim Omans, Bob Parrott, Ken Quigley, Larry Rannals, Gilberto Ruiz, Russell Sanna, Carl Steinitz, John R. Stilgoe, Jeff Tayman, LtCol Scott Thomas, Bill Tippets, LtCol Brian Tucker, Diane Walsh, Chris White, Robin Wills, and Susan Wynn. Photographs of military maneuver activities were provided by AC/S Environmental Security, MCB Camp Pendleton. All other photographs were taken by members of the research team or by participants in the 1996 study *Biodiversity and Landscape Planning: Alternative Futures for the Region of Camp Pendleton, California.*

2008

Authors

Allan W. Shearer is an Assistant Professor of Landscape Architecture at Rutgers—The State University of New Jersey. His work focuses on methods for creating scenarios of possible futures. He participated on this project while a Research Fellow at the Harvard Graduate School of Design and previously contributed to several projects relating to regional growth around military bases and written on the consequences of climate change on environmental and human security. He has been a Visiting Scholar at the Pell Center for International Relations and Pubic Policy at Salve Regina University and a Donald D. Harrington Fellow at The University of Texas at Austin. He graduated from Princeton University and received a Master of Landscape Architecture and Ph.D. from Harvard University.

David A. Mouat is an Associate Research Professor in the Division of Earth and Ecosystem Sciences at the Desert Research Institute, Reno, Nevada. His research interests include relating landscape characteristics to issues of ecosystem health and land degradation. He has a particular interest in global arid ecosystems and served as the Chairman of the Group of Experts of the United Nations Convention to Combat Desertification. Recently, he managed the interagency Biodiversity Research Consortium (BRC) investigation on possible consequences of growth in the California Mojave Desert. He holds a B.A. in Physical Geography from the University of California at Berkeley and received his Ph.D. in Geoecology from Oregon State University.

Scott D. Bassett is an Assistant Professor of Planning at the University of Nevada at Reno. His research focus is on the computer simulation of spatial patterns and processes in natural resources with an emphasis on the integration of spatial information from multiple disciplines. He has been an active participant within the Gap Analysis program for Utah and Nevada and recently developed spatial models to address how land use decisions could impact natural resources within the Upper San Pedro River Watershed, Arizona and Sonora, Mexico. He received his Doctor of Design at Harvard University. He holds a B.S. in Geography and Anthropology and a M.S. in Fisheries and Wildlife Ecology from Utah State University.

Michael W. Binford is a Professor of Geography at the University of Florida and holds the University of Florida Foundation Research Professor 2002-2005 chair. His work investigates the relationships between human settlement and ecological systems and has a particular focus on land–water interactions. Sites of recent studies have ranged from the plateau of Lake Titicaca to inland and coastal sub-basins along the Amazon to the hills of Thailand and to the island of Haiti. He received his B.S. in Biology from Kansas State University, M.S. in Fishery Biology and Experimental Statistics from Louisiana State University, and Ph.D. in Zoology and Geology from the University of Indiana.

Craig W. Johnson is a Professor Emeritus of Landscape Architecture and Environmental Planning at Utah State University. His research areas are land reclamation, urban forestry, and urban wildlife planning. A licensed landscape architect in Idaho, Minnesota, and Utah, he is the Director of the Visual Resource Assessment section of Ecotone, a Logan, Utah based consulting firm. He received his Bachelor of Landscape Architecture from Michigan State University and Master of Landscape Architecture from the University of Illinois. He earned an M.S. in Fisheries and Wildlife Biology from South Dakota State University.

Justin A. Saarinen worked as a Research Assistant under Michael Binford and was responsible for implementing the hydrologic model that allowed stream flow analysis of Santa Margarita River under current and possible future conditions. He received a B.A. and M.S. in Geography from the University of Florida.

Alan W. Gertler is a Research Professor in the Division of Atmospheric Sciences at the Desert Research Institute, Reno, Nevada. His research includes both laboratory and field studies of atmospheric chemistry with particular emphasis on the impact of mobile sources on the environment, urban air quality in megacities, the development of new methods to attribute ambient pollutants to specific sources, and the use and impact of alternative fuels for transportation and power generation. Currently he is the President of the International Union of Air of Pollution Prevention and Environmental Protection Associations (IUAPPA). He received a B.S. in Chemistry from SUNY Albany and Ph.D. in Chemistry from UCLA.

Jülide Kahyaoğlu-Koračin is an Assistant Research Professor in the Division of Atmospheric Sciences at the Desert Research Institute, Reno, Nevada. Her research interests include numerical simulations and transport and dispersion studies of atmospheric pollutants, emissions inventory development and emissions modeling of air pollutants, numerical weather predictions, data assimilation and forecasting, and climate change and its interactions with air quality. She earned her Ph.D. in Atmospheric Sciences from the University of Nevada, Reno, and her B.S. and M.S. in Physics from the University of Marmara, Istanbul, Turkey.

Chapter 1

Scenario-Based Studies for Landscape Planning

Allan W. Shearer

Introduction

Changes to the environment are often considered in terms of "project scale" decisions—to matters that are precisely identified, that are readily controlled, and that are easily contained within a delineated boundary. But as is well recognized by landscape planners, designers, and managers, natural processes—and an ever increasing number of globalizing socio-cultural processes—are not influenced by parcel maps or socially constructed political boundaries. Therefore to some degree the ultimate success of these same projects will be determined by future situations that are only vaguely glimpsed, that are beyond immediate influence, and that take place far beyond the horizon. These concerns may include the macro-level performance of the economy, the introduction of new technologies, changes in population and demographics, the refocus and redirection of social priorities, the enactment of new laws and new regulations, the effects of natural disasters, and, perhaps most significantly, the actions of neighboring stakeholders. While many of the general effects of marginal changes to the built environment are typically understood, anticipating the cumulative consequences of multiple decisions over a large area is complicated by uncertainties about the location, timing, and use of new development. Herein lies an acute and seemingly overwhelming problem: Because the future has not yet happened, it offers no facts on which we can preemptively act; that is, there are no pieces of information that can be verified by physical evidence or eyewitness testimony. Instead, there are only assumptions about how the world might develop and what that change might mean.

An approach to help manage the inherent uncertainty of assumptions about the future is to examine the potential consequences of an action or set of actions against a set of plausible, yet different contexts. By doing so, a person, organization, company, or government can make decisions that are more resilient to the possible conditions of tomorrow. In general, these alternative views of the future are referred to as *scenarios*.[1]

Dictionaries typically define the word "scenario" as a "summary of the events of a play, film, or novel." But in the 1950s Herman Kahn appropriated the term for long-range visions of the future that were used to speculate on contingent relationships and the dynamics of change.[2] More formally, the scenario came to be understood as a "hypothetical sequence of events constructed for the purpose of focusing attention on causal processes and decision points."[3] Since then, many have contributed to the development of techniques for crafting and for using scenarios.[4] While the ever expanding understanding of "scenario" resists a universal definition, most conceptions of the term share four principles:

1. Scenarios are fictional (where fictional is understood to mean unverifiable but plausible, not fanciful) accounts which represent a process of change over some duration.
2. Scenarios describe situations, actions, and consequences which are contingently related.
3. Scenarios are understood to be predictive judgments which describe what could happen, not predictions which describe what will happen, or even what is likely to happen.

4. Scenarios organize information within explicitly defined frameworks.

There are several benefits of scenario-based approaches to decision making.[5] First, scenarios provide an aid, or more formally a heuristic, for explicitly defining events within their contexts and for assessing consequences. Facts and data are, in themselves, largely meaningless until they are placed within a framework of understanding which includes the interaction of social, economic, political, and technological factors. Scenarios provide a means to relate and comprehend isolated pieces of information within a single framework and—equally important—to compare meanings across different frameworks. Second, the level of specificity required for the description of a scenario—the who, what, where, when, and why of an action—can lead a decision maker to consider implications that would have been missed had the representation of the subject matter been limited to abstract principles and general statements. That is, details matter. For example, an approach to mass transit in one part of the country may not work in another because of differing demographics, access to resources, physical conditions (such as the weather), regional customs, and local law. Third, related to the issue of specificity, the kind of detail offered in a story illustrates the implications of a theory, rather than abstractly espouses it. By doing so, the information is often more readily understood by a broader audience as applicable to concrete and, often, current needs.[6] Fourth, because scenarios are fictional, they can serve as artificial case histories which illustrate the implications of policies which might be ignored if examples only from the past "real" world are considered. To the degree that the future will offer unprecedented situations, basing decisions only on the evidence of the past may be shortsighted. Finally, scenarios provide a means to facilitate the discussion of planning options across stakeholder groups, professional disciplines, and levels of management.

Scenarios have been used to augment discussion on a wide variety of topics—including agriculture,[7] communications,[8] economics,[9] energy,[10] the environment,[11] health policy,[12] legal systems,[13] security studies,[14] technology policy,[15] transportation,[16] and water use.[17] But, since each decision takes place in a specific context characterized by available information, values and perceptions of stakeholders, and constraints of time, not all scenario studies are (or can be) the same. Therefore, the scenario development process should be tailored to address a given decision-making situation. To provide that custom fit, one needs to understand the kinds of assumptions that inform a vision of the future. This chapter summarizes a series of analytic frameworks that can allow these assumptions to be rendered explicitly, and thereby shared and discussed with others in a scenario-development and use process.

Scenarios and Alternative Futures

Figure 1.1 provides a representation of several general aspects concerning thoughts about the future.[18] The circle on the center grid represents a mapping of the present conditions relative to two concerns, one charted on the x axis and one charted on the y axis. For example, one issue might be the density of new development; the other the availability of mass transit. Of course, reducing the understanding of the world to two issues is a simplification; however, representing the infinite number of interests that comprise contemporary society on a two-dimensional page is complicated, if not confusing, and so for graphic convenience only two are used to illustrate the point. Piercing the center of the grid, a double-headed arrow represents time. To the left is the past, along which are other grids that mark specific moments of history. Again, circles represent the conditions of those times mapped relative to the two axes of concern. Connecting the circles are solid lines representing the sequence of events which led from one period to the next. Multiple lines reflect the situation that the evolution from one state to the next may be explained in different ways, depending on one's point of view and the emphasis given to different influences. Some understanding may be found in each of the alternative explanations and, moreover, the advancement of historical knowledge will undoubtedly add still more perspectives. To the right of the center grid lies the future. As with the past, grids mark specific times and the multiple

Figure 1.1
Scenarios and Alternative Futures

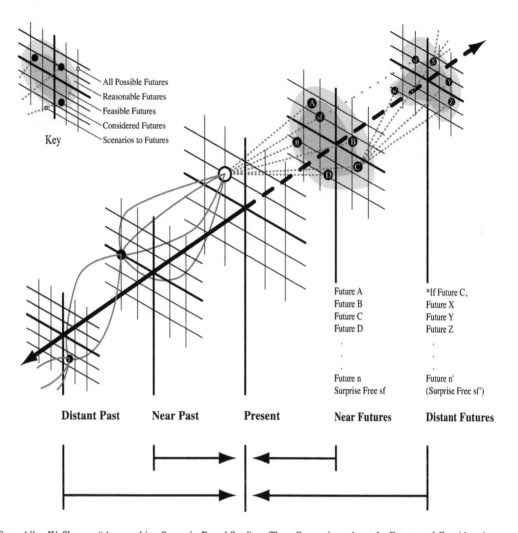

Key

All Possible Futures
Reasonable Futures
Feasible Futures
Considered Futures
Scenarios to Futures

Future A
Future B
Future C
Future D

.
.
.

Future n
Surprise Free sf

*If Future C,
Future X
Future Y
Future Z

.
.
.

Future n'
(Surprise Free sf')

Distant Past   Near Past   Present   Near Futures   Distant Futures

From Allan W. Shearer, "Approaching Scenario-Based Studies: Three Perceptions about the Future and Considerations for Landscape Planning," *Environment and Planning B: Planning and Design* 32:1 (January 2005), pp. 67-87. Republished with permission of Pion Limited, London.

dots represent alternative future conditions. The dashed lines which connect periods of time are the scenarios. Like histories, scenarios explain how the world could move from one state to another; however, as noted earlier, unlike histories which are based on verifiable facts, scenarios are based on assumptions.

A point that should be emphasized is the distinction between the terms "scenario"

and "alternative future." Although they are sometimes used interchangeably in the literature, there is considerable methodological benefit in differentiating the expressions. An understanding of the future requires both a comprehension of the forces of change and an awareness of possible significant moments, and, as such, there needs to be a distinction between scenarios (the means of change) and alternative futures (the ends of change). Hence, an alternative future is defined as a possible state; and a scenario is defined as a means to achieve that state. While the scenario and alternative futures are clearly related, they inform the decision making process differently. The scenarios serve three purposes: (1) In describing the means of change, each scenario establishes a sense of credibility to its respective alternative future. While a given future might result from different courses of action,[19] the articulation of at least one course of events provides plausibility, an element which cannot be underestimated in policy debate when uncertainties are at issue. (2) The specifics of each scenario can serve as mile markers by which decision makers could note the emergence of and progress along a path that leads to a specific future. (3) The specifics of each scenario can also serve as the basis to consider related changes. Social trends do not emerge in isolation and elected officials or government agencies do not act by idiosyncratic whim. Hence, just as the plot of a novel may prompt the reader to consider the lives of characters and conditions beyond the printed page, the plot of a scenario may prompt the decision maker to consider related spin-off concerns. In turn, the recognition of these concerns can contribute to a richer sense of what may come. The alternative futures serve two different purposes: (1) They quantify the consequences associated with a given path of change. Development related impacts on the environment are often incremental and cumulative and the snapshots offered by alternative futures can help decision makers to take stock. (2) The alterative futures can serve to compare the effects of different paths of change.

Several additional notes to the diagram should be considered. The "surprise free" alternative future, which is marked "sf" on the diagram, is sometimes referred to as "the trend" or "business as usual" expectation. It is the future that can be anticipated if there are no significant changes in the social, political, economic, technical, or environmental aspects of the world. Within each grid of the future are shaded areas that classify the alternatives as possible, reasonable, and feasible. Distinguishing scenarios at this admittedly generalized level of potential realization is a pragmatic compromise between the need to assess the potential viability of a future and the desire to avoid erroneous claims of

precision. The arrows at the bottom of the diagram represent the important qualification that scenario-based studies, or planning studies in general, use the information of the recorded past and the assumptions of the possible future to meet the perceived need of the present better.

Finally, the more removed a time period is from the present, the smaller it is drawn. This convention reflects the broad presumption that the more distant an era is from our own, the more difficult it can often be to imagine and understand its consequences. However, this generalization should not be mistaken for a universal theorem that all aspects of the future become equally more difficult to anticipate as the time horizon is extended. Looking backward at history, some parts of the distant past are known better than more recent events. For example, because of written records and other archaeological artifacts, we know more about the Roman Empire circa AD 100 than we do about Native American settlements circa AD 1000. Similarly, assumptions about some aspects of the distant future may be more apparent and taken with greater confidence than others of the near future.

Change

Given the notion that scenarios describe change, an important set of assumptions follow from the kinds of change that are to be considered. Ken Boulding provides a useful set of metaphors for distinguishing four general kinds of change that could be the focus of a scenario-based study.[20]

Planets are elements of the future that are part of very stable systems. The metaphor follows from the workings of the solar system. For example, it is possible to predict the time of the next solar eclipse and the location from which it can best be viewed. Because uncertainty is minimal, even measurable planet-like elements allow for singular states futures to be forecast with great confidence.

Plagues are elements of the future that are inherently difficult—if not impossible—to predict, but that can cause great harm to people and places. These include floods, fires, volcanic eruptions, tsunamis, and rapidly spreading infectious disease. Scenarios built around plagues are often used for worst-case, acute crisis management; however, they might also be used to examine competing assumptions about how a system might recover or change from a plague-like shock, such as extended drought (e.g., how plant communities might respond to a prolonged drought). The concept in its extreme form can be more generally understood as a trigger to a system-wide change from one state of (more or less) equilibrium to another.

Plants are elements of the future that are in the process of becoming evident. The general outcome can be anticipated, but the precise form may not be known. Hence an acorn from an oak tree will produce an oak, not a maple tree; but, if the sapling does not receive sufficient water, the mature tree may be undersized. Demographic bubbles are also examples of plants: The baby boom generation was born after World War II putting in place a succession of societal opportunities and demands as they grew older. These phenomena have also been referred to as "predetermined elements."[21] In Boulding's framework, plans (such as land use plans) are a kind of plant. Since most community plans have (by design) excess capacity, a scenario study might examine the variations of growth within a plan for the next 20,000 new residents. A version of a plant-like scenario for environmental management is Plans Build-Out, which simulates the complete implementation of a land use plan.

Plays are elements of the future that result from the interactions of people (or institutions). While it is often easy to assume that all (or most) people will behave the same way in the same circumstance (the logic of the situation)[22] or that individuals will stick to established patterns of behavior (the logic of the dispositional traits),[23] the future may offer possibilities that never before existed. Scenarios that focus play-like uncertainty are useful when no strategic plan exists or is operational, or when there is no precedent for action.[24]

Depending on the decisions that are pending, the uncertainties of plagues, plants, and plays may all become what are called in this book (and elsewhere in the literature) "critical uncertainties"—those aspects of the future that are most difficult to predict, but that could have profound impacts on the ultimate success of a decision.

Attitudes toward Change and the Decision Environments

Distinguishable from kinds of change is one's position regarding the prospect of change. Russell Ackoff identifies four such attitudes.[25] It should be recognized that all of these attitudes can (and do) coexist, and a person's or organization's attitude may evolve not only with time, but also with the specific decision that is under consideration.

Reactive—This attitude is typified by a belief that the conditions of the past were better than those of the present, and that steps should be taken to restore the remembered "golden age."

Inactive—This attitude is characterized by a sound satisfaction with the present and a vested interest in maintaining the status quo; hence, any change is resisted.

Proactive (also called interactive)— This attitude is distinguished by a sense of little satisfaction with the present and also an unwillingness either to recreate the past (which was also imperfect) or to accept an inevitable future. The proactive attitude captures the belief that the future is the result of actions taken in the present. Implicit in this attitude is the need for positive visions of what society might achieve.

Preactive—This attitude is based on the presumption that the future cannot significantly be shaped by individuals or organizations, but instead will emerge from forces that are not readily controlled. Embedded in the preactive attitude is a desire for a forecast of what may happen in the hopes of exploiting new opportunities and avoiding pitfalls.

Closely related to one's attitude toward change is the degree to which the future can purposefully be created. To qualify one's position about assumptions of agency, it is useful to take explicit consideration of what T.F. Emery and E.L. Trist have called the "decision environment"—the social-physical space in which the decision is to be made and subsequent actions taken.[26] They distinguish three kinds of environments.

Internal Environment—The decision maker has complete control and he/she/they can optimize a solution to a desired goal.

Transactional Environment—The decision maker does not have complete control, but he/she/they can influence outcomes and events.

Contextual Environment—The decision maker has no influence and can, at best, pre-position oneself for anticipated change.

The attitude toward change and the decision environment often come together in one of two kinds of studies. "Normative" studies are typified by a proactive attitude operating in a (minimally) transactional or (maximally) internal decision environment. In contrast, "descriptive" studies are typified by a preactive attitude operating in a

contextual decision environment. This split between normative and preactive attitudes and assumptions about decision environments is reflected in what is, arguably, the broadest distinction among kinds of scenario-based studies: normative studies, which seek to identify preferable futures; and descriptive studies, which aim to identify possible futures without regard for preference. This distinction has been in evidence since scenario techniques were first developed in the years following World War II. Some, such as the French philosopher and bureaucrat Gaston Berger, saw opportunities for inventing a better world.[27] Others, such as Kahn, saw uncertainties and, although they accepted that their tomorrows would be different from their yesterdays, they doubted that the future could successfully be engineered in all aspects. It should be noted that, somewhat confusingly, many studies which use the expression "alternative futures" in the title concern normative futures,[28] whereas those that use scenarios usually consider descriptive futures.[29]

Scenarios for Landscape Planning

How might these general concerns about scenario-based studies be applied to environmental change? Here, the framework for landscape planning devised by Carl Steinitz provides one approach to consider the possible consequences of regional change for landscape planning and management decisions.[30] As described in the scenario-based study, *Biodiversity and Landscape Planning: Alternative Futures for the Region of Camp Pendleton, California*[31] (and in other studies) the framework asks six questions, each of which is related to a type of model. Over the course of a study, the set of questions is asked three times: the first time to define the context and scope of the investigation, the second time to specify the methods of inquiry and analysis, and the third time to carry the project forward to a set of conclusions. Figure 1.2 illustrates the generalized framework.

The six questions, listed in the usual order for defining the context of a study, are:

I—How should the state of the landscape be described: in content, space, and time?

Figure 1.2
Framework for Landscape Planning

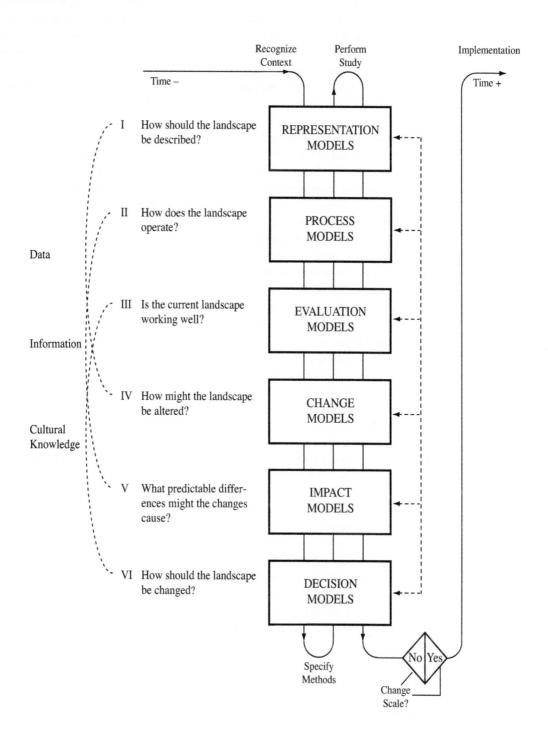

Redrawn from Carl Steinitz, "A Framework for Theory Applicable to the Education of Landscape Architects," *Landscape Journal* 9:2 (January 1990), pp. 136–143. Copyright 1990 by the Board of Regents of the University of Wisconsin System. Reproduced by permission of the University of Wisconsin Press.

Answering this question requires defining a vocabulary and a syntax to identify those characteristics of a place relevant to a particular study. To describe the physical, biological, and social components of a very large study area, a computer-based Geographic Information System, or GIS, is typically used to organize spatially explicit data on the region.

II—How does the landscape operate? What are the functional and structural relationships among its elements? Once the pertinent components of the landscape have been identified and defined, relationships between the parts are established. These processes can be physical, such as flooding or soil moisture loss; biological, such as potential California gnatcatcher habitat; or cultural, such as land management and protection status or visual preference. In most cases, these relationships are modeled using the available data in the GIS. Creating process models requires specialized scientific or technical knowledge, and so this part of the process often requires collaboration with relevant topical experts.

III—Is the current landscape working well? The initial evaluation of the landscape is made by operating the process models on the data that represent the baseline state of the study (in principle, these are current conditions; operationally, they are the most recent usable data). Given the complexity of social and ecological processes and given the managerial need to accommodate an array of demands, multiple assessments are typically needed. Further, some issues may be assessed in terms of multiple impact models. For example, indexes of potential biodiversity may be assessed by species richness, a suite of single-species indicator models, or by the landscape ecological pattern. While the set of models are related, each individual model may be based on a different premise (as in the case of the biodiversity models). Therefore, each

model may present different implications for landscape planning and management. Also note that as used in the framework, the term, "well" can be defined several ways. For example, it might refer to a statutory or regulatory requirement (e.g., part-per-billion of a contaminate). For pragmatic reasons, the condition of "well" is often interpreted simply as the *status quo*.

IV—How might the landscape be altered by what actions, where, and when? The issue of change is multifaceted. Depending on the needs of a given study, the considered changes can include those brought about by exogenous forces (such as broad social, economic, or political pressures) or by endogenous actions (such as the implementation of plans, investments, or regulations by local stakeholders). Change may be projected or simulated by a variety of means. Commonly considered future change conditions include the full implementation of the existing plans (called Plan or Plans Build-Out) and a Surprise Free future in which current trends continue for the examined period of time.

V—What predictable differences might the changes cause? Operating the process models on the change scenarios and comparing the results with the baseline evaluations (question III) yields impact assessments.

VI—How should the landscape be changed? Each impact assessment reveals one aspect of how the alternative scenarios are predicted to affect the landscape. With this information, decision makers are better prepared to decide how the landscape should be changed, or how to prepare for changes that are beyond their control. It should be recognized that "change" of the landscape includes (the obvious) instances of building *something* (houses, roads, etc.), but can also include instances of establishing protection measures that preclude future modification.

Initially asking the six questions in the order presented above provides decision makers with a general sense of what might readily be done in a study. After this preliminary assessment, decision makers ask if the kinds of data (i.e. raw data), information (i.e. classified data), and knowledge (i.e. assessed data) they have at the present time is sufficient to make a decision. Often, there are needs for additional details, greater accuracy, or increased precision. For example, a community may be considering options for locating new subdivisions in a large watershed. The decision criteria includes considering the potential increase in flooding for the 20-year storm event. A model for quantifying flood volume and timing is available, but it requires soils data that are more accurate and precise than those currently available. Hence, moving forward with the decision requires improved soils data. Or, a community whose economy is based on tourism is concerned about the impact of urbanization on visual quality along scenic roads. Hence, moving forward with the process requires identifying criteria for what contributes to "scenic quality," including specifying what roads have these qualities, the extent of views along these roads, and what features of the landscape tourists find preferential (objects such as barns, sunflower fields not corn fields, etc.). The need to address these kinds of issues underscores this framework as being decision-driven, rather than data-driven. That is, data must be found to support the decision-making process; the decision-making process is not *a priori* structured around the available data.

Scenarios can be used in two parts of the framework. Most commonly (and perhaps most obviously) they can be used in Step III—Change. In the design and planning professions, normative plans are proposed and then evaluated in terms of some set of criteria. The decision maker's task is then to select the plan with the best performance metrics. It is also possible to use scenarios in Step IV—Decision. In this case, Step III—Change is (as before) defined in terms of a local plan. Its impact is assessed and then that impact is considered in light of several contextual scenarios. For example, suppose the change is the construction of a new sea wall. Impact models show that it will protect a town from currently rare 10-foot waves. The decision process might include scenarios that consider rising sea levels due to climate change. When used in Step IV, the scenarios can be viewed as a kind of geographic vulnerability analysis. The scenarios presented in this book for the study of Southern California were of this type.

Hypothetical Example

The influence of these perceptions about the future on the framing of scenario studies can be suggested by a relatively simple example of a hypothetical town, which is home to a few thousand people and governed by an all-powerful (and, let us say, well-respected and admired) mayor. Within the town boundaries are the expected houses, apartment buildings, schools, offices, stores, workshops, factories, roads and sidewalks, police and fire departments, utility facilities for power, communications, and waste treatment, and open space. The mayor has the funds to build a single new road that is intended to increase cross-town commercial traffic. It is for him or her to decide where the road should be routed.

At one level of understanding, such a town could be viewed as a closed system. As such, the town is viewed metaphorically as the whole of the universe and the internal environment is conceived as the same as the contextual environment. Here, the mayor is in a position to adopt a proactive attitude to change and can follow a strategy to create a set of normative scenarios. Each of the scenarios would represent a different road plan. The plans would be compared with each other, and the one deemed best in terms of the stated goal of increasing commercial traffic would then be implemented. Importantly, and so far only implicit in this example, the evaluation of the plans assumes that the users of the new road are known and that the users' needs can be accommodated. That is, the modes of transportation are either assumed to be stable with the present or are assumed to be forecast with high degrees of accuracy and precision. In many cases this is a safe assumption, as modes of transportation have not changed frequently in the past; but, they have changed: from cart, to railroad, to automobile. This point is made not to criticize the first city planners for not anticipating later technologies, nor to suggest that more recent planners should have anticipated

Segway scooters; however, it does serve to highlight that decisions are made on assumptions of stability which may prove to be incorrect.

At another level of understanding, the fictional town of the example is also a part of a larger geographic region. To the degree that the town and its surrounding communities can be seen to influence each other, there is a transactional environment. Possible elements of this decision environment might include the construction of regional schools, the alignment of roads which interconnect the towns, and the development of commonly shared economic resources. In the case of a strongly defined, multifaceted transactional environment, the mayor could still adopt a partly proactive attitude toward future change, but may want to temper his or her zeal knowing that not all things may go according to his or her hopes. Scenarios developed under these circumstances may best be constructed as partly normative and partly descriptive. Relating to the example, the mayor can control the placement of the new road but may only be able to influence the placement of the other new roads created by the other towns which may connect to the town. Implicit in this variation of the example, as so far described, is the assumption that the entire contextual environment is generally stable. In particular, although the various towns have some existing relationships, they will continue to operate as individual political entities. Here, analogies might be made to the evolving structure of Europe.

At still another level, the town also exists within a still larger region, nation, and world. Elements such as the national economy and the development of new and transferable technologies would factor into a consideration of the contextual environment. To extend the road example further, the mayor controls the placement of roads in his or her town and influences the placement of roads in adjoining towns. But, what if a railroad with transcontinental connections is built through the region to the west of our study area and siphons off all import and export traffic? Is the town's and the region's investment in new roads wasted? Should such uncertain contingencies be factored into the decision to build the new roads? More generally, if contextual matters seem to dominate local concerns, or if the transactional environment is weak (that is, the towns are isolated or hostile dukedoms and do

not collaborate with each other), then the mayor may want to adopt a preactive attitude, follow the business model of strategy, and develop descriptive scenarios that would allow him or her to position the town in such a way as to reap the potential benefits of changes he or she cannot control.

Challenges for Land Use Scenarios

Physical Scale and Temporal Scale

The above example of a hypothetical town is useful for understanding some of the basic principles for designing a scenario-based study; however, in approaching real world decisions, several challenges must be considered.

Achieving a match between decision and data is complicated by two factors of scale. First, in most parts of the world, the area over which a given natural process operates (e.g., a watershed or the globe) is different, and typically larger, than the area which a given social process associated with land use is managed (e.g., a town or metropolitan region). Second, the time over which natural processes operate are often different—sometimes longer, sometimes shorter—than the time horizon of planning activities, which are dependant on a variety of social constructions including election cycles and financial transactions (e.g., bond maturity periods). As a general rule, the more aggregated the geographical scale (i.e., the larger it is), the more slowly a system's dynamics unfold; the less aggregated, the more quickly it responds to immediate events.[32]

Commensurately, most studies explicitly delimit spatial and temporal scales. In general, scenario-based studies that focus on natural processes that operate at the global or hemispheric scale use time horizons on the order of 50–100 years; studies that focus on social management of cities and regions use time horizons on the order of 10–20 years.

With the scale-related needs of the typical investigation noted, the call to consider human-ecological relationships at multiple scales would seem to be self-evident given ever increasing awareness about the interconnectedness of natural processes and the desire for sustainable development. But while multi-scale studies have been done in the past, they are relatively rare, a situation that likely results from several inherent challenges. From a computational or technical perspective, there can be definitional problems in matching the levels of abstraction (data accuracy and precision) used at different scales. As a recent review paper of multi-scale scenarios noted: different goals and methods used at different scales may produce incompatible results; "linking is difficult when the relevance of issues and processes change with scale . . . [and] credibility is often sacrificed at one scale or another."[33] Further, from a social perspective, there may be resistance from local stakeholders to validate broad assumptions given perceptions of local specificities or to accept the portrayals of themselves or other stakeholders. [34]

Manufactured Uncertainties

The work of some historians has presented a case that a significant aspect of civilization-making has been the transformation of the environment to meet basic needs and express societal values.[35] More narrowly, it has been argued that a significant element of modern western civilizations has been the continuing effort to minimize the uncertainties that stem from natural processes and social dynamics.[36] The phrase, "manufactured uncertainties" captures the situation that new aspects of uncertainty arise from efforts to constrain or eliminate previously recognized insecurities.

There are two aspects of this (largely but not exclusively western) societal development. The first might be considered a matter of our attentiveness. As we scientifically and technologically advance, our attitude toward risk can be seen to escalate as we become more and more aware of new uncertainties. As we control one hazard, a new one becomes evident. The second aspect of manufactured uncertainties concerns how our interconnected social world operates. The work of civilization to manage risk has been accomplished by the application of an expanded knowledge base through ever larger and increasingly abstract expert solutions. In doing so, our solutions have transformed what were once local—local in terms of both time and space—and largely idiosyncratic risk into regional and global and systematic risk. This circumstance means that if (or when) dangers overcome protective measures,

the impacts can register over a larger area, affect more people, or otherwise be more severe. Thus, the dilemma presented by manufactured uncertainties is cyclic and seemingly feeds on itself: our increasing perception risk propels us to do more and more to cope with ever new uncertainty, but our modern answers bring about situations for which our tools have limited effect.[37]

An example of systematic risk brought on by the management of a specific uncertainty is United States flood control policy, which is the largest and arguably most complex flood control system in the world.[38] It is an amalgam of federal, state, and local organizations that work to maintain river traffic and protect certain economic interests. Its "plan" ("plant-like" element in Boulding's scheme) resulted from the "play" of various interest groups; and, once in place, the plan (like all operational plans) focused future action for a limited set of goals (as opposed to other hypothetical goals). Along the length of the Mississippi River and its tributaries, there are a series of dams, levees, and side channels that have been built to manage stream flow. In Louisiana, the course of the river is artificially maintained to assist trade and industry. Much of the engineering works were put in place to accommodate river flow from the north. The "plague" of Hurricane Katrina hit from the south. And once the walls of the levees were either breached or crossed by the storm, there was no place for the water to go. The system built to protect the city contributed to the damage of its infrastructure and to the harm of its residents. To the degree that the strength of Katrina was amplified by anthropogenic activity (e.g., "global warming"), there is a positive feedback loop in the plays-plagues system. The design of the river, the (specified and actual) quality of construction, and (possible) effects of climate change all provided manufactured uncertainties.

A consideration of intersections among planets, plagues, plants (plans), and plays can help to explore aspects of systematic risk and conditions of extreme or radical uncertainty—under which problems become ill-defined and the possible outcomes of our actions are unknown.[39] Nassim Nicholas Taleb has recently described very surprising events that can materialize from such relationships and associated feedback loops, "black swans," after the bird that European scientists said could not exist given their

Continental observations, but that they could not deny after visiting Australia.[40]

Concluding Comments

Larry Gregory and Anne Duran reviewed some of the literature on experiments aimed at understanding the psychology associated with the use of scenarios.[41] In examples ranging from the likely future subscription to cable television, to perceptions about the risk of contracting a disease, to anticipating the outcome of presidential elections, they found consistent evidence that scenarios not only make people more aware of contingencies, but that scenarios also alter expectations about possible future events. But, unless the underlying assumptions about a study are carefully proscribed, the representation of the future may, in turn, be ill defined and decisions based in whole or in part on such thinking may be unintentionally and unfortunately biased. The perceptions about the future that have been discussed in this chapter each provide a useful, if not necessary, caveat to a given scenario process. Explicitly addressing these considerations as part of a study will lead to more enlightening processes and better decisions.

If the future is perceived as open, then the notion of a "good decision" might best be understood as a decision that is robust to a set of possible future contexts. In terms of biological sciences and environmental management, robust decisions go hand-in hand with maintaining resilient ecosystems.[42] In thinking about human occupied (and increasingly human dominated) landscapes, it is also important to consider society's needs for and uses of land. Each of the four scenarios developed for the study that will be described in this book present unique potential challenges for environmental conservation in Southern California.

NOTES:

[1] This chapter draws on and expands ideas presented in Allan W. Shearer, "Approaching Scenario–Based Studies: Three Perceptions about the Future and Considerations for Landscape Planning," *Environment and Planning B: Planning & Design* 32:1 (January 2005). pp. 67–87; and Allan W. Shearer, David A. Mouat, Scott D. Bassett, Michael W. Binford, Craig W. Johnson, and Justin A. Saarinen, "Examining Development–Related Uncertainties for Environmental Management: Strategic Planning Scenarios in Southern California," *Landscape and Urban Planning* 77 (2006), pp . 359–381.

[2] Art Kleiner, *The Age of Heretics: Heroes, Outlaws, and the Forerunners of Corporate Change* (New York: Currency Doubleday, 1996), pp. 150, 368.

[3] Herman Kahn and Anthony J. Wiener, *The Year 2000: A Framework for Speculation on the Next Thirty Years* (New York: Macmillan, 1967), p. 6.

[4] For example: Edward Cornish, *Futuring: The Exploration of the Future* (Washington, D.C.: The World Future Society, 2004); Wendell Bell, *Foundations of Future Studies: Human Science for a New Era, volume 1—History, Purposes, and Knowledge* (New Brunswick, NJ: Transaction Books, 1997); N.C. Georgantzas and William Acar, *Scenario–Driven Planning: Learning to Manage Uncertainty* (Westport, Connecticut: Quorum Books, 1995); R.J. Lempert, S.W. Popper, and S.C. Bankes, *Shaping the Next One Hundred Years: New Methods for Quantitative, Long–Term Policy Analysis*, RAND report MR–1626 (Santa Monica, California: RAND Corporation, 2003); Gill Ringland, *Scenario Planning: Managing for the Future* (New York: John Wiley & Sons, 1998); Peter Schwartz, *The Art of the Long View: Planning for the Future in an Uncertain World* (New York: Currency–Doubleday, 1991); Kees van der Heijdn, *Scenarios: The Art of Strategic Conversation* (New York: John Wiley & Sons, 1996); U. von Reibnitz, *Scenario Techniques*, translated by P.A.W. Rosenthal (New York: McGraw–Hill, 1987).

[5] Herman Kahn, *On Alternative World Futures: Issues and Themes—Draft*, Hudson Institute Report # HI–525–D/3 (Croton–on–Hudson, New York: The Hudson Institute, April 14, 1966), pp. 23–27; Kees van der Heijdn, p. 51; Liam Fahey and Robert M. Randall, "What is Scenario Learning?" in *Learning from the Future: Competitive Foresight Scenarios*, Liam Fahey and Robert M. Randall, eds. (New York: John Wiley & Sons, 1998), pp. 3–21, this note pp. 12–14.

[6] van der Heijden, pp. 116–120.

[7] Robert Costanza, Alexey Voinov, Roelof Boumans, Thomas Maxwell, Ferdinando Vila, Lisa Wainger, Helena Voinov, "Integrated Ecological Economic Modeling of the Patuxent River Watershed, Maryland," *Ecological Monographs* 72:2 (May 2002), pp. 203–231.

[8] Tony Stevenson, "Netweaving Alternative Futures—Information Technocracy or Communicative Community?," *Futures* 30:2–3 (March – April 1998), pp. 189–198.

[9] Aart R. de Lange, "A Dynamic Input–Output Model for Investigating Alternative Futures— Applications to the South–African Economy," *Technological Forecasting and Social Change* 18:3 (1980), pp. 235–245.

[10] Andrii Gritesevskyi and Nebojsa Nakicenovic, "Modeling Uncertainty of Induced Technological Change," *Energy Policy* 28:13 (November 2000), pp. 907–921.

[11] Jari Kaivo–oja, "Alternative Scenarios of Social Development: Is Analytical Sustainability Policy Analysis Possible? How?," *Sustainable Development* 7 (1999), pp. 140–150.

[12] Robert F. Rushmer, *Humanizing Health Care—Alternative Futures for Medicine* (Cambridge, Massachusetts: Massachusetts Institute of Technology Press, 1975).

[13] Jim A. Dator and Sharon J. Rodgers, *Alternative Futures for the State Courts of 2020* (Washington, D.C.: State Justice Institute; Chicago, Illinois: American Judiciary Society, 1991).

[14] Pinar Bilgin, "Alternative Futures for the Middle East," *Futures* 33:5 (June 2001), pp. 423–436.

[15] Victor Basiuk, "Technology, Western Europe's Alternative Futures, and American Policy," *Orbis—A Journal of World Affairs* 15:2 (Summer 1971), pp. 485–506.

[16] Thomas W. Bonnett and Robert L. Olson, *Scenarios of State Government in the Year 2010* (Washington, D.C.: Council of Governors' Policy

Advisors, 1993).

[17] Department of Water Resources, *California Water Plan Update 2005*, Report No. B–160–05 (Sacramento, California: Department of Water Resources, 2005).

[18] Shearer (2005).

[19] Kahn and Weiner, 1967.

[20] Kenneth E. Boulding, "World Society: The Range of Possible Futures," in Elise and Kenneth E. Boulding, *The Future: Images and Processes* (Thousand Oaks, California: Sage Publications, 1995), pp. 39–56.

[21] Pierre Wack, "Scenarios: Uncharted Waters Ahead," *Harvard Business Review* 63:5 (September – October 1985), pp. 72–89; Pierre, Wack, "Scenarios: Shooting the Rapids," *Harvard Business Review* 63:6 (November – December 1985), pp. 139–150.

[22] Clayton Roberts, *The Logic of Historical Explanation* (University Park, Pennsylvania: Pennsylvania State University Press, 1996), pp. 160–198.

[23] ibid.

[24] For example, William M. Jones, "Fractional Debates and National Commitments: The Multidimensional Scenario," RAND Memorandum RM–5259–ISA (Santa Monica, California: The RAND Corporation, March 1967).

[25] Russell Ackoff, *Creating the Corporate Future: Plan or Be Planned For* (New York: John Wiley & Sons, 1981).

[26] F.E. Emery and E.L. Trist, "The Causal Texture of Organizational Environments" *Human Relations* 18 (1965), pp. 21–32; F.E. Emery and E.L. Trist, *Towards a Social Ecology: Contextual Appreciations of the Future in the Present* (New York: Plenum Press, 1973).

[27] A. Cournand and M. Levy, (Eds), *Shaping the Future: Gaston Berger and the Concept of Prospective* (New York: Gordon and Breach, 1973).

[28] For example, L.R. Beres, H.R. Targ, *Constructing Alternative World Futures: Reordering the Planet* (Cambridge, Massachusetts: Schenkman, 1977).

[29] For example, Ringland (1998).

[30] Carl Steinitz, "A Framework for the Theory Applicable to the Education of Landscape Architects (and Other Environmental Design Professionals)," *Landscape Journal* 9:2 (Fall 1990), pp. 136–143.

[31] Carl Steinitz, Michael Binford, Paul Cote, Thomas Edwards, Jr., Stephen Ervin, Craig Johnson, Ross Kiester, David Mouat, Douglas Olson, Allan Shearer, Richard Toth, and Robin Wills, *Biodiversity and Landscape Planning: Alternative Futures for the Region of Camp Pendleton, California* (Cambridge, Massachusetts: Harvard University Graduate School of Design, 1996).

[32] *Millennium Assessment, Ecosystems & Human Well–Being: Synthesis* (Washington, D.C.: Island Press, 2005).

[33] Reinnette Biggs, Ciara Raudsepp–Hearne, Carol Atkinson–Palombo, Erin Bohensky, Emily Boyd, Georgina Cundill, Helen Fox, Scott Ingram, Kasper Kok, Stephanie Spehar, Maria Tengo, Dagmar Timer, and Monika Zurek, "Linking Futures across Scales: A Dialog on Multiscale Scenarios," *Ecology and Society* 12:1 [Article 17] (2007) <http://www.ecologyandsociety.org/vol12/iss1/art17>.

[34] R.J. Scholes and R. Biggs, Ecosystem Services in Southern Africa: A Regional Assessment (Pretoria, South Africa: Council for Scientific and Industrial Research, 2004); K. Kok, R. Biggs, and M. Zurek, "Methods for Developing Multiscale Participatory Scenarios: Insights from Southern Africa and Europe," *Ecology and Society* 12:1 [Article 8] (2007) <http://www.ecologyandsociety.org/vol12/iss1/art8>.

[35] For example: Felipe Fernandez-Armesto: *Civilizations: Culture, Ambition, and the Transformation of Nature* (New York: Free Press, 2001); William Cronon, *Nature's Metropolis: Chicago and the Great West* (New York: Norton, 1991); John R. Stilgoe, *Common Landscape of North America, 1580 to 1845* (New Haven: Yale University Press, 1982);

[36] Anthony Giddens, *The Consequences of Modernity*, (Palo Alto, California: Stanford University Press, 1990); Ulrich Beck, *Risk Society: Towards a New Modernity,* trans. Mark Ritter, (London: Sage Publications, 1992); Ulrich Beck and Johannes Willms, *Conversations with Ulrich Beck,* trans. Michael Pollak, (Cambridge: Polity, 2004).

[37] P.H. Liotta and Allan W. Shearer, "Zombie Concepts and Boomerang Effects: Uncertainty, Risk, and Security Intersection though the Lens of Environmental Change," in *Environmental Change and Human Security: Recognizing and*

Acting on Hazard Impacts, P.H. Liotta, David A. Mouat,William G. Kepner, and Judith Lancaster (Eds.), (Dordecht, The Netherlands: Springer–Verlag, in press, expected 2009).

[38] Karen O'Neill, *Rivers by Design: State Power and the Origins of U.S. Flood Control* (Durham, North Carolina: Duke University Press, 2006).

[39] John Maynard Keynes, *The General Theory of Employment, Interest and Money* (New York: Harcourt Brace, 1936 [reprint 1964]; John Maynard Keynes, "The General Theory of Employment," *Quarterly Journal of Economics*, 51 (1937), pp.

209–223.

[40] Nassim Nicholas Taleb, *The Black Swan: The Impact of the Highly Improbable* (New York: Random House, 2007).

[41] W.L. Gregory and A. Duran, "Scenarios and Acceptance of Forecasts," in J.S. Armstrong, ed., *Principles of Forecasting: A Handbook for Researchers and Practitioners* (Boston: Kluwer Academic, 2001), pp. 519–540.

[42] Stephen R. Carpenter, "Ecological Futures: Building an Ecology of the Long Now," *Ecology* 83:8 (2002), pp. 2069–2083.

Chapter 2

The Uncertainties of Regional Development and Their Possible Effects on Natural Resources Management

Allan W. Shearer and David A. Mouat

Growth beyond the borders of towns and cities has had a mixed reputation in America. On the one hand, since the mid-nineteenth century the ability to afford a suburban or semi-rural home has marked a degree of economic prosperity.[1] The draw of a quiet home with a landscaped yard to escape the pressures of the city was, and still is, strong. Enabled by new modes of transportation—first trolleys, then commuter trains, and finally highways—urbanites seized the opportunity to relocate to more natural environs.[2] Today opportunities afforded by advances in telecommunications continue to extend the range of physical distance between the home and the main office. Equally important in understanding the demands for a suburban or rural lifestyle is that in addition to the attractions offered, would-be residents were (and in some cases still are) pushed away from cities by higher taxes, the quality of schools, and notable crime rates.[3] Yet, on the other hand, development beyond municipal borders has also been associated with social problems and environmental degradation. Commonly and often pejoratively known as *sprawl*, such growth is characterized by leapfrog or scattered development of low-density houses punctuated by commercial strips.[4] From a social standpoint, it has been criticized for contributing to inefficient outlays for infrastructure, unaffordable housing, and the decline of inner cities—which, of course, feeds back on the system, reinforcing some of the reasons that make suburban life attractive.[5] Environmental impacts of sprawl include the loss of productive agricultural land and natural vegetation, ecosystem fragmentation, increased runoff of storm water and the commensurate increased risk of flooding, and greater amounts of air pollution associated with increased automobile usage.[6]

Regional development that reduces native vegetation makes the amount that remains relatively more valuable for the maintenance of ecological systems, including the conservation of local flora and fauna. As fewer and fewer lands become more and more biologically significant, stricter management regimens may become necessary to support and conserve the natural environment. These increased requirements can, in turn, reduce the availability of these lands for other uses including recreation, resource harvesting, and, in the case of military lands, training exercises.

One of the many regions where there are conflicts between conserving natural resources and accommodating development is Southern California. The region boasts some of the richest biodiversity in the nation and is home to several federally listed Threatened and Endangered species including the California gnatcatcher (*Polioptila californica*), cactus wren (*Campylorhynchus brunneicapillus*), and least Bell's vireo (*Vireo bellii pusillus*). It supports a variety of vegetation communities including coastal lagoons, vernal pools, coastal scrub, chaparral, grassland, oak woodland, coniferous forests, and at its eastern extent, desert scrub. The region is also one of the most attractive places to live and faces tremendous development pressures. Broadly looking toward the future, metropolitan areas in Southern California rank high on the U.S. Department of Commerce's list of places facing the greatest population growth over the next ten years.[7] Orange County is forecast to see an increase of 377,000 people and San Diego could see 471,000 new residents.

This situation presents a difficult question: Can each of the region's land stewards—which include federal, state, and local agencies and non-

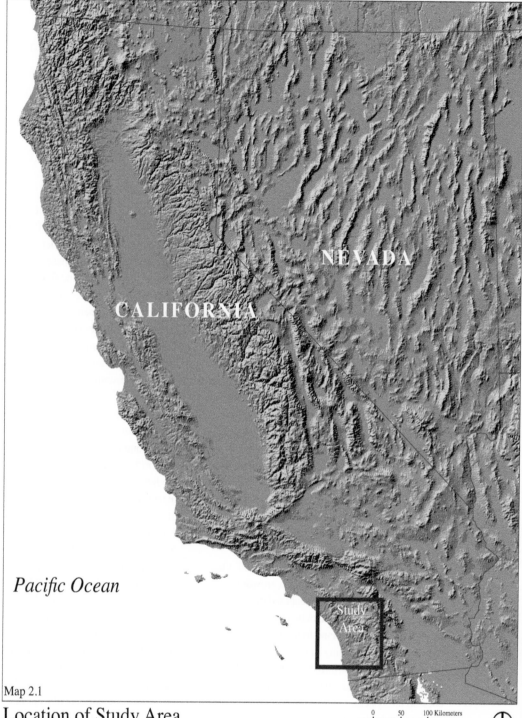

Map 2.1

Location of Study Area

governmental organizations—successfully manage the natural resources in their care if they, acting individually or as a group, only control a small portion of the larger ecosystem?

This study seeks to provide a framework to explore issues associated with growth in the area between Los Angeles and San Diego and to consider the possible consequences on environmental management. The study area, shown in Map 2.1 and in detail in Map 2.2, is anchored by three large military installations: Marine Corps Base (MCB) Camp Pendleton—which for the purposes of this research project includes Marine Corps Air Station (MCAS) Camp Pendleton, MCAS Miramar, and Naval Weapons Station (NWS) Fallbrook. The extent of analysis includes the land between the San Juan River to the north and the San Diego River to the south. This area is encompassed by a bounding box 68 miles north–south and 73 miles east–west.

The impetus to study the relationships between military bases and their surrounding regions warrants comment. Implicit in the "train the way you fight" principle is the availability of realistic training environments. Maintaining the health and viability of ecological systems that provide these conditions can hence be understood as a necessary component to successful long-term training operations. Since the substantial expansion of military bases at the start of World War II, the Department of Defense (DoD) has taken increasing steps to maintain the natural resources on the lands that it administers.[8] To cite several examples of such efforts: In 1942, initiatives with the Soil Conservation Service worked to protect erosion problems that resulted from mechanized vehicles. Post-war natural resources management activities included the planting of trees and ground covers and the development of techniques for controlling fires on maneuver lands. Through the 1950s, professional natural resources managers were hired to assist at the installation level and most bases developed land management plans. In the 1960s, DoD joined most other federal agencies in adopting a multiple-use strategy of land use management. During this period the leasing of lands for agricultural use and timber harvesting became more extensive. The Sikes Act of 1960 further institutionalized the management of natural resources by authorizing cooperative plans with the Fish and Wildlife Service.[9] Through the

1970s and 1980s, there was increased emphasis on concerns over the clean-up of hazardous waste materials and restoration of damaged ecosystems. This period also saw the introduction of environmental legislation, including the National Environmental Policy Act (NEPA)[10] and the Endangered Species Act (ESA).[11] These laws added new requirements to all federal agencies and marked a turning point in environmental regulation, which had formerly been left largely to state governments.

In 1990, the federal government identified habitat modification and loss of *biodiversity* as at the highest level of risk to the country.[12] The term biodiversity, a shorthand expression for biological diversity, is generally understood to incorporate three aspects of the ecosystem: habitat, species variety, and the genetic variability within a species.[13] An area with high or rich biodiversity has an assortment of all three components. During the Clinton administration, the protection of biodiversity was identified as part of a larger initiative to adopt an ecosystem-based approach to federal land management.[14] This new focus became operational through the application of the principles of conservation biology to environmental policy.[15] A relatively new discipline, conservation biology incorporates elements of ecology, wildlife biology, geology, and other natural sciences to provide a basis to conserve plant and animal species in fragmented habitats.[16]

The adoption of ecosystem management has gradually evolved as the science that supports practicable plans has developed. In 1994, a Department of Defense memorandum was circulated to "ensure that ecosystem management becomes the basis for the future of DoD lands and waters."[17] In 1996, a DoD Instruction stated, "the conservation of biodiversity is a component of overall ecosystem integrity and sustainability, which in turn supports the military mission."[18] Also in 1996, *Conserving Biodiversity on Military Lands* was issued to help introduce DoD personnel to the principles of this new approach to land stewardship.[19] The report was prepared by The Nature Conservancy and sponsored by the Department of Defense "Biodiversity Initiative." On the national scene, the expected benefits of ecosystem management as a means to achieve better land stewardship became a "settled" issue with broad public support.[20] Currently, the

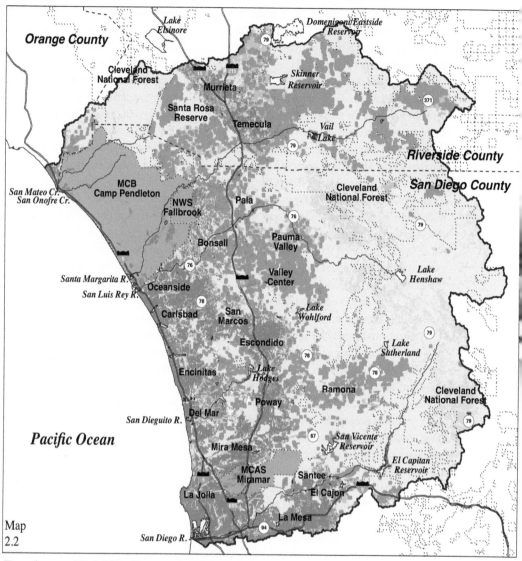

Map
2.2

Region of MCB Camp Pendleton & MCAS Miramar
Existing Conditions 2000
Study Area Locations

Grassland and Shrubland 853476 acres (49.2%)		Urban and Suburban 260191 acres (15%)	
Forest and Riparian 106218 acres (6.1%)		Interstates	
Agriculture and Orchards 57611 acres (3.3%)		Rivers	
Military Maneuver and Impact 145720 acres (8.4%)		Federal Lands	
Rural and Ex-Urban 294005 acres (17%)		Counties	

— **See color insert following page 204** —

implementation of ecosystem management continues to evolve as federal land managers grapple with the responsibility of considering their decisions in the context of large-scale environmental structures while retaining only the authority to act within agency and installation boundaries.

The evolution of conservation techniques has provided DoD with improving means to address increasing concerns over natural resources conservation. It could almost go-without-noting that armed forces personnel are, of course, not the only occupants of the large training bases. As substantially undeveloped areas, the installations are host to a wide variety of plant and animal species. By some measures, DoD lands are disproportionately important for the maintenance of habitat. In terms of acreage controlled, DoD ranks fifth among other federal agencies, behind the Bureau of Land Management (approximately 272 million acres), the Forest Service (~193 million acres), Fish and Wildlife Service (~91 million acres), and the National Park Service (~76 million acres). However, DoD ranks highest in terms of administering land that has been identified as supporting federally listed Threatened and Endangered species. In 1996, it was reported that military lands provide habitat for over 200 listed species.[21] A slightly earlier Forest Service report, which was based on published records of known species locations, placed DoD as the federal agency responsible for maintaining the highest number of Threatened and Endangered species.[22]

Several factors contribute to the ecological importance of military lands:[23] DoD has installations in a wide variety of geographical conditions and so it is likely to have a greater total variety of species and possibly a greater number of species than agencies which manage fewer kinds of ecosystems. More specifically, DoD properties include types of ecosystems that are either underrepresented or unrepresented in the land holdings of other federal agencies.[24] Indeed, DoD maintains some of the few large unbuilt parcels in the relatively densely populated eastern United States. Also, access to DoD lands is heavily restricted and many training activities result in relatively little environmental impact when compared to the commodity production that occurs on other federal lands. And finally, on the whole, DoD lands have been more closely surveyed than those of other agencies.

This research project is not the only investigation to examine the ecological interrelationships between military bases and their surrounding areas; nor, is it even the first to investigate these issues in this part of the country. In 1996, the study *Biodiversity and Landscape Planning: Alternative Futures for the Region of Camp Pendleton, California* explored how the cumulative effects of urban growth might influence the biodiversity of the region.[25] Conducted by a team of researchers from Harvard University, Utah State University, the U.S. Environmental Protection Agency, the U.S. Forest Service, and The Nature Conservancy, the project sought to investigate some basic questions on the possible relationships between development and biodiversity. These questions included: How might biodiversity be measured? What are the technical costs and benefits of a computer simulated modeling approach to landscape planning for biodiversity? Can such an approach serve as the basis for regional planning and negotiation? What are potentially effective strategies for the conservation of biodiversity?

As part of the 1996 research, three scales of change were considered. At what is commonly called "site scale," the restoration of no-longer-active percolation ponds on Camp Pendleton and possible wildlife corridors between the base and The Nature Conservancy's Ecological Reserve on the Santa Rosa Plateau were investigated. At a larger scale, development schemes for the area around Oak Grove, a sub-watershed within the Santa Margarita River Basin were compared. Finally, at the regional scale, several alternatives for growth were assessed in terms of a set of environmental process models including landscape ecological pattern, potential habitat for a suite of indicator species, and total species richness. The first alternative was a *Build-Out* (that is, a complete implementation) of the then existing municipal plans as collected by the regional planning agencies. Several other alternatives, which sought to achieve the dual goals of 1) accommodating the expected population forecast and 2) managing for biodiversity, were also developed and compared. The Private Conservation Future encouraged large-lot ownership with strict development controls to buffer and connect important areas of habitat. The Multi-Centers Future

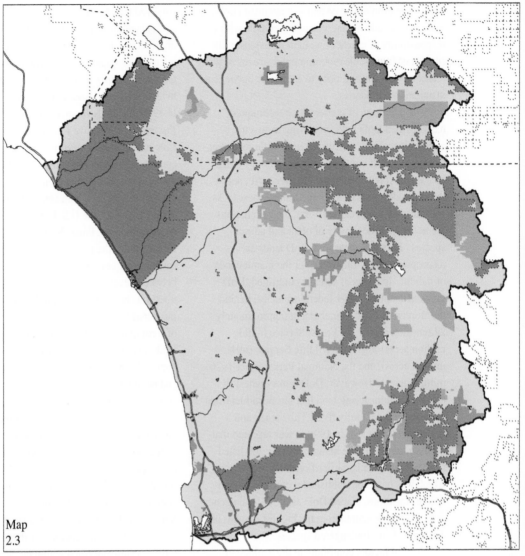

Map
2.3

Region of MCB Camp Pendleton & MCAS Miramar
Existing Conditions 2000
Land Ownership

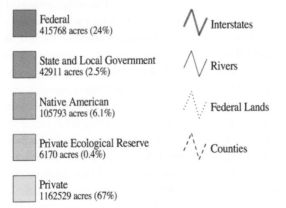

Federal
415768 acres (24%)

State and Local Government
42911 acres (2.5%)

Native American
105793 acres (6.1%)

Private Ecological Reserve
6170 acres (0.4%)

Private
1162529 acres (67%)

Interstates

Rivers

Federal Lands

Counties

focused on the creation of several new small towns, leaving a considerable amount of land undisturbed. The New City Future concentrated growth into a single community away from some of the more ecologically sensitive lands. The Spread Future served as a test of the need to do any planning to protect regional biodiversity. It propagated a pattern of uncontrolled, low-density housing across the region without regard to any ecological needs. Finally, a variant of the Spread Future introduced an aggressive conservation plan after ten years of unchecked growth.

The study documented in this book extends the area of analysis south to include much of the rapidly developing part of west-central San Diego County, including MCAS Miramar. Additionally, this work takes a different approach. Rather than identify potential means to achieve strategic goals which are presumed to be commonly shared and politically viable, it is organized to examine critical uncertainties—those issues over which individual stakeholders have (or at least seem to have) little or no control, but which will significantly determine the ultimate success of any local actions taken in the near future.

Some of the critical uncertainties that are perceived as salient to the area at this time include:

- Will there be sufficient water to satisfy all perceived needs?
- Will there be sufficient energy to satisfy all perceived needs?
- Will mass transit become widely accepted?
- Will higher densities of development be acceptable?
- Will future environmental regulations be different than they are today?
- Will conservation of the environment become more important to the public-at-large?

These and other questions served as the basis for an extended exploration of what might happen in the region. Combinations of *yes* and *no* answers reflected different possible future states and prompted questions of how these conditions might arise and what might follow. The details of who,

what, where, when, and how provided, in turn, a more complete mapping of how the future might unfold. While the number of possible combinations of even a short list of critical uncertainties can grow very large very quickly, four scenarios captured many of the sets of regional concerns and led to distinct patterns of development. These scenarios became the Coastal, Northern, Regional Low-Density, and Three-Centers Futures.

The approach used in this study requires a comment about its intentions and possible interpretations. By focusing on critical uncertainties rather than goals, the possible futures that are considered in this report reflect concerns about what tomorrow *might* become, not preferences for what it *should* become. In each of the scenarios, there may be what some would call winners and losers. Similarly, it is also possible that a person reading this report would very much want to live in some of the described alternative futures or, equally possible, not live in any of them. While such assessments could mark the success or failure of a comprehensive regional planning initiative, they are incidental to the goals of this current study. Instead, the four alternative futures developed for analysis are best understood as different contextual models against which the risks and opportunities of local actions might be considered. For example, as asked above, can the various actions taken to manage the lands of a military base succeed given different possible changes to the larger landscape? Such questions about the management of natural lands are not unique to the military and many of the underlying concerns are common to other organizations, agencies, and institutions throughout the region. It is the hope of the researchers that the approach taken in this study and its results will also be of benefit to all of the area's stakeholders as they look to make their own decisions.

This report is organized to state the methods of analysis that have been used to examine specific pressures that the region's land managers could face. Building on the overview of the principles of scenario-based studies provided in Chapter 1, this book develops and explores a consideration of the future to better understand the consequences of change on natural systems. Chapter 3 summarizes other scenario-based studies that have specifically explored development around military installations and which have contributed to a better understanding of the relationships between bases, their neighboring communities, and the surrounding regions. Chapter 4 gives an introduction to the study area, its history, its physical and biological features, and describes how these are represented for the purpose of analysis. Chapter 5 provides an overview of the mapping techniques used for these analyses and provides photographs of representative land cover types. Chapter 6 reviews the process used to understand the dynamics of the region and to identify the critical uncertainties. Chapter 7 presents four scenarios that play out the logic of these uncertainties and related spatially explicit alternative futures. Chapters 8, 9, and 10 consider the impacts of these changes on stream flow, air quality, and a suite of biologic models including vegetation change, landscape ecological pattern, potential habitat for selected indicator species, and total species richness. Chapter 11 summarizes and discusses the results.

Readers should consider some important caveats to this report. First, the investigators are conducting independent research and are not providing planning services to the United States Marine Corps, the Department of the Navy, or to any of the other regional stakeholders. Second, the investigators have made assumptions about the area based on publicly available documents. Similarly, the research models are based on published information. To the best of the investigators' knowledge, these sources are accurate. Third, while distinctions are made between pubic and private lands (Map 2.3), private property boundaries (i.e. lines between individual parcels) are not considered. Similarly, local government jurisdictions are not considered in the alternative futures. Instead, the study area was treated as a single, albeit very complex, geographic entity. Fourth, concurrent to this research, several local governments, including Western Riverside and San Diego Counties, initiated efforts to revise the land use plans for their respective jurisdictions. At the time of this project's conclusion, these government studies were still in development and were not considered within the scope of work. Finally, this research has a selective and limited focus on environmental issues that may contribute to the management of unbuilt lands with a geographic boundary determined by the location of several large military installations. As such, this work is not intended to represent a comprehensive inventory or analysis of the issues facing the region.

NOTES:

[1] Anthony Downs, *Opening Up the Suburbs: An Urban Strategy for America* (New Haven, Connecticut: Yale University Press, 1973); John R. Stilgoe, *Borderland: Origins of the American Suburb, 1820 – 1939* (New Haven, Connecticut: Yale University Press, 1988).

[2] Peter Mieszkowski and Edwin S. Mills, "The Causes of Metropolitan Suburbanization," *Journal of Economic Perspectives* 7:3 (Summer 1993), pp. 135–147.

[3] Charles F. Adams, Howard B. Fleeter, Yul Kim, Mark Freeman, and Imgon Cho, "Flight from Blight and Metropolitan Suburbanization Revisited," *Urban Affairs Review* 31:4 (March 1996), pp. 529–543.

[4] Reid Ewing, "Is Los Angeles–Style Sprawl Desirable?" *Journal of the American Planning Association* 63:1 (Winter 1997), pp. 107–126, this note p. 108.

[5] Robert H. Freilich and Bruce G. Peshoff, "The Social Costs of Sprawl," *Urban Lawyer* 29:2 (Spring 1997), pp. 183–198, this note, p. 184; Ewing, pp. 113–118; Joseph Persky and Wim Wiewel, *When Corporations Leave Town: The Costs and Benefits of Metropolitan Job Sprawl* (Detroit, Michigan: Wayne State University Press, 2000).

[6] Michael P. Johnson, "Environmental Impacts of Urban Sprawl: A Survey of the Literature and Proposed Research Agenda," *Environment and Planning A* 33:4 (April 2001), pp. 717–735.

[7] Published in National Association of Home Builders, *Smart Growth: Building Better Places to Live, Work and Play* (Washington, D.C.: National Association of Home Builders, 2000), p. 5.

[8] For a history of land use and natural resources management on Department of Defense holdings, see J.R. Arnold and R. Wiener, *The U.S. Army Corps of Engineers and Natural Resources Management on Army Installations 1941 – 1987* (Fort Belvoir, Virginia: Engineering and Housing Support Center and Office of History, U.S. Army Corps of Engineers, 1989).

[9] The Sikes Act of 1960 as amended, Public Law 86–797, 16 United States Code 670(a)–670(o), 74 Statute 1052 (September 15, 1960); subsequently amended.

[10] The National Environmental Policy Act of 1969 as amended, Public Law 91-190, 42 United States Code 4321-4347 (January 1, 1970); subsequently amended.

[11] The Endangered Species Act of 1973 as amended, Public Law 93-205, 16 United States Code 1531–1544, 87 Statute 884 (December 28, 1973).

[12] U.S. Environmental Protection Agency–Science Advisory Board, *Reducing Risk: Setting Priorities and Strategies for Environmental Protection*, Environmental Protection Agency Report # SAB-EC-90-021 (Washington, D.C.: U.S. Environmental Protection Agency, 1990), p. 12.

[13] John C. Ryan, *Life Support: Conserving Biological Diversity*, Worldwatch Paper 108 (Washington, D.C.: Worldwatch Institute, April 1992), p. 7. For other definitions of biodiversity which are both more expansive and more restrictive, see David Takacs, *The Idea of Biodiversity: Philosophies of Paradise* (Baltimore, MD: The Johns Hopkins University Press, 1996), pp. 46–50.

[14] For an overview of ecosystem management principles and applications within the federal government, see *Ecological Applications* 6:3 (August 1996), pp. 692–974; for applications of ecosystem management within government agencies, see Congressional Research Service, *Ecosystem Management and Federal Agencies*, Report # 94–339–ENR (Washington, D.C.: Library of Congress, April 19, 1994).

[15] Takacs, pp. 41–99. A review of Takacs' book by John Horgan, "It's Not Easy Being Green," *New York Times Book Review* 147:50,670 (January 12, 1997), p. 8 emphasizes the political use of the term.

[16] For background in conservation biology see, Graeme Caughley, *Conservation Biology in Theory and Practice* (Cambridge, Massachusetts: Blackwell Science, 1996); Richard B. Primack, *Essentials of Conservation Biology*, 2nd edition (Sunderland, Massachusetts: Sinauer Associates, 1998); Malcolm L. Hunter, *Fundamentals of Conservation Biology*, 2nd edition (Malden, Massachusetts, Blackwell Science, 2002).

[17] Office of the Under Secretary of Defense (Acquisition and Technology), "To Deputy Under Secretary of Defense (Environmental Security) and others," Subject: Implementation of Ecosystem Management in DoD (August 8, 1994).

[18] U.S. Department of Defense, *Instruction 4715.3*, "Environmental Conservation Program,"

(May 3, 1996), p. 16.

[19] Sherri W. Goodman, in Michele Leslie, Gary K. Meffe, Jeffrey L. Hardesty, and Diane L. Adams, *Conserving Biodiversity on Military Lands: A Handbook for Natural Resources Managers* (Arlinton, Virginia: The Nature Conservancy, 1996), Foreword.

[20] David N. Bengston, George Xu, David P. Fan, "Attitudes toward Ecosystem Management in the United States, 1992 – 1988," *Society and Natural Resources* 14:6 (July 2001), pp. 471–487.

[21] Leslie, et al., p. 9.

[22] Curtis H. Flather, Linda A. Joyce, and Carol A. Bloomgarden, *Species Endangerment Patterns in the United States*, U.S. Forest Service General Technical Report # RM–241 (Fort Collins, Colorado: U.S. Department of Agriculture, Forest Service, Rocky Mountain Forest and Range Experiment Station, 1994), p. 12.

[23] David Rubenson, Marc Dean Millot, Gwen Farnsworth, and Jerry Aroesty, *More Than 25 Million Acres? DoD As a Federal, Natural, and Cultural Resource Manager*, RAND Corporation Report # MR–715–OSD (Santa Monica, California: RAND Corporation, 1996), pp. 14–15.

[24] Leslie, et al, pp. 9–10.

[25] Carl Steinitz, Michael Binford, Paul Cote, Thomas Edwards, Jr., Stephen Ervin, Craig Johnson, Ross Kiester, David Mouat, Douglas Olson, Allan Shearer, Richard Toth, Robin Wills, *Biodiversity and Landscape Planning: Alternative Futures for the Region of Camp Pendleton, California* (Cambridge, Massachusetts: Harvard University Graduate School of Design, 1996).

Chapter 3

Scenario-Based Studies of Military Installations and Their Regions

Allan W. Shearer

Three previous scenario-based studies have contributed to better understanding the present and possible future environmental interrelationships between military installations and surrounding regions. The purpose of this chapter is to summarize and compare the scope of work and the methods used in each of these investigations. As will be demonstrated, the use of scenarios for military land use management is a multi-faceted endeavor.

The projects are *Biodiversity and Landscape Planning: Alternative Futures for the Region of Camp Pendleton, California* (1994 - 1996),[1] *Analysis and Assessment of Military and Non-Military Impacts on Biodiversity: Mojave Desert Case Study* (1998 - 2002),[2] and *Alternative Futures and Changing Landscapes: The Upper San Pedro Basin, Arizona and Sonora* (1998 - 2001).[3] The underlying organization of each of these studies can be understood in terms of Carl Steinitz's "Framework for Landscape Planning" which asks six questions to understand environmental change: "How should the landscape be described? How does the landscape operate? Is the current landscape working well? How might the landscape be altered, and by whom? What predictable impacts might the [planned and/or anticipated] changes cause? And, should the landscape be changed; if so how?"[4] The first three questions provide a means for describing and evaluating current conditions; the second three questions provide a means for understanding possible changes relative to the present. In a typical study, the questions are asked three times. The first pass proceeds in the sequence presented above and is done quickly to establish the needs of the study, identify which resources are readily available, and

prescribe those that must be developed. The second pass is done in reverse order, beginning with the nature of the decision. It is at this stage when the design of the study is specified. It is important to note that this design process begins with the decision or decisions that need to be made, not with the data. That is, the data are selected or acquired given the needs of the questions at hand; the questions are not asked given the data that are available. The third pass completes the work and is approached, once more, from the question of landscape representation. At each question, planners and/or stakeholders are required to address issues of methodology. For example, the third question—How might the landscape change?—might be addressed through the perpetuation of current trends or be the result of a willful act to change those trends (or some combination of the two).

A characteristic shared by all three of the studies is a focus on the relationships between regional biodiversity and development. Toward this end, each study evaluates the present conditions and potential impacts of the future against a set of mathematical models that emulate natural systems and are tailored to the conditions of each site. It should also be noted that, beyond a general methodological approach and a focus on the conservation of biodiversity, the projects also share considerable overlap in research personnel. However, despite the similarities of scope, design, and staff, these three projects differ in terms of the circumstances in which the work was initiated, the degree of interaction between the researchers and area stakeholders, and the types of scenarios that were developed.

Alternative Futures for the Region of Camp Pendleton, California

The first study on the region of Camp Pendleton was conducted by researchers at the Harvard University Graduate School of Design, Utah State University, the U.S. Forest Service, the U.S. Environmental Protection Agency, and The Nature Conservancy. It was primarily funded by the Strategic Environmental Research and Development Program (SERDP), a joint operation between the Department of Defense, the Department of Energy, and the Environmental Protection Agency. Additional funds were provided by the U.S. Forest Service. The aim of the project was to assess Camp Pendleton's (then) current role in supporting the region's biodiversity and to investigate how that role might change in the future given different patterns of urban development and conservation strategies. Throughout the 1990s, the importance of biodiversity assumed an increasingly prominent place in the debate of environmental policy, not only within DoD but also across the nation.[5] One locus of biodiversity-related investigation was the Biodiversity Research Consortium (BRC), which was comprised of researchers from numerous federal agencies working in cooperation with scientists, planners, and policy experts in academia and non-governmental organizations (NGOs). The central goals of the BRC were to identify means of analyzing biological diversity nationwide; to search for correlations between species diversity and environmental diversity; to evaluate the comparative risk to biodiversity using species diversity, environmental diversity, and stressor information; and to begin developing approaches to managing environmental diversity in order to achieve species diversity goals.[6] Several of the researchers on the first Camp Pendleton project were associated with the BRC and this study was envisioned to contribute to the understanding of how changes in urban development stress habitats of native species. Development was seen as contributing to habitat loss and degradation in two ways. First, it caused what can be called *direct stress* through the construction of roads, homes, and businesses and the corresponding loss of natural vegetation. Second, development caused *indirect stress* through habitat fragmentation and other changes to natural

systems, such as accelerated storm water runoff and consequent flooding. Given this agenda, the researchers chose the area around Camp Pendleton because it had both high biodiversity and faced significant urban growth pressures.

The study area was defined by the drainage basins of the rivers that flow through or are adjacent to the base and included parts of San Diego, Riverside, and Orange Counties. A rectangular bounding box drawn around the irregular basin boundaries measures approximately seventy-two miles east–west, by forty miles north–south. While short of spanning entire ecoregions, the area did encompass a variety of habitat types ranging from the Pacific Coast to foothills to mountains, to inland valleys, and—at the eastern extent—to desert. Three scales of change were examined: regional, sub-regional (defined by a third-order stream drainage basin), and local (a wildlife corridor near Camp Pendleton and the restoration of on-base wastewater percolation ponds for riparian habitat). Regional scale environmental assessment included changes in visual preference, acreage of soils valued for agriculture, risk of fire and subsequent vegetation change, and volume of discharge of the Santa Margarita River. Possible change in biodiversity was examined through three models. First, change in the landscape ecological pattern (which considers habitat patch size, geometry, and connectivity and is one way to approach the issue of fragmentation);[7] Second, a set of detailed single species potential habitat models of selected reptiles, birds, and mammals; And third, total species richness (variety of amphibians, reptiles, birds, and mammals) as a function of potential habitat.[8]

The regional scenarios were developed by advanced students in the fields of architecture, landscape architecture, and urban planning at the Harvard Graduate School of Design in a semester-long studio course. Students had access to the research team's Geographic Information System (GIS) files that used data from 1990 to 1993 as "current conditions" and were referred to as 1990+. These files included land use, land ownership, elevation, slope, soils, streams, and vegetation. The students also had access to the results of the research team's computer models that evaluated the 1990+ conditions, in terms of biodiversity and stream flow in major drainage basins. The scenarios were developed, in part, through GIS modeling and, in part, through more traditional methods of design practice (e.g., drawing lines on maps). The students were asked to propose options that would accommodate the expected twenty-year population forecast of 500,000 people and manage for biodiversity. Working in groups of three to six people, six futures were developed: 1) The *Private Conservation* future conserves biodiversity through large-lot ownership and management of important habitat areas. 2) The *Multi-Centers* future concentrates growth in a set of small towns. 3) The *New City* future concentrates growth in a single new urban area. 4) One student team was asked to investigate two possible futures that would meet the projected change in population, but dismiss (or, at least, ignore) the explicit planning goal of managing for biodiversity and to instead maximize development revenue. The first of these alternatives was referred to as *Spread* growth. 5) A variant of the Spread future was *Spread with Conservation*, which allowed unchecked growth for ten years and then implemented a substantial conservation effort and directed growth for the next ten. The two Spread scenarios were created to test if the other regional designs would be necessary to conserve biodiversity. The assessment of this alternative future was that if no conservation plans were implemented within the first ten years *and* if the economy remained very robust—allowing the widespread construction of "rural villas" with extensive gardens and orchards—then the remaining unbuilt land would be significantly fragmented into small, isolated areas with negligible value as habitat. As such, implementing even a strong conservation plan after that time would be ineffective for maintaining the region's biodiversity. 6) Finally, the *Plans Build-Out* alternative fully implemented the existing municipal and county plans. An important note for comparing the alternatives is that the *Plans Build-Out* alternative did not utilize the population projection, but simply filled all available land with its planned use. Given simplifying assumptions about the density of development within the generalized land use categories used for the study, *Plans Build-Out* could accommodate well over one million additional people.[9]

In 1997, a second studio-course at the Harvard Graduate School of Design conducted a follow-up to the project and used the alternative futures and associated impact assessments from the then completed research project to propose a single best plan which would manage for biodiversity and accommodate growth in the region around Camp Pendleton.[10] Again students worked in teams of three to six people; however, the organization of the work was structured differently. In the first (1996) studio, the teams worked in parallel with one another, each pursuing the same strategic objectives at a regional scale (i.e., accommodating population growth and managing for biodiversity), albeit through different means. In the second (1997) studio, the teams likewise engaged the same question of balancing development and conservation goals, but did so at different spatial scales. One group considered—or, perhaps it would be better to say, reconsidered—the region as a whole, a second group approached the sub-regional scale in and around the town of Temecula, a third group examined the scale of a large institutional investor-stakeholder, and, lastly, a fourth group looked at the scale of the individual homeowner. The final exercises of the studio required the groups to come together and identify which actions could be made at the federal, state, regional, county, local, and private levels of decision making. Part of this process included noting conflicts that resulted from approaching similar problems at these different scales. These problems ranged from questions about who (individually or collectively) is ultimately responsible for addressing certain problems to discussions of how to best provide solutions. This follow-up study was funded by the Graduate School of Design.

Mojave Desert Case Study

Following the study on the region of Camp Pendleton, researchers from the U.S. E.P.A., the Desert Research Institute, Oregon State University, Utah State University, and the U.S. Forest Service investigated possible growth and consequent environmental impacts in California's portion of the Mojave Desert. This project was initiated and funded through SERDP. The Mojave spans more than 77,000 square miles over parts of California, Nevada, Arizona, and Utah. Three-quarters of the area is owned by the federal government, primarily through the Bureau of Land Management (BLM) and the National Park Service (NPS). Thirty of California's cities and towns are located in the Mojave, as are four major military installations: Marine Corps Air Ground Combat Center at Twenty-Nine Palms, Marine Corps Logistics Base Barstow, Naval Weapons Testing Station China Lake, Edwards Air Force Base, and Fort Irwin which is home to the Army's National Training Center (NTC). Concerns over the relationships between military operations and environmental stewardship needs increased in the region throughout the 1980s and 1990s as new training initiatives—particularly at the NTC—were planned. Much of the debate was centered on the protection of the Desert Tortoise (*Gopherus agassizii*), which had been listed as a Threatened Species by the U.S. Fish and Wildlife Service in April 1990. However, during the time of the study there were also more reaching efforts to craft a conservation plan for the larger region around the NTC.[11]

The study area for this project was roughly a triangle, bounded by the mountains that border the Mojave to the west and south, and by the diagonal north-west to south-east California-Nevada state line. Within California, parts of Inyo, San Bernardino, Los Angeles, and Riverside Counties were included. While the study area did not comprise an entire ecoregion, biogeographic boundaries were partially used to delimit the area of analysis area. Again, concerns about potential impacts on biodiversity played an important role in defining the scope of work. Toward understanding the influences of development on the natural systems, three levels of species impact modeling were considered. First, multi-species analysis included total species richness of terrestrial vertebrates, species richness for federally listed Threatened and Endangered Species, and species endemic to the Mojave. Second, focal species models examined habitat needs of selected species in greater depth. And third, a small number of single species models, which employed an ever greater degree of detail and which incorporated field-based research, were developed.

A *Build-Out* scenario, which fully implemented the local plans, and six alternatives that allocated the projected population from 2000 to 2020, were considered.[12] The alternative scenarios were created through a modeling process based on trends of past development as well as assumptions about future population increases. Variations of the trend were created by changing the distribution of land that was available for development. The allocation of houses within each of the alternative futures was based on a *logit* model of development potential. Logit models allow probabilities to be assigned to distinct states of categorical data based on a number of non-random, independent observable factors.[13] In this case, the issue of concern was private land use and the possible states were unbuilt land and built land; the independent factors contributing to the likelihood that a given piece of land would be developed were distance to existing development, percentage of surrounding development, distance to primary roads, distance to secondary roads, presence within an incorporated city, and percent slope of the terrain. The probabilities for these factors were calculated by comparing the location of development that occurred between 1970 and 1990 to a 1970 base line. The probabilities were then extended to post-1990 conditions. In each scenario, county population forecasts were applied through a statistically derived average population density of 3.76 people per hectare (2.47 acres) and new houses were sited at the locations that had the greatest probability of development.

The *Trend* scenario plotted the future most likely to occur if the pattern established by the recent past continues for another twenty years. *New Roads* directed the locations of future growth by the construction of several primary roads. Four scenarios examined land trades and the establishment of buffer zones. *Exchange 1* traded private land with low development probability and high biodiversity value for public land with high development probability and low-biodiversity value. *Exchange 2* established a three-mile buffer around all of the military installations. The buffer was created by trading the privately owned land for a comparable amount of public land located farther away from the installations. *Exchange 3* traded private "in-holdings" which are adjacent to lands that are specifically managed for the conservation of biodiversity for other non-military public lands that have a high probability of development. (NB: An in-holding is a private parcel of land that is within the boundaries of public lands.) *Exchange 4* established a five-mile buffer along each side of the ten flight paths that comprise the R-2508 Complex (a training and testing area which is jointly managed by China Lake, Edwards, and Fort Irwin). Similar to the other scenarios, private land within the buffer was acquired through an exchange with non-military public lands. Each of the six alternative scenarios was also varied through three additional permutations: 1) allocating the expected population growth to a housing density of 20 people per hectare; 2) allocating 1.5 times the expected number of people at the current population density of 3.76 people per hectare; 3) allocating 1.5 times the expected number of people at a density of 20 people per hectare. Counting the *Build-Out* alternative and the permutations of the six alternative scenarios, twenty five alternative futures were initially compared, although not all of these were used in the project's final report.

Alternative Futures for the San Pedro River Basin, Arizona and Sonora

The third alternative futures study was conducted on the region around the U.S. Army's Fort Huachuca, which has been in operation since 1877. The work was conducted by researchers from Harvard, the Desert Research Institute, the University of Arizona, and the U.S. Army Construction Engineering Research Laboratory (CERL). It was initiated by the U.S. Army Training and Doctrine Command (TRADOC) and funded by the Department of Defense Legacy Resource Management Program (usually referred to simply as Legacy) which supports projects designed to preserve, restore, and manage natural and cultural resources.

As is the case with many installations, Fort Huachuca's role as a steward for the natural landscape has become more prominent over the last twenty-five years as the nearby communities have grown.[14] As a large, mostly undeveloped piece of land, the Post faces a variety of natural resources management issues. Among these concerns is maintaining nesting and forage habitat for the migratory lesser long-nosed bat (*Leptonycteris curasoae*), which was listed as

an Endangered Species in 1988.[15] However, the greatest environmental concern in this region is not the continued conservation of on-base habitats, but the conservation of the nearby Upper San Pedro River that provides critical habitat for neo-tropical migrating birds. The San Pedro is one of the few free-flowing rivers in the Colorado River system. It originates in Mexico and flows northward over one hundred and thirty five miles into the Gila River. Its ecological importance was given special status in 1988, when the U.S. Congress created the San Pedro Riparian National Conservation Area (SPRNCA).[16] SPRNCA, which is administered by the Bureau of Land Management, spans roughly forty miles along the river's course and covers 56,000 acres of riparian habitat. Throughout much of the year, little or no water can be seen in the stream channel, but flow beneath the surface supports riparian vegetation. Groundwater is drawn for several activities including copper mining operations in Cananea, Mexico, agricultural irrigation—primarily, but not exclusively—in Arizona, and residential development on both sides of the international border.[17] Groundwater pumping has resulted is a cone of depression in the aquifer near the town of Sierra Vista and Fort Huachuca. The cone acts as a sink, drawing water from the surface and subsurface flows. As more water is diverted for use (residential, commercial, agricultural, etc.), the cone gets larger and greater stress is placed on the vegetation along the river.

The fate of the river has garnered a significant amount of attention from environmental groups including The Nature Conservancy, which listed the San Pedro as among its "top ten" places of critical concern.[18] Similarly, American Rivers called it one of the most endangered rivers of 1999.[19] Some of this interest has resulted in legal actions taken on behalf of the ecosystem. In 1996, the Southwest Center for Biodiversity (now the Center for Biodiversity) alleged that expansions of Fort Huachuca's operations were harming the river by causing urban growth within the basin.[20] The argument was predicated on the assertion that the base was the underlying factor driving new development, and with that new development came increasing groundwater extraction that was incrementally reducing stream flow. The Southwest Center believed that given this impact, the Post was required to file an environmental impact statement under the National Environmental Policy Act (NEPA).[21] Its complaint was filed in a petition to the Commission for Environmental Cooperation (CEC), a tri-national agency that was created with the North American Free Trade Act (NAFTA). The CEC certified the legitimacy of the petition and requested a response from the U.S. Environmental Protection Agency (EPA), which is responsible for the enforcement of environmental regulations associated with NEPA. In addition, the CEC proceeded with its own investigation under powers provided by its charter. Contemporaneously, but not apparently contingently, the Southwest Center withdrew its complaint when the CEC investigation was initiated. The focus of the CEC report was to identify possible means to achieve a sustainable balance between development needs and conservation objectives in the area.[22] The document proved to be controversial among residents of the area and was seen as an internationalist intervention into a local problem.[23] And so, while the development pressures in this region were (and are) not as great as those in Southern California if measured by sheer population increase, the relationship between development and environmental impact is, in the case of groundwater reserves, very acute. Moreover, the political tensions in the area are quite high, intensified by national and international attention. The regional water situation would make it difficult for the Army to expand current activities or introduce new operations that would perceptibly increase its water usage.

The study area for this DoD sponsored study of the region was defined by the Upper San Pedro River Basin plus an extension to the west in order to include grassland habitat which has become less common in the region due to fire suppression.[24] A bounding box around the area is approximately thirty miles east–west by seventy miles north–south. The set of biologic impact models was similar to that used in the other alternative futures studies mentioned and included landscape ecological pattern, a suite of single species potential habitat models, and species richness. Additionally, visual preference, vegetation change, and groundwater flow were modeled.

The possible futures considered for the study were based on a "Scenario Guide" that identified a series of issues relating to population and demographics, water management, and land management. For each issue, several policy options were identified. A draft version of the Guide was distributed to local residents for comments regarding the scope of issues that had been identified and the range of policy options listed for each issue. A final version, which was slightly shorter than the draft, was then distributed to the environmental management staff at Fort Huachuca, the County Planning office, and members of the community at a public meeting. Guides were also made available via the Internet. Over 200 Guides were distributed and approximately 50 were returned. A sample question from the guide is provided in Figure 3.1.[25]

Each answer specified a scenario parameter. In the case of the example given in the figure, the response distributed the expected new residents into different housing types. Other questions called for restrictions or accelerations of development and for the implementation of different conservation initiatives. Each scenario made use of the same housing allocation model to site new homes. It was based on the probability of development as reflected by criteria identified in a survey of local developers and real estate agents. The model sub-divided the region into smaller markets and allocated new homes in five year intervals; however, due to computational limits, the probabilities were calculated only at the start of the process, and not revised after each iteration. New development in Mexico was more simply modeled by expanding existing population centers.

Based on the responses to the Guide, three primary scenarios were identified. Additionally, several variations to these possible futures were also modeled. However, unlike the permutations of the scenarios created for the Mojave project, the variations for this study were not uniformly and systematically applied. Instead, the variations reflected concerns that were specific to a larger pattern of responses. The *Plans* scenario was based on the current local plans and the local population forecasts. *Plans-1* doubled the expected population increase in Arizona. *Plans-2* used the expected population increase in Arizona, but doubled the population in the Sonoran towns and increased mining operations in Cananea. *Plans-3* used expected population forecasts, but limited growth

Figure 3.1
Example Question from the Alternative Futures for the Upper San Pedro River Basin Scenario Guide

Distribution of New Residents

The location of the new homes in the Upper San Pedro River Basin will impact the future of the region in terms of the siting of new schools and the construction or improvement of roads. For the purpose of this study, residential areas are classified into four categories: Urban, Suburban, Rural, and Ex-Urban.

Urban homes are within an incorporated municipal boundary and are on lots smaller than 1 acre. This category includes multi-family homes. Urban homes are connected to the municipal sewer system and use the municipal water supply.

Suburban homes are built on 1–4 acre lots and are within 2 miles of an incorporated municipal area. Suburban homes use septic tanks and are connected to the municipal water supply.

Rural homes are built on lots of 4 acres or more. They are built on generally open areas (such as ranching or agricultural areas). They have private, on-site wells, and use septic tanks.

Ex-Urban homes are also built in generally open areas (such as ranching or agricultural areas) but are built on lots of 1–2 acres. Typically Ex-Urban homes are clustered in groups of five or more. Ex-Urban homes typically use septic tanks and use a common water supply (usually a private water company). Ex-Urban homes are usually accessed by dirt roads or low-volume paved roads.

In the alternative future you envision . . .

___% of the new population should live in Rural homes
___% of the new population should live in Suburban homes
___% of the new population should live in Urban homes
___% of the new population should live in Ex-Urban homes
====
100% Total new population

in Arizona to currently urbanized areas. The *Constrained* scenario assumed a lower than forecast population increase in Arizona and concentrated new growth to existing urban areas. *Constrained-1* also assumed the reduced population increase within Arizona, but doubled the on-base population of Fort Huachuca. *Constrained-2* used the lower population forecasts and closed Fort Huachuca. The *Open* scenario assumed a fifty percent higher than forecast population in Arizona and fewer controls on development than are presently used. *Open-1* closed Fort Huachuca and maintained current controls on rural residential development. *Open-2* doubled the current population of Fort Huachuca and the population of towns in Sonora; Mexican mining operations were also expanded.

Comparison of the Three Alternative Futures Studies

Considered as a set, the three alternative futures projects share a focus on finding broad solutions to counter pressures on biodiversity, but display a range of methodological differences. Referring back to the discussion in the first chapter of *normative* and *exploratory* scenarios—that is, futures that are preferred vs. futures that could happen—each of these studies investigates futures that are generally normative. This normative orientation is most readily apparent in the Camp Pendleton and Mojave studies which both explicitly state their objectives as finding ways to accommodate development and conserve biodiversity. The scenarios for the San Pedro Basin are also generally normative, albeit in a complex way. The Scenario Guide that served as the basis for generating the alternative futures stated the process of collecting opinions on the future was not "a vote" and the researchers were looking for distinct patterns of response, not necessarily popular ones. Further, no measure of the strength of preference—either the force of individual conviction or the sum of collective opinion—was considered in the final array of alternative futures. Yet, while the set of scenarios for this study were not solely designed to examine policy options toward the predetermined goal of protecting biodiversity, the concerted aims and objectives of the respondents were embedded in each of the questions, and

therefore, in each of the aggregate alternative visions of the future. As exemplified by the Scenario Guide question quoted above, issues were discussed in terms of what *should* happen in the region, not what theoretically could happen or what was perceived as most likely to happen. As a result, the final set of scenarios can be understood as representative of a range of regional preferences, some of which place a greater value on biodiversity, some of which place a lesser value on biodiversity.

Within the general classification of normative-based studies, several similarities and differences among the three projects can be distinguished. First, there is the implicit assumption in all of the studies that the stakeholders of the respective regions can, to a very large degree, control their own fates—although none of the studies makes a claim about the political resolve to do so. Only two factors were presented as beyond local control. First, each study treats population increase as an exogenous variable. Second, in the San Pedro study two of the futures include the closure of Fort Huachuca. Should such an event transpire, the decision would be made by Congress and the President, not the local residents. It should be noted that, appropriately for this specific issue, the Scenario Guide phrased the option as "Fort Huachuca closes," not "Fort Huachuca *should* close." In terms already discussed (again see Chapter 1), the assumption that the local residents can largely shape their region's future, these studies project what Russell Ackoff would call a *proactive* attitude toward change.[26]

A related characterization, also discussed in the previous chapter, concerns the degree to which stakeholder influence can be successfully exerted in the larger social environment. In each of these three studies the multiple stakeholders have considerable ability to influence their respective regions. If their efforts were combined, it is assumed that they would be able to significantly effect desired changes. Given these characteristics, all of the projects reflect what could be called largely self-contained or internal environments (following F.E. Emery and E.L. Trist).[27] Within this categorization, a more precise distinction can be seen in the kinds of agency and stakeholder interactions that are prescribed in each of the studies. The Camp Pendleton studies assume the possibility that a very large number of stakeholders—acting through three levels of government—can successfully work together. Moreover, it suggests that because of the threat posed by the *Sprawl* alternatives, that they do so sooner, rather than later. Similarly, the San Pedro study also calls for a multitude of very specific actions taken at all levels of regional participation, from the individual private citizen to the collective federal government. More modestly, the Mojave study investigates possible regional growth, but limits the areas considered for conservation to those on or near federal lands. While the establishment of buffer zones around installations might be achieved through changes in land use planning policy at the state level and/or local level, most of the actions could be accommodated by federal agencies working in cooperation with only a relatively small number of private citizens.

The three projects are also marked by differences of how the scenarios were constructed, and these differences, in turn, may have implications on the results. The approach used in the Camp Pendleton study was one of *deductive inference*. That is, the alternative futures were, by and large, developed first as fully rendered designs or visions; the scenarios, which describe how they could be realized, were then identified, or "reverse engineered," from the point-of-view of the future back to the present. The Mojave and San Pedro studies used processes based on *inductive inference*.[28] That is, each alternative future resulted from scenarios that charted a course of action that started in the present and moved forward toward the future. It has been argued that deductive and inductive approaches to scenario construction each have their benefits and limitations.[29] Deductive approaches tend to result in scenarios which, taken as a set, are more diverse and often surprising, but also risk the possibility that they are too far removed from reality to be of substantial use in a decision-making process. Inductive approaches tend to result in scenarios which, again taken as a set, are more similar to each other, but risk being too conservatively grounded in the politics, economics, and social conditions of the present to provide any conceptual breakthroughs. Inductively generated scenarios are also likely to emphasize futures with dramatic short-term effects and discount those futures with slowly developing, but incrementally significant effects. Although judging the degree of surprise offered by a scenario

is inherently subjective, the Camp Pendleton scenarios do seem to offer a relatively wider array of possibilities. For example, given the longstanding trend of low-density suburban development in that part of the country, how many residents or planning professionals would envision a new densely populated city? Also, despite the *trillions* of possible unique responses to the Scenario Guide used for the San Pedro study, the pattern of submitted replies and the consequent scenarios was quite narrow.

A second methodological issue associated with these studies is the logic used within the scenarios. Implicit in both the Mojave and San Pedro studies is the notion that while the land available for development in the future might change, the logic underlying the processes of development will not. This assumption is embedded in the use of single predictive models to place houses within the alternative patterns of land available for construction. A question that begs asking is could changing the location of available land also change the structural relationships of development? That is, in these new conditions, do the same assumptions about the preferences for building in a given location hold? For example, the *Exchange-2* set of scenarios in the Mojave study takes land near the installations out of development consideration and replaces the lost acreage with land owned by other federal agencies. Among the factors apparently not considered in the statistically derived development preference model for the Mojave study was proximity to specific kinds of military development (for example, proximity to military flight lines vs. ground maneuver area vs. cantonment area, etc.). Speculatively, the expected noise generated from these facilities might serve to slow growth in these areas. Trading these lands for those that abut a national park and may be characterized by relative quiet (since they are not sites of military maneuvers) could change development priorities and, perhaps, spark a development boom that brings more people and different densities of development. Similarly, in the San Pedro study, the *Constrained-2* and *Open-1* scenarios close Fort Huachuca and make parts of the post available for development. Given that much of the installation is relatively flat and has highly valued views of the nearby mountains, this land would seem prime for immediate development. Adding additional complexity to this possibility is Fort Huachuca's role as the largest employer in the county. Would the loss of jobs resulting from its closure lower expectations about the number of people expected to move to the area, or, at least, sufficiently influence the demographic profiles of the new residents that the preference for housing types would change? On the one hand, developing scenarios with the assumption that the structural components of the region will remain largely unaffected by changes in the constituent parts of the local social, economic, and political norms can be considered pragmatic and may even be a prerequisite for scenarios that assume that the stakeholders can affect the future. However, on the other hand, this step also places the scenarios in a very conservative framework and may render them blind to significant consequences of actions taken in the near and distant futures. By comparison to the other projects, the scenarios for the first Camp Pendleton study lack detail and seem underdeveloped. Given the constraints of time, the students focused their efforts on the details of the alternative futures rather than on the means to achieve them. That is, they focused on the ends of the future, rather than the means of the future. This situation makes it difficult, if not impossible, to identify all of the assumptions that underlie development in each alternative and so analyzing the scenario logic to look for strengths and weaknesses is not a particularly fruitful endeavor.

Beyond matters of method, the three projects are also marked by the differences of context in which they were undertaken. Perhaps the most evident transformation is the degree of interaction between the researchers and the installations in crafting the scenarios. As noted, the regional scenarios for the study of Camp Pendleton were created by design students at Harvard. Similarly, scenarios for the Mojave study were created at Utah State, led by researchers in the Department of Landscape Architecture and Environmental Planning. In most respects, these situations are not remarkable. Many—if not most—sponsored research projects utilize specialized competence from outside the funding agency to provide critical assessments and expert judgments. The scenarios for the study of Fort Huachuca and the San Pedro Basin however, took an important step toward involving base personnel and area residents to a greater degree through the use of the Scenario Guide. One potential

benefit of this process was that by including a greater degree of local input in the creation of the scenarios, the alternative futures better reflect not only an array of abstractly preferred land use values for the region, but the actual preferred strategies of the local stakeholders. Describing options in the same kind of language and comparing each of them through a set of assessments can be a way to select the best, or at least most politically viable, alternative. Although the San Pedro study was not designed to be an exercise in conflict resolution, the methods used for the research could allow its product to contribute to such a use. While this possibility is notable, it should be kept in mind that while the opinions of the local stakeholders provided the specifics of the scenarios which were mapped and compared, the selection of the issues and the phrasing of the policy options were limited to a specific set of issues. More significantly, the selected issues were chosen by the researchers. Thus, the degree of input was considerable, but constrained.

A second difference among the studies, which is related to the use of scenario-based studies, concerns organizational positioning of the work. The first study on the region of Camp Pendleton and the Mojave study were both funded and administered by SERDP, which, while funded by the Department of Defense, is outside the command structure of the installations. The San Pedro study was funded by Legacy, another operation outside the installation command structure, but was initiated by TRADOC, which is the command responsible for the Post. Without putting too fine a point on the issue, the initiation of the San Pedro study by its branch command could be seen as suggesting that scenarios are becoming better integrated into the institutional thinking of Department of Defense land and natural resources management. As noted in Pierre Wack's work with scenarios for Royal/Dutch Shell Oil, developing scenarios is an empty exercise if the work is not integrated into the decision-making process at all levels of administration.[30] Although only time will tell, the perception within the command structure that scenario-based land use studies can contribute to supporting the military mission and the inclusion of base personnel may prove to be significant in ensuring that the scenarios address the perceived needs of line managers and may hence be better used in their deliberations.

Forecasting and Mitigating Future Urban Encroachment Adjacent to California Military Installations

Although not an alternative futures project *per se*—since it only considers one future—the study *Forecasting and Mitigating Future Urban Encroachment Adjacent to California Military Installations: A Spatial Approach*,[31] provides what is perhaps the most explicit consideration of development-driven pressures on military lands that has been undertaken outside federal government sponsorship. The work was conducted by researchers at the Berkeley Center for Environmental Design Research and was commissioned by the California Technology, Trade, and Commerce Agency (CTTCA) as part of its mandate from the state government to develop strategies for the retention or reuse of DoD facilities.[32]

Two levels of analysis were considered in the project. The first was a macro-scale investigation of potential urban growth around twenty-three of the state's twenty-six active military installations. County population forecasts for 2010 and 2020 were applied to a statistical probability model of urban growth that was calibrated on development data from 1972 to 1996. New development within a ten kilometer (slightly over six miles) buffer zone around the facilities was then projected. Additionally, the research identified the loss of potential critical habitat for species listed as Threatened or Endangered by U.S. Department of Interior[33] within five-, ten-, and twenty-kilometers of all twenty-six active bases. The possible future conditions were then compared to a baseline of 1996 conditions to assess the potential magnitude of lost habitat. The conclusion of this analysis was that while more people will live near a base in 2020 than in 1996, most of the projected growth will be directed away from the bases. As a result, the study concludes that the most significant encroachment-related impacts will likely be limited to only several bases: Marine Corps Air Station (MCAS) Miramar, Marine Corps Logistics Base Barstow, Travis Air Force Base (AFB), Navy Base Coronado, Marine Corps Base (MCB) Camp Pendleton, and Naval Weapons Station (NWS) Fallbrook.[34]

The second level of analysis in this study provided a more detailed examination of potential growth impacts around Camp Pendleton, Miramar, Travis, and Edwards, four of the larger facilities located in the state. Three sets of issues were considered: the likely extent of future urban growth adjacent to each installation; how and whether future urban growth in the general vicinity of each installation will affect its role as a critical habitat reserve; and whether and how projected urban growth will increase the number of nearby residents impacted by noise from base operations."[35] This part of the CTTCA report is useful in that it attempts to specify unique encroachment concerns facing each facility. For example, it recognizes that MCB Camp Pendleton is the only large area of high-quality habitat on the Southern California coast, and that as the surrounding area develops, Pendleton's role as a habitat for threatened and endangered species will become increasingly more important. Another example of base-specific analysis concerned the orientation of the flight lines at Miramar, Edwards, and Travis and the prospect of increased development in areas that are subjected to high levels of noise during take-offs, landings, and low-altitude maneuvers. Edwards is unique in that it maintains the nation's only low-altitude supersonic flight corridor. Miramar's location in southern San Diego County makes it the most likely facility to be completely surrounded by development. (NB: Camp Pendleton also conducts flight operations; however, perhaps because its air station is located near the center of the base and far away from civilian development, its air operations were not—apparently—considered in the analysis of flight line noise.) It should be noted that the specified concerns were identified by the researchers and not by the installation staffs or others.

Beyond the spatial analysis developed by this research, arguably the most far-reaching element of the Berkeley report is the list of legislative and regulatory options that could allow military mission needs to be better considered in local land use planning activities.[36] The authors note that with the exception of a very small number of special districts, land use decisions within the state are made at the county or municipal levels of government. Local land use control is not unique to California and is, indeed, the norm throughout the nation. Possible options cited to better consider the use of military lands include state review and/

or appeal of local planning decisions, the creation of multi-jurisdictional planning boards, and new state permitting powers relating to development near installations. While none of these options explicitly call for the inclusion of *training land* as an officially recognized category of land use, as suggested in testimony before the House of Representatives,[37] each provides a means to consider military mission requirements within the contexts of local land use needs and broader environmental goals. Given that not all states enable or administer land use planning in the same way as California, the options listed in the report might not be practical (or even legal) across the country, and so general statements on the ability to transfer these ideas to other jurisdictions cannot easily be made. Regardless, as a set the options span a wide range of alternative approaches which could minimize or mitigate development-driven pressures on military lands.

The authors of the Berkeley report make two significant qualifications to the scope of their research. The first is that the future that is considered for their analysis is based on statistically identified trends.[38] If the trends do not hold—that is, if the future is different than the recent past—then the pressures of development on the military training activities and on the environment may also be different, perhaps less severe or perhaps far more acute. The second is that in order to better understand the environmental impacts associated with urban growth, an area that encompasses "the entire ecological region" (or ecoregion[39]) would need to be considered and issues such as habitat fragmentation should be included.[40]

Conclusions

In conclusion, two important limitations to these alternative futures projects need to be discussed. First, there is a question of just how much the local stakeholders are prepared to work together to shape the future of their regions. By presenting normative visions of the future, each study can be seen as a contribution toward improving the circumstances in which people live. As noted by Robert Costanza, such visions are not only laudable pursuits, but are necessary first steps in moving toward sustainable living conditions.[41] But toward meeting needs in

the not-distant-future, a question must be asked about the prospects for implementing any of these futures on a timely basis. The regions considered in these studies are large and diverse, and the successful pursuit of a single, shared vision is daunting. As those who have attended meetings of local governments or even watched them on local cable television can attest, a petition for a single residential zoning variance can be a complicated and protracted matter. Regional planning efforts are, of course, much, much more problematic in that they require not only consensus among neighbors, but also among municipal, county, state, and federal government agencies on both the desired ends and the means to achieve them. It must be asked, what if mutual agreements cannot be identified and acted upon? Raising this point may seem pessimistic, but should be considered. In such situations, adopting what Ackoff calls a preactive approach could be beneficial. As discussed in Chapter 1, with such an approach individual stakeholders attempt to understand the range of possibilities that are beyond their control in order to better pre-position themselves to take advantage of opportunities and avoid hazards.

Second, there is a question of the degree to which the effects of larger social-political-economic environments will impact the region being studied. As noted, population is considered an exogenous factor in all of these studies, but the success of any of these plans is also dependent on issues of culture, economics, politics, and technology which are determined—or at least worked-out—at state, national, and international levels of interaction. Just as the region will change, so too will the nation and local plans made on assumptions of widespread stability, and predictability may prove to be misguided. Given the likelihood of change, these broader considerations deserve greater attention. In order to consider the possible influence of outside effects and the possible effects of non- or limited-cooperation, a more descriptive scenario process is (or is also) needed. Such a process of investigation would not only make different kinds of scenarios, but would also use the scenarios in a different way.

Another way to consider these differences is in terms of Steinitz's six-part framework for landscape planning (also discussed in Chapter 1). In normative studies, the scenarios are used to define landscape change, which is the third step of the process. The changes are regional in scope and scale. They are then compared *against each other* in the sixth framework step in which a decision is rendered. That is, at the decision step, one asks: Which scenario should be created, which alternative future should be developed? In descriptive studies, the question of change is not regional, but local. The scenarios are used in the sixth step as a means to test the potential success of the local change. That is, at the decision step one asks: Does the change make sense under the conditions of each scenario? Such a process will be described in a following chapter on the development of scenarios for the region of MCB Camp Pendleton and MCAS Miramar.

NOTES:

[1] Carl Steinitz, Michael Binford, Paul Cote, Thomas Edwards, Jr., Stephen Ervin, Craig Johnson, Ross Kiester, David Mouat, Douglas Olson, Allan Shearer, Richard Toth, Robin Wills, *Biodiversity and Landscape Planning: Alternative Futures for the Region of Camp Pendleton, California* (Cambridge, Massachusetts: Harvard Graduate School of Design, 1996).

[2] Mary E. Cablk, Jamie DeNormandie, Thomas C. Edwards, Jr., Robert Fisher, Manuel Gonzalez, Jill S. Heaton, Lori Hunter, Kimberly Karish, A. Ross Kiester, Robert Lilieholm, S. Mark Meyers, David Mouat, Natalie Robbins, Matt Stevenson, Richard Toth, *Analysis and Assessment of Military and Non-Military Impacts on Biodiversity in the California Mojave Desert* (Reno, Nevada: Desert Research Institute, 2002).

[3] Carl Steinitz, Hector Arias, Scott Bassett, Michael Flaxman, Tomas Goode, Thomas Maddock, III, David Mouat, Richard Peiser, Allan Shearer, *Alternative Futures for Changing Landscapes: The Upper San Pedro River Basin in Arizona and Sonora, Mexico* (Washington, D.C.: Island Press, January 2003).

[4] Carl Steinitz, "A Framework for the Theory Applicable to the Education of Landscape Architects (and Other Environmental Design Professionals)," *Landscape Journal* 9:2 (Fall 1990), pp. 136–143.

[5] David Takacs, *The Idea of Biodiversity: Philosophies of Paradise* (Baltimore, MD: The Johns Hopkins University Press, 1996), pp. 41–99.

[6] Biodiversity Research Consortium, *Objectives* (May 21, 1997). Available on-line at <http://bufo. geo.orst.edu/brc/objectives.html>.

[7] For a description of the principles of landscape ecology, see Richard T.T. Forman and Michel Godron, *Landscape Ecology* (New York: John Wiley & Sons, 1986); Richard T.T. Forman, *Land Mosaics: The Ecology of Landscapes and Regions* (Cambridge, England: Cambridge University Press, 1995).

[8] Definitions of potential habitat were based on existing Wildlife Habitat Relations models. See, Kenneth E. Mayer and William F. Laudenslayer, Jr., *A Guide to Wildlife Habitats of California* (Sacramento, California: Department of Forestry and Fire Protection, 1988).

[9] Based on unbuilt, but planned private land with: 3 people per household; Rural Residential density of 1 house per 4 acres; Single Family density of 1 house per 1/4 acre; Multi-Family density of 2 houses per 1/8 acre. Also note that the Steinitz, et al. (1996) document contains a typographic error in the reported map statistics: spatial units for each map are reported as hectares; however, the figures actually provide the number of 30 meter x 30 meter cells—the resolution of LANDSAT data. To convert cells to hectares, the reported number should be divided by 11.1.

[10] Carl Steinitz, Chad Adams, Lauren Alexander, James De Normandie, Ruth Durant, Lois Eberhart, John Felkner, Kathleen Hickey, Andrew Mellinger, Risa Narita, Timothy Slattery, Clotilde Viellard, Yu-Feng Wang, E. Mitchell Wright, *An Alternative Future for the Region of Camp Pendleton, California* (Cambridge, Massachusetts: Harvard Graduate School of Design, 1997).

[11] Anne W. Chapman, *The National Training Center Matures: 1985 – 1993* (Fort Monroe, Virginia: United States Army Training and Doctrine Command, 1997), pp. 65–126.

[12] Matthew R. Stevenson, Richard E. Toth, Thomas C. Edwards, Jr., Lori Hunter, Robert J. Lilieholm, Kimberly S. Karish, James DeNormandie, Manuel Gonzalez, and Mary Cablk, "What If . . . ? Alternative Futures for the California Mojave Desert," (2002). Available on-line at <http://www01.giscafe.com/TechPapers/Papers/paper036/p192.htm>.

[13] For additional description of logit models, see J.S. Cramer, *The Logit Model: An Introduction for Economists* (London: Edward Arnold, 1991).

[14] William G. Kepner, C.M Edmonds, and Christopher J. Watts, *Remote Sensing and Geographic Information Systems for Decision Analysis in Public Resource Administration: A Case Study of 25 Years of Landscape Change in a Southwestern Watershed*, EPA Report # EPA/600/R-02/039 (Las Vegas, Nevada: U.S. Environmental Protection Agency, Office of Research and Development, National Exposure Research Laboratory, June 2002).

[15] Ronnie Sidner, "A Bat Boom at Fort Huachuca," *Endangered Species Bulletin* 25:6 (November – December 2000), pp. 12–13. Available

on-line at <http://endangered.fws.gov/esb/2000/11–12–13.pdf>.

[16] An Original Bill to Provide for the Designation and Conservation of Certain Lands in the States of Arizona and Idaho, and For Other Purposes. Public Law 100–696 (November 18, 1988), 16 United States Code § 460xx(a) (1994). Also known as Arizona–Idaho Conservation Act of 1988.

[17] For an overview of the physical characteristics of the basin and the development pressures, see U.S. Department of the Interior, Bureau of Land Management, *The Upper San Pedro River Basin of the United States and Mexico: A Resource Directory and an Overview of Natural Resource Issues Confronting Decision–Makers and Natural Resources Managers*, BLM Report # BLM/AZ/PT–98/021 (Phoenix, Arizona: Bureau of Land Management, Arizona State Office, May 1998).

[18] "The Nature Conservancy's Protection Initiative for the 1990s," *Nature Conservancy* 41:28 (1991), pp. 28–29.

[19] American Rivers, Press Release, "San Pedro River Named One of the Nation's Most Endangered Rivers" (Washington, D.C.: April 12, 1999). Available on-line at <http://www.amrivers.org/pressrelease/pressmersanpeddro1999.htm>.

[20] Robert G. Varaday, Margaret Ann Moote, and Robert Merideth, "Water Management Options for the Upper San Pedro Basin: Assessing the Social and Institutional Landscape," *Natural Resources Journal* 40:2 (Spring 2000), pp. 223–235, this note pp. 225–226; Diana M. Liverman, Robert G. Varaday, Octavio Chávez, and Roberto Sánchez, "Environmental Issues Along the United States–Mexico Border: Drivers of Change and Responses of Citizens and Institutions," *Annual Review of Energy and the Environment* 24 (1999), pp. 607–643, this note p. 637, available on-line at <http://las.arizona.edu/liverman/acree.pdf>.

[21] The National Environmental Policy Act of 1969, as amended, Public Law 91–190, 42 United States Code 4321–4327 (January 1, 1970), subsequently amended.

[22] Commission for Environmental Cooperation, *Ribbon of Life: An Agenda for Preserving Transboundary Migratory Bird Habitat on the Upper San Pedro River* (Montreal, Canada: Communications and Public Outreach Department of the CEC Secretariat, 1999). Available on-line at <http://www.cec.org/files/PDF/sp–engl_EN.pdf>.

[23] See Bill Hess, "Environmental Drive–By Study," *Sierra Vista Herald/Bisbee Daily Review* (June 15, 1997), p. 1A; Bill Hess, "Nazism, Communism, Now Environmentalism," *Sierra Vista Herald/Bisbee Daily Review* (June 24, 1997), p. 1A

[24] Conrad J. Bahre, "Human Impacts on the Grasslands of Southeastern Arizona" in Mitchel P. McClaren and Thomas R. Van Devender, eds., *The Desert Grassland* (Tucson, Arizona: University of Arizona Press, 1995), pp. 230–254; Guy R. McPherson, "The Role of Fire in Desert Grasslands" in McClaren and Van Devender, pp. 130–151.

[25] Steinitz, et al., (2002), Appendix.

[26] Russell Ackoff, *Creating the Corporate Future: Plan or Be Planned For* (New York: John Wiley & Sons, 1981), pp. 52–65.

[27] F.E. Emery and E.L. Trist, "The Causal Texture of Organizational Environments," *Human Relations* 18:1 (February 1965), pp. 21–32; F.E. Emery and E.L. Trist, *Towards a Social Ecology: Contextual Appreciations of the Future in the Present* (New York: Plenum Press, 1973), pp. 43f. The relationship of transactional and contextual environments in regards to scenarios is made in Kees van der Heijden, *Scenarios: The Art of Strategic Conversation* (New York: John Wiley & Sons, 1996), p. 6.

[28] C. Ducot and G.J. Lubben, "A Typology for Scenarios," *Futures* 12:1 (February 1980), pp. 51–57.

[29] Helmut Jungermann, "Inferential Processes in the Construction of Scenarios," *Journal of Forecasting* 4:4 (1985), pp. 321–327.

[30] Pierre Wack, "Scenarios, Shooting the Rapids," *Harvard Business Review* 63:6 (November–December, 1985), pp. 131–142.

[31] John Landis, Michael Reilly, Robert Twiss, Howard Foster, and Patricia Frontiera, *Forecasting and Mitigating Future Urban Encroachment Adjacent to California Military Installations: A Spatial Approach* (Sacramento, California: California Technology, Trade, and Commerce Agency, June 20, 2001). Available online at <http://www.regis.berkeley.edu/cttca/finaldocs/cttca_report062101.pdf>.

[32] California Defense Retention and Conversion Act of 1999. California Statute § 425 (September 16, 1999). Also known as the Knight Bill.

[33] For its assessment of potential habitat, the study matched vegetation data from the California Gap Analysis Program with species habitat needs data from the California Fish and Game's Wildlife Habitat Relations model. Habitat quality was also scored for each species on a scale of 1 (lowest quality) to 5 (highest quality). An index for each spatial analysis cell was then developed that summed the potential species present by the respective quality of habitat. Scores for each cell ranged from 0 to 66. Landis, et al., pp. 8–9.

[34] Landis, et al., pp. v–vi.

[35] Landis, et al., p. 37.

[36] Landis, et al., pp. 51–61.

[37] Edward Hanlon, Jr., testimony in U.S. House of Representatives, Committee on Government Reform, *Hearing on Challenges to National Security: Constraints on Military Training*, 107th Congress, 1st session (Washington, D.C.: U.S. Government Printing Office, May 9, 2001), p. 124. Available on-line at <http://frwebgate.access. gpo.gov/cig–bin/getdoc.cgi?dbname=107_house_ hearings&docid=f:75041.pdf>.

[38] Landis, et al., p. 35.

[39] For competing definitions of ecoregions of the United States, see J.M. Omernik, "Map Supplement: Ecoregions of the Coterminous United States," *Annals of the Association of American Geographers* 77 (1987), pp. 118 – 125; Robert G. Bailey, *Description of the Ecoregions of the United States*, U.S. Forest Service Miscellaneous Publication No. 1391 (Ogden, Utah: U.S. Department of Agriculture, 1995).

[40] Landis, et al., p. 41.

[41] Robert Costanza, "Visions of Alternative (Unpredictable) Futures and Their Use in Policy Analysis," *Conservation Ecology* (On-line Edition) 4:1, Article No. 5; also in Robert Costanza, "Four Visions of the Century Ahead: Will It be Star Trek, Ecotopia, Big Government, or Mad Max?" *The Futurist* 33:2 (February 1999), pp. 23–28, this note p. 23.

Chapter 4

The Regional Context of MCB Camp Pendleton and MCAS Miramar

David A. Mouat

The study area for this project is the context region for Marine Corps Base (MCB) Camp Pendleton and Marine Corps Air Station (MCAS) Miramar. Naval Weapons Station (NWS) Fallbrook is also located within the boundaries of investigation. The region comprises parts of southern Orange, southwestern Riverside, and western San Diego Counties (approximately 75% of the study area lies in San Diego County).

There are few places in the world where, on a typical March day, a person can have a choice of surfing, skiing, or enjoying wildflowers.[1] While the region's rugged terrain and associated seismic activity are dramatic, these features do not give the region its specific character. The landscape's response to strongly seasonal climatic and hydrologic forces, reflected in its droughts, fires, floods, and erosion make the region distinctly Mediterranean.[2] These forces combined with its unique physiography, provide conditions that support a spectacular diversity of native flora and fauna and also attract residents and visitors to the area.

Physiography and Geology

The study area is characterized by a narrow coastal strip abutting against coastal marine terraces. These surfaces are flat top mesas, mesa slopes, dissected canyons, arroyos, or ravines. Occasionally, the canyon bottoms are wide enough to have floodplains and terraces. Four perennial streams and rivers cross the study area and are characterized by having those features. They include the Santa Margarita River which drains much of the study area within Riverside County and flows through MCB

Camp Pendleton, the San Luis Rey River which drains most of Palomar Mountain and northern San Diego County, the San Dieguito River which drains much of west central San Diego County, and the San Diego River which forms the southern boundary of the study area. In addition, some ephemeral streams have significant floodplains and terraces. These include the San Juan Creek in southern Orange County, the San Mateo Creek that flows through MCB Camp Pendleton, and the San Marcos Creek that flows through the city of Escondido. Farther inland, typically east of I-15, the topography is hilly to mountainous with occasional alluvial plains, such as at Lake Henshaw, or plateaus, such as the Santa Rosa.

The study area is divisible into two large physiographic provinces on the basis of surface geology and relief (Map 4.1). The *Peninsular Range Province* is east of the western boundary of plutonic and metavolcanic surface rocks of Mesozoic age.[3] This province contains the higher land in the region. Within San Diego County, Palomar Mountain, the San Ysidro, Volcan, Cuyamaca, and Laguna Mountains dominate the landscape with several peaks over 6,000 feet (Map 4.2). In Riverside County, the San Jacinto Mountains dominate the northeast landscape of the study area. In the northwest part of the study area, straddling parts of Orange, Riverside, and San Diego Counties, the Santa Ana Mountains dominate. The Peninsular Ranges are also characterized by large westward tilting fault blocks with typically steeper slopes on the eastern sides (Map 4.3). Palomar Mountain is typical of this type of structure. One of the most spectacular tilting faults is the San Jacinto Mountains, just north and east of the study area, with a relief in excess of 10,000 feet. The eastern flank of

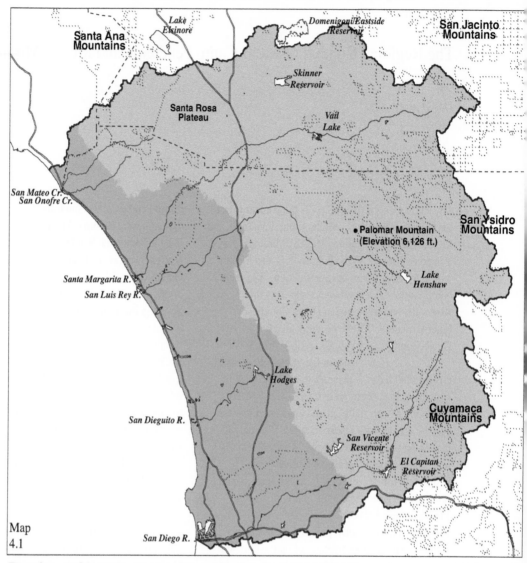

Map
4.1

Region of MCB Camp Pendleton & MCAS Miramar
Existing Conditions 2000
Physiography of the Study Region

 Peninsular Range Province
1220827 acres (70.4%)

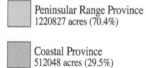 Counties

Coastal Province
512048 acres (29.5%)

 Interstates

 Rivers

Federal Lands

the Santa Ana Mountains is bounded by the Elsinore Fault, which has a strong topographic influence as evidenced by the steep scarp west of Lake Elsinore.

The generally lower lying *Coastal Province* is largely comprised of sedimentary surface rocks of Tertiary age. The dominant relief of the Coastal Province is a series of marine terraces, locally referred to as *mesas*. These mesas were formed or abraded by coastal marine action associated with sea level fluctuation. All of the various levels of marine terraces are dissected by canyon systems, and the degree of dissection increases with age of the terrace.

Like the rest of Southern California, San Diego County has a number of active earthquake faults. These faults generally run in a northwest–southeast direction and are the product of crustal stresses associated with movement of the Pacific and North American lithospheric plates. In modern times the strongest recorded quake in coastal San Diego County was the M5.3 temblor that occurred on 13 July 1986 on the Coronado Bank Fault, 25 miles offshore of Solana Beach, north of Del Mar.[4] Historic documents record that a very strong earthquake struck San Diego on 27 May 1862, which damaged buildings in Old Town and opened cracks in the earth near the mouth of the San Diego River. This destructive temblor was centered on either the Rose Canyon or Coronado Bank faults and descriptions of damage suggest that it had a magnitude of about 6.0 on the Richter scale.[5]

The soils of the region are a reflection of the geology, slope, climate, and vegetation. The soils near the coast developed within the sandy marine sediments of the Coastal Province and tend to be loose and sandy, an ideal medium for agriculture and landscaping. The soils developed within the Peninsular Ranges tend to be thin and well-drained. In both provinces, there may be areas of instability and landslides may occur. Landslides rarely result from seismic activity, but rather are due to the intrinsic low shear strength of the soils. When these soils are saturated with water, gravitational stress can greatly exceed the shear strength and result in slope failure. This process becomes exacerbated by construction on steep slopes that undercuts the soils' stability and results in increased stress. Removal of vegetation by fire may also result in the soils becoming saturated.

Climate

The climate of the study area is considered to be largely Mediterranean with desert influences on the lee sides of mountain ranges away from the coast. The area is characterized by warm, dry summers and cool, wet winters. Variability within the region reflects latitude, elevation, aspect, and proximity to the ocean. In general, precipitation increases from south to north and from the ocean to the mountains. The coast is affected by fog generated by advection across the cool California current.[6] The climate is influenced by the proximity of the subtropical high pressure of the North Pacific that is located in the north during summer and in the south during winter.[7] High pressure systems are characterized by dry subsiding air that results in fair weather systems. When the high pressure shifts eastward, as it occasionally does during fall and winter, the dry subsiding winds come from the east and experience additional heating. These east winds are known locally as *Santa Ana* winds and are responsible for very hot and dry conditions with attendant drying of vegetation. *Santa Ana* conditions occasionally result in catastrophic fires.[8]

The effect of the cold ocean current, in addition to producing the ubiquitous summer fog, is to cool the coast. Daytime high temperatures along the coast during summer months are typically in the 70 degree (Fahrenheit) range, while just a few miles inland and away from the coastal influence, temperatures may exceed 100 degrees. Given these characteristics, the climate of the coastal strip, including the cities of San Diego and Oceanside, is classified according to the Köppen classification as "semiarid steppe" and not Mediterranean as the cool ocean current keeps the coastal zone quite arid.[9]

Vegetation

The vegetation of the study area is of interest within the scope of this project for three reasons: 1) it is an important component of the landscape which is so attractive to the residents of the region, 2) it is readily consumed by fire posing a threat to both biodiversity and to human habitation, and 3) it is a critical part of the habitat. Vegetation in Southwest California is a function of the predominantly Mediterranean climate, geology,

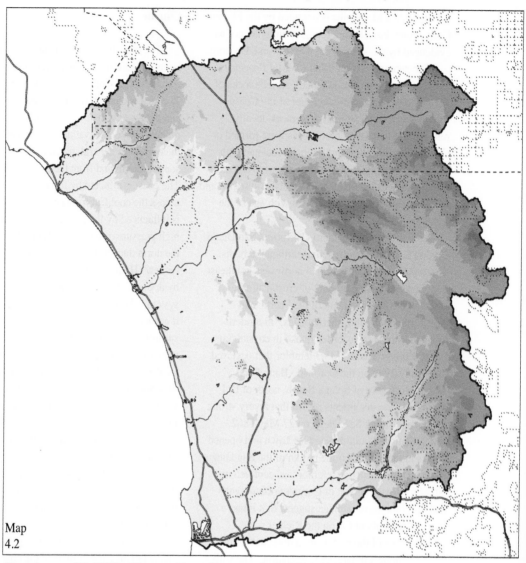

Map
4.2

Region of MCB Camp Pendleton & MCAS Miramar
Existing Conditions 2000
Elevation

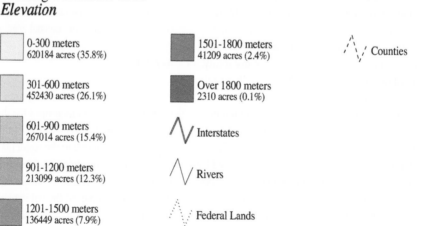

0-300 meters 620184 acres (35.8%)	1501-1800 meters 41209 acres (2.4%)	Counties
301-600 meters 452430 acres (26.1%)	Over 1800 meters 2310 acres (0.1%)	
601-900 meters 267014 acres (15.4%)	Interstates	
901-1200 meters 213099 acres (12.3%)	Rivers	
1201-1500 meters 136449 acres (7.9%)	Federal Lands	

soils, rugged terrain, fire, and human activities.

As stated previously, the Mediterranean climate is characterized by warm dry summers and cool moist winters. Much of the coast of California and Baja California is influenced by this climate. San Diego receives on average about 11 inches of precipitation annually, although it was closer to 3 inches in the 2001–2002 season. Over 90% of the annual precipitation falls during the winter half of the year. The elevation of the study area slopes gradually upward from the coast inland, with moderately high mountain systems located only 10 to 30 miles from the coast. Precipitation increases dramatically with elevation (to over 40 inches on Palomar Mountain).[10] Temperature decreases as elevation increases. On the lee slopes of the mountains, a rain shadow effect lowers precipitation to less than 4 inches. A result of this highly varied climate is a concomitantly varied mix of vegetation.

The vegetation, thus, is not only affected by a strong summer drought but also by low annual precipitation. Precipitation variability also affects regional vegetation. Thus, the key to understanding the vegetation of the study area is to understand the mechanisms for how plants adapt to a combination of low annual precipitation, high seasonality in which virtually no precipitation falls during the six hottest months of the year, and high variability as witnessed by the investigators during the 2001–2002 period, when the annual precipitation fell to less than 30% of normal.

Adaptive strategies of plants to a climate with summer drought include: the evolution of sclerophyllous leaves (which have a low surface to volume ratio and waxlike surfaces which reduce transpiration), the ability of leaves to open and close stomata (thereby reducing transpiration), heat evading foliage (which might include leaves which are oriented vertically to reduce reception of sunlight), summer deciduousness and senescence, subsurface food and water storage (which might include the presence of a bulb, long tap roots for obtaining deep seated soil moisture, shallow rootlets for rapidly extracting brief ephemeral rains), and germination response to wildfires.[11]

Sclerophylly in vegetation is common in many of the plant communities of the study area including mixed evergreen forest, live oak woodland, chaparral, and others, as a mechanism for adapting to the long summer drought. Both annual and perennial grasslands are adapted to the long summer drought. Most of the vegetation of the region, then, is either sclerophyllous, graminoid, or of another physiognomy (such as annual or perennial forbs) which senesce or otherwise reduce photosynthesis during the long summer drought.

Mapping Vegetation

There are many ways in which vegetation can be categorized and mapped. Phytosociologists (those who study plant associations) note mosaics of vegetation that seem to repeat themselves.[12] Vegetation itself consists of various more or less mappable units, possibly arranged in definite patterns across the landscape.[13] For the purposes of this study, vegetation patterns or types that can be useful for habitat identification and discrimination (assuming that habitat is, in part, a function of vegetation as well as other factors) have been mapped. The intent of the maps is to communicate a highly complex set of information about vegetation in a simplified spatially referenced form and to provide spatially referenced numerical data about vegetation that can be used for analytical purposes.[14]

Küchler asserts that classification is the first step in the mapping process and that mapping expresses the classification spatially or cartographically.[15] Sawyer and Keeler-Wolf are in agreement with earlier phytosociologists in stating that a vegetation map is a symbolic representation of visually distinct groupings of plants.[16] There are two basic ways to initiate a mapping process. First, following an *a priori* system of organization, the earth as a whole can be subdivided into smaller and smaller units; second, with freedom from preconceived notions, what is *seen* on the ground can be noted and sorted, *a posteriori*. Cartographic considerations are ever present and include the limitation of mapping scale and the precision and accuracy of classification schemes. For example, a classification can be more descriptive and detailed and can involve more floristic and structural information than could be perceived on an aerial photo or depicted on a map.

Typically, one seeks to map one of two different kinds of vegetation. *Potential* vegetation maps show what would occur in the area if activities

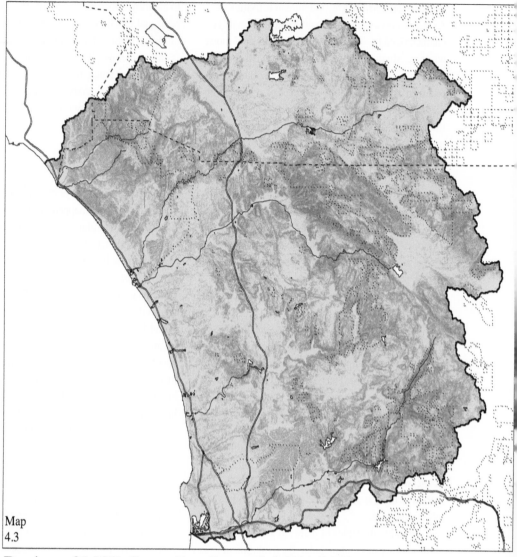

Map
4.3

Region of MCB Camp Pendleton & MCAS Miramar
Existing Conditions 2000
Slope

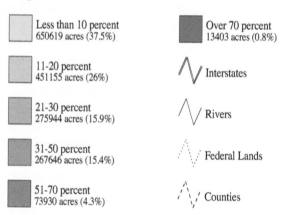

Less than 10 percent 650619 acres (37.5%)		Over 70 percent 13403 acres (0.8%)
11-20 percent 451155 acres (26%)		Interstates
21-30 percent 275944 acres (15.9%)		Rivers
31-50 percent 267646 acres (15.4%)		Federal Lands
51-70 percent 73930 acres (4.3%)		Counties

associated with European influence did not exist. *Actual* vegetation maps show what presently exists in the area. In Southern California, the tremendous amount of urbanization as well as vegetation change brought about by human influence—especially those associated with changes to the natural fire regime and the introduction of exotic species—results in a completely different actual pattern of vegetation existing today than would have occurred without European influence. Mapping potential vegetation in Southern California might presuppose that the landscape *could* or even *might* return to a condition it would be in without European influence. However, the presence of urbanization, the reality of exotic invasive species, and the influence that people exert, make it highly unlikely that the pre-European vegetation will return. In this project, *actual* vegetation is mapped. The reader should also note that *potential* habitat, which is analyzed as part of this study, is discriminated on the basis of the actual vegetation and other terrain factors that occur in the landscape.

Vegetation Classification

For this project, the primary purpose of the vegetation mapping exercise was to develop habitat relationships for selected species. Thus, vegetation was mapped at what might be considered at the level of major biotic communities.[17] Throughout the process, the researchers recognized vegetation types that, in a hierarchical sense, are referred to as "series."[18] A series is one of the distinct plant–animal communities found within a biome. In this hierarchical classification system, a series occurs within a major biotic community, which in turn is a subset of a formation that is characterized by a specific climate. Thus, an oak series containing an Engelmann oak association (a unit finer than a series) might be delineated within the California Evergreen Woodland biotic community, which, in turn is found within Warm Temperate Forests and Woodlands of Brown's classification.[19] The following table illustrates this hierarchical scheme (excerpted and adapted from Brown).[20] The categories printed in bold type are typical of the types mapped for this investigation:

123 Temperate Forests (**Mixed Forest**) and Woodlands
   123.4 Californian Evergreen Woodland
      123.41 Encinal Oak Series (**Oak Woodland**)
         123.414 *Quercus engelmannii*
            (Engelmann oak) Association
   123.42 Walnut Series

While in some instances, a specific association—such as the Engelmann oak woodland—could have been mapped, an inconsistency of mapping would have occurred. It was felt that the needs for habitat mapping at the scale of the project necessitated the mapping at the specified levels. As such, vegetation was not mapped hierarchically and types representing rather broad classes (such as Mixed Forest) were mapped alongside more specific classes (such as Coastal Sage Scrub). The type of names used in this study represent those previously described by Holland[21] and referenced in *The Manual of California Vegetation* within the California vegetation mapping program.[22] *The Manual of California Vegetation* was the result of a program developed by the California Native Plant Society and the California Department of Fish and Game to adopt a uniform vegetation classification system that would be agreed upon by a variety of users in order to facilitate communication. The goal was to protect rare, threatened, and endangered plant communities across jurisdictional boundaries. In order to determine the extent of these communities, the juxtaposition of those with all other communities in the state was considered essential.[23] While not strictly hierarchical, this classification describes the vegetation types at the "series" level. *The Manual of California Vegetation* may not be the final word on vegetation classifications of California, but it provides the most recent effort in a long history of vegetation classification efforts within the state and it certainly represents the views of many people and organizations. Yet, as the authors of that work state, a classification serves a particular purpose; one classification will not serve all purposes.[24] Thus, it is most likely that additional classifications will emerge to help address future needs.

Map
4.4

Region of MCB Camp Pendleton & MCAS Miramar
Satellite Image from November 18, 2000
Composite of Landsat ETM Bands 7, 4, and 2

— See color insert following page 204 —

Mapping Vegetation of the Study Area

Map 4.5

Region in 2000
Satellite Image from Nov. 18, 2000
Landsat ETM Band 4

The 1996 alternative futures-based study of the Camp Pendleton region utilized publicly available data from a variety of sources. A close inspection of that project's vegetation map will reveal seams where different sources meet.[25] For example, in Map 4.7, one can see lines where two different sources of data abut. While the use of these data was methodologically viable and operationally expedient, there were concerns over the effects of such a heterogeneous mix on the quality of the associated biodiversity assessments. Yet, in approaching the current study, it was also recognized that a detailed vegetation mapping effort would be beyond the scope of the investigation. Balancing means and ends, existing information on vegetation described in the categories and scales needed for habitat assessment was used to delineate *actual* vegetation (as opposed to *potential* vegetation) on Landsat Enhanced Thematic Mapper (ETM) imagery. Considerable field verification was performed in order to ensure some degree of accuracy.

One of the most useful vegetation base maps was produced by the San Diego Association of Governments.[26] The vegetation databases were created as part of regional conservation planning efforts in the San Diego area and came from a number of sources (including the City and County of San Diego, the Multiple Species Conservation Program, the Multiple Habitat Conservation Program, the Marine Corps, and others). The classification used in the vegetation mapping was a modified Holland system. The 135 habitat types were collapsed to 17 categories. Most of these were applied for the vegetation classification of this study and relabeled. For example: Woodlands was re-classified Oak Woodland and Montane Coniferous Forest became Mixed Forest. The vegetation databases from Western Riverside County came from the University of California, Riverside.[27]

A November 18, 2000 Landsat ETM image[28] was acquired for the study area region (Maps 4.4 and 4.5. That particular date was chosen for a number of reasons: 1) it was very near the start date of the project, 2) it was a very clear image, and 3) it accentuated differences among land cover types. With image processing and geographic information systems (GIS), a comprehensive land use/land cover

map was created using the Landsat ETM satellite image as the major component for identifying vegetative cover. The Landsat data are comprised of seven spectral bands and a single high-resolution panchromatic image. For the identification of vegetation, six of the spectral bands—bands 1, 2, 3, 4, 5, and 7—were used. Band 6, delivered as two separate measurements, was not used because of its portrayal of values in the thermal realm. The pixel (or cell or raster) size of band 6 is 90 meters by 90 meters (roughly 295 feet by 295 feet), while the pixel size of the other seven spectral bands is approximately 30 meters by 30 meters (roughly 100 feet by 100 feet). This difference in resolution provided a second reason for not using band 6. The panchromatic image, with a pixel size of 10 meters (almost 33 feet), was not used due to the difference in resolution and because the electromagnetic spectrum was included in the other spectral bands.

Identification of vegetation cover types began by using an unsupervised statistical technique that classified the six bands of Landsat ETM imagery into 120 discrete classes. That is, the spectral signatures of each pixel were compared against all other pixels in the image. Pixels with similar signatures were grouped together. Through this process, 120 potential classes were identified. This grouping provided an initial look at the possible vegetation types occurring at a diversity of

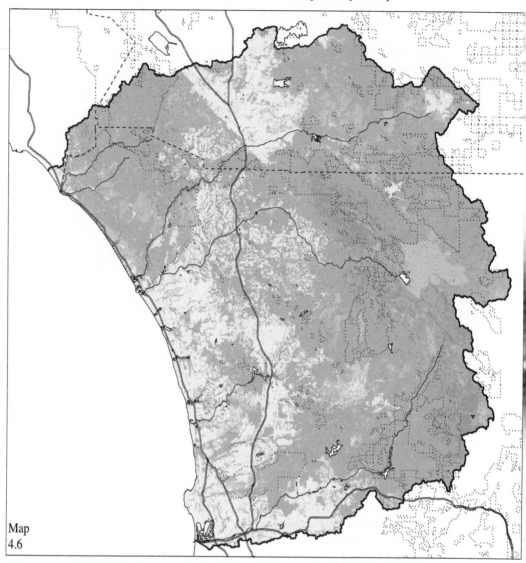

Map
4.6

Region of MCB Camp Pendleton & MCAS Miramar
Existing Conditions 2000
Vegetation

0   2.5   5 Kilometers
0   2.5   5 Miles

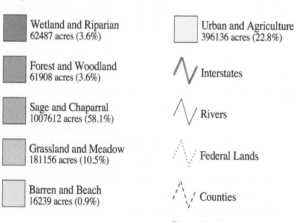

Wetland and Riparian
62487 acres (3.6%)

Forest and Woodland
61908 acres (3.6%)

Sage and Chaparral
1007612 acres (58.1%)

Grassland and Meadow
181156 acres (10.5%)

Barren and Beach
16239 acres (0.9%)

Urban and Agriculture
396136 acres (22.8%)

Interstates

Rivers

Federal Lands

Counties

— See color insert following page 204 —

elevations and on a variety of slopes and aspects. Slope and aspect caused shadowing on a number of different vegetation types; thereby causing the same vegetation to appear spectrally different on the Landsat imagery. Vegetation dominated by an individual species often appeared spectrally dissimilar when located on different soil types or when associated with different vegetation species. The associated vegetation and soil types in the region vary by elevation;[29] thus, elevation was indirectly used to account for some of the spectral dissimilarity caused by these two factors.

With the compilation of three key pieces of thematic information—existing vegetation maps, 120 discrete classes from the Landsat imagery, and elevation—a land cover map of existing conditions was created with a resolution of 30 meters. To create this map, 120 classes were overlaid on the elevation and the existing vegetation maps. Each class within the 120 classes was summarized by existing vegetation and by elevation. The frequency of existing vegetation type by elevation was plotted class-by-class and a land cover type was assigned. Assignment of a class to a land cover type was done based on the vegetation type with the highest frequency for a specified elevation. Ground verification of some classes was done where confusion over the proper land cover type persisted. If land cover type classification errors occurred, as verified through field checks, they were corrected by reclassification of the class into the correct land cover type. Once completed, a land cover map was created for utilization by other models.

Land use was assigned through analysis of housing densities and by overlaying existing land use maps. Data obtained from MCB Camp Pendleton[30] and MCAS Miramar[31] were used to identify Military Maneuver and Impact Areas. Commercial, industrial, transportation, and housing locations were identified by overlaying the urban lands identified in the land cover map with land use information from SANDAG[32] and from the prior study of the region.[33] Running a moving window filter over the entire land cover map assessed housing density and allowed identification of the Ex-urban and Rural residential land uses. Areas with 1 house on 1 to 5 acres were identified as Rural residential. Areas with 1 house on 5 to 20 acres were identified as Ex-urban.

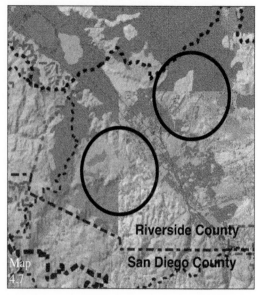

Detail from the Vegetation Map Used in the 1996 Study *Biodiversity and Landscape Planning: Alternative Futures for the Region of Camp Pendleton, California.* The circles highlight seams between two different data sources.

The Vegetation Types

The vegetation classification is shown in Map 4.6. As stated previously, the vegetation is influenced by several factors including short- and long-term precipitation patterns and terrain factors. As might be expected, there are zones of vegetation that reflect the gradients of climate and are partially affected by the other parameters. For the purposes of this description, the vegetation classification needs to be associated with patterns of biodiversity and, as such, is broad rather than fine.

Mixed Forest. Found at the highest elevations, the Mixed Forest is comprised of conifer forest dominated by various species of Pine (*Pinus* spp.) including Ponderosa, Coulter, and Jeffrey (*P. ponderosa, P. coulteri,* and *P. jeffreyi*, respectively), among others. Big Cone Douglas Fir and Incense Cedar (*Pseudotsuga macrocarpa* and *Calocedrus decurrens*, respectively) are also common. Approximately 2% of the study area is in Mixed Forest. The higher elevations of Palomar Mountain have well-developed coniferous forest communities. Conifer forest may occur from as low as 4,250 feet to over 6,500 feet. The forest is often interrupted with mountain meadow grasslands where perennial species dominate. Meadows comprise considerably less than 1% of the study area.

Oak Woodland. Below the forest and sometimes interfingering with it, especially on south facing slopes, is a woodland biome dominated by oaks (*Quercus* spp.). The often complete oak canopy may have a number of species including the deciduous Black Oak (*Q. kelloggii*) as well as the evergreen Interior Live Oak and Canyon Live Oak (*Q. wislizenii* and *Q. chrysolepis*, respectively). Engelmann and Coast Live Oak (*Q. engelmannii* and *Q.* agrifolia, respectively) may occur at lower elevations. The evergreen Madrone (*Arbutus menziesii*), and the evergreen California Bay Laurel (*Umbellularia californica*) may also be important canopy species. Many species common in the lower elevation chaparral may be found in the understory. More common species might include Toyon (*Heteromeles arbutifolia*), manzanita (*Arctostaphylos* spp.), vines (*Vitex* spp.), and poison oak (*Toxicodendron diversiloba*). Oak woodland comprises less than 2% of the study area and occurs from about 2,000 feet to over 5,000 feet. It typically occurs on deeper clayey soils than does chaparral.

Below the forests and woodland are several vegetation types that are mapped for the purpose of habitat characterization. These include riparian communities, coastal sage scrub, chaparral, grassland, and desert scrub. In addition, a coastal sage scrub/chaparral mix and a very localized wetland were also delineated.

Riparian Vegetation. Riparian vegetation occurs in association with the numerous, largely ephemeral, drainage systems that traverse parts of the study area. Where the riparian systems are large and perennial (such as the Santa Margarita River), a number of vegetation types might be found. These may include short-lived plants occupying the annual floodplain, vegetation occurring in marshy areas, early successional vegetation occupying lower terraces, and gallery forests occupying higher terraces. While covering just 3.5% of the study area, the riparian zones provide inordinately high quality habitat for wildlife and are greatly valued for recreational uses. Native wetland vegetation consisting of cattails, rushes, reeds, and sedges is at risk of being converted to the highly invasive Giant Reed (*Arundo donax*), which provides little or no

habitat for native fauna. Prominent in numerous parts of the study area, this exotic is especially prevalent along the lower Santa Margarita River floodplain within MCB Camp Pendleton. The Marine Corps has devoted a considerable amount of effort to eliminate *Arundo*. On the lower terraces of the major drainage systems, relatively short-lived willow-dominated vegetation often flourishes. This vegetation, like the Giant Reed and other marshy vegetation, is often scoured and removed by infrequent floods (such as would occur once in ten or twenty-five years). At higher terraces, riparian vegetation may be dominated by a tree overstory. California Sycamore (*Platanus racemosa*), various oak species (*Quercus* spp.), Fremont Cottonwood (*Populus fremontii*), and California Walnut (*Juglans californica*) are typically found in these communities. Two or more of these tree species often occur together and comprise the gallery forest. In spite of the very high diversity of riparian vegetation, it is simply mapped as one type. However, it should be understood that the implications of the composition are considerable for wildlife habitat.

Coastal Sage Scrub. Coastal Sage Scrub, also referred to as *soft chaparral*, occupies approximately 17% of the study area. It is found on foothills and mesas of western San Diego County, on MCB Camp Pendleton, and west of Interstate-15. The vegetation consists of low (less than 6-feet tall) shallow rooted, multiple stemmed shrubs with soft bendable leaves—hence the name *soft chaparral*. The phenology of coastal sage scrub species differs substantially from that of chaparral shrubs. The evergreen shrubs of chaparral produce new stem growth principally during the spring.[34] In contrast, the coastal sage scrub species initiate new stem growth soon after the commencement of the fall rains during the coldest parts of the year. One possible explanation for this difference may be due to the fact that chaparral species tend to have much longer tap roots than sage species. These long roots are able to tap deeper water reserves. In comparison with chaparral, the lower growing usually more open coastal sage scrub occupies drier sites and is composed of dominants whose principal adaptive mode is exploitation of soil moisture in upper soil horizons during the cool winter season.[35] Unlike evergreen chaparral species, coastal sage species shed their leaves, affording them greater summer drought tolerance. The coastal sage shrub species are also characterized by having foliage that may be aromatic and gray in appearance. Many species may be drought deciduous with leaves that desiccate and fall off as the summer drought progresses. Most species sprout after burning with stand recovery in as little as 10 years.[36] A large number of species may be present, but a few species dominate. These include Black Sage (*Salvia mellifera*), White Sage (*Salvia apiana*), Sugarbush (*Rhus ovata*), California or Wild Buckwheat (*Eriogonum fasciculatum*), Laurel Sumac (*Malosma laurina*), and California Sagebrush (*Artemisia californica*, which is also a sage, but from an entirely different family than the other two sages).

Chaparral. Chaparral, often referred to as *hard chaparral* to distinguish it from coastal sage scrub, is characterized by evergreen, stiffly branched, needle or broad leathery-leafed (sclerophyllous) shrubs mostly 3- to 10-feet tall. Chaparral is the most extensive vegetation type of the study area (and of the state) covering approximately 40% of the region. In fact, chaparral communities cover approximately 10% of California, occurring in scattered locations on the foothills of the mountains in both Northern and Southern California. In Southern California, chaparral communities are a dominant feature of the landscape and cover much of the lower elevation of the region's mountains; they are also found in the areas of lower elevation near the coast. In the coastal regions of Central and Southern California, chaparral communities are closely associated with coastal sage scrub communities. Chaparral communities are often subject to large seasonal temperature fluctuations with high summer temperatures and low winter temperatures.

Chaparral is especially prominent north of MCB Camp Pendleton in the Santa Ana Mountains, east of Temecula in western Riverside County, and in interior San Diego County both west and east from Palomar Mountain, and southeast of Ramona. Chaparral communities are normally extremely dense and can form an almost impenetrable vegetation with little understory. Some stands are composed of more open stands of shrubs that sometimes gain the stature of small, bushy trees. The composition of chaparral is quite variable and is often composed of a diversity species. As a result,

chaparral communities have been subdivided into several different types depending on location and dominant species. Within the study area, several types are readily differentiated and include Chamise Chaparral (*Adenostoma fasciculatum*), Manzanita Chaparral (several species of manzanita, e.g., *Arctostaphylos pungens*), Scrub Oak Chaparral (several species of oak, e.g., *Quercus berberidifolia*), and mixed chaparral. These different types are quite prominent in the study area, but are mapped under the single category of chaparral. Chamise chaparral tends to be somewhat monospecific (that is, mixed with few other species) with a concomitant relative lack of biodiversity while the other three have quite a bit of diversity. It is especially common on hot, dry sites (especially on south and west facing slopes) forming extensive stands. Regrowth is slow after fires due to poor site conditions. Manzanita chaparral generally occurs on deeper soils and at higher elevations than other chaparral types. Scrub oak chaparral is a mesic type occurring on north facing slopes below 3,000 feet and on all aspects above 3,000 feet. Red shanks (*Adenostoma parvifolium*) chaparral also occurs in the study area in very limited distribution east of Temecula and east of El Cajon.

Chaparral is replaced by grasslands (typically annual grasses) on frequently burned sites especially on the more arid borders and at low elevations and by oak woodland on more mesic sites, especially where fires are less frequent and intense.[37] Coastal Sage Scrub and Chaparral could not be differentiated in approximately 1.5% of the study area, and was mapped as a combined unit, Mixed Coastal Sage Scrub and Chaparral.

Grassland. Grassland vegetation covers approximately 10% of the study area, occurring as small to large patches throughout the region. It is especially prevalent on MCB Camp Pendleton, east of Temecula, the east side of Palomar Mountain including the Warner Valley, and surrounding the town of Ramona. This vegetation type is also a composite of numerous plant communities. Some annual forb vegetation such as Russian thistle (*Salsola* spp.) with little annual or perennial grass component has been mapped as Grassland. Most of the grassland mapped in the study area consists of exotic annual species such as wild oat and brome (*Avena* spp. and *Bromus* spp., respectively) and forbs such as filaree (*Erodium cicutarium*), bur clover (*Medicago hispida*), and cheeseweed (*Malva parviflora*). Typical of the grasslands dominated by introduced species is the Warner Valley, adjacent to Lake Henshaw. The introductions have largely come about since the 1850s.[38] In fact, nearly the entire extent of grassland found in California today[39] is that of a landscape dominated by introduced, largely annual, species. Brown, in reporting on the works of others, notes that more than 400 alien species account for 50% to 90% of the vegetation cover.[40] One notable exception occurs on the Santa Rosa Plateau east of MCB Camp Pendleton. There, as a result of careful management led by The Nature Conservancy, over 7,400 acres of perennial bunchgrass prairie has been preserved. The flora of the Santa Rosa Plateau is truly impressive (with 583 species and subspecies), but it is Purple Needlegrass (*Nassella pulchra*, formerly *Stipa pulchra*) that dominates the plateau. Both annual and perennial grasslands, particularly under grazing, green soon after the first rains and stay green into spring, especially if the winter rains have been normal or above normal. Annual grasslands start to *senesce* (brown off) by late April or early May, with perennial grasses senescing slightly later.

Desert Scrub. Desert Scrub occurs along the eastern border of the study area, but with considerable area adjacent to the Eastside/ Domenigoni Reservoir in the north central part of the region. The vegetation tends to be much more open (often less than 10% cover) than the chaparral into which it grades. In the eastern part of the study area, its boundary with chaparral tends to be on the east facing and much drier slopes of the mountain ranges of the region. The vegetation composition has more in common with the dry Colorado Desert to the east than the chaparral to the west. Typical species include Creosotebush (*Larrea tridentata*), White Bursage (*Encelia farinosa*), Smoketree (*Psorothamnus spinosus*), Brittlebush (*Encelia farinosa*), Mormon Tea (*Ephedra* spp.), and Saltbush (*Atriplex* spp.).

Vegetation Change and Fire

Whether vegetation is viewed as a continuum (i.e., Gleasonian) or as an almost organismic unit

(i.e., Clementsian),[41] there is little debate that it changes. In the complete absence of human influence, vegetation is seen to change with short and long term variations in climate or with catastrophic occurrences such as natural fires, floods, landslides, and insect infestations. The concept of change is embodied within the concept of succession. That is, vegetation changes from one type to another, with time, human impact, or natural cause. Different successional stages of vegetation are dominated by species that either reproduce quickly and with little competition from other species or by species that outcompete other species.

In the early twentieth century, one of the premier proponents of landscape study in the United States, Carl Sauer, stated his view on the vegetation of the inhabited world "as resulting not only from natural evolutionary processes but also of historic human perceptions and uses of the land."[42] The writings of early pioneers[43] are replete with commentary and accounts about vegetation that differ markedly from that seen today. Although often anecdotal, these accounts permit an understanding of the changes that have occurred. Numerous historians, geographers, plant ecologists, and others have written about changes in the landscape of the American West.[44]

For example, one area that has received considerable scholarly attention is southern Arizona. It once had vast grasslands that dominated the landscape. According to Hastings and Turner, these ecosystems were perpetuated by fires initiated by Native Americans.[45] European settlement brought changes in land use and these vegetation communities have subsequently been degraded or diminished. As part of their investigations, Hastings and Turner compared photographs from the late 19th and early 20th centuries with those from the 1950s and 1960s and revealed what are often striking changes in the composition and structure of the vegetation. Bahre notes that the most apparent directional vegetation changes in southeastern Arizona have been the introduction and expansion of exotics; the removal of native plant cover for settlement; an overall decline in [perennial] native grasses; an increase in woody plants; and the general degradation of vegetation cover due to various human activities.[46] Those who have studied this change in the western United States attribute overall degradation to changes in fire frequency and intensity usually due to suppression, and the results of intensive grazing. McPherson notes that considerable evidence suggests that widespread livestock grazing reduced fine fuel, and therefore fire frequency, in desert grasslands after 1880.[47] Invasive native species such as mesquite and exotics were introduced to the grasslands by livestock and flourished in the absence of fire. Agricultural irrigation and livestock watering lowered water tables and exacerbated the problem and degraded riparian habitat. Grazing and shrub removal by ranchers to encourage the growth of grasses maintained a grassland albeit in a degraded state (NOTE: observations by the author). In proximity to increasing urbanization, grazing has been removed and the degraded grasslands have been altered to an even more degraded shrub vegetation.[48]

The vegetation of southwest California has been extensively altered by human activity especially in the post-Mexican period. In a similar manner to that of southern Arizona, cattle grazing introduced exotic species, and thereby altered native habitat. Human occupation suppressed fire. Water was pumped from the ground or diverted from perennial sources. Even more dramatically than in southern Arizona, human settlement has been accompanied by the use of exotic plants for landscaping. Afro-Australasian species such as *Eucalyptus* spp. and iceplant (a genus of the *Mesembryanthyaceae* family) were planted as ornamentals and escaped to colonize natural habitats and to replace native species. Numerous annual grasses and forbs were accidentally introduced and escaped.

The vegetation of the region is closely related to and shaped by the processes of fire and has been studied for decades. Chaparral communities are known to be very closely associated with fire. Chaparral shrubs are very flammable due to the resinous foliage, woody stems, accumulated litter, and standing dead branches. Wildfires play a key ecological role in the development and perpetuation of these communities. Chaparral shrubs are highly adapted to fire by either re-sprouting from an underground burl (lignotuber) or by having seeds that are stimulated to germinate by the heat treatment of the fire and by chemicals in the charcoal left after the fire. According to Callaway and Davis, when fire is excluded from hard chaparral, the vegetation transitions to oak woodland.[49] When chaparral is either grazed or burned (with or without grazing), its transition rates to other vegetation types are low. According to Keeley, soft chaparral (or coastal sage scrub) on frequently burned sites is replaced by grassland and on infrequently burned sites by oak woodland.[50] Robin Wills, a fire manager with the National Park Service, also notes that frequently burned coastal sage scrub does not result in its maintenance, but causes a transition to grassland. He further refers to this grassland as being comprised largely of annual grasses and forbs. As urban expansion continues, Wills notes, there is likely to be an increase in ignition opportunities. These fires will have a negative effect on the preservation of coastal sage scrub.[51]

The effect of the clearing chaparral and coastal sage scrub for development seems to have a similar result to excessive burning. The analysis of this study shows that the clearing results in a replacement either by barren land or by grassland. Furthermore, the grassland is not a native perennial grassland, but is comprised of annual grasses and forbs. The difference is dramatic in terms of biodiversity value. Native perennial grasslands have a high intrinsic biodiversity value, whereas annual grasses and forbs have relatively low biodiversity values.

The grazing of land by livestock (primarily horses, cattle, and sheep) will result in a gradual shifting of grass species from perennial to annual. It is likely that such change occurred in the 19th and early 20th centuries. Livestock also has the effect of clearing brush within the oak woodland and forest vegetation types. In this respect, livestock serves to mimic fire. Unfortunately, it also introduces exotic species. With the suppression of fire and an increasing removal of livestock grazing, the oak woodland and forest types are changing their appearance by having a more closed understory. This understory was sparse in the pre-European period and has since become dense. The understory is also highly burnable and are likely to be much more intense as a result of this fuel buildup than in the past when fire suppression was not as prevalent.

NOTES:

[1] Philip R. Pryde, "Introduction" in Philip R. Pryde, ed., *San Diego: An Introduction to the Region*, (Dubuque, Iowa: Kendall Hunt Publishing Company, 1984), pp. 1–12, this note, p. 2.

[2] Amalie J. Orme, "The Mediterranean Environment of Greater California," in Antony R. Orme, ed., *The Physical Geography of North America* (Oxford, England: Oxford University Press, 2002), pp. 402 – 424.

[3] David S. McArthur, "Geomorphology of San Diego County" in Philip R. Pryde, ed., *San Diego: An Introduction to the Region* (Dubuque, Iowa: Kendall Hunt Publishing Company, 1984), pp. 13 – 30.

[4] Thomas A. Deméré, "Faults and Earthquakes in San Diego County." Available on-line at <http://www.sdnhm.org/research/paleontology/sdfaults.html>.

[5] ibid.

[6] Orme (2002), pp. 402 – 424.

[7] Philip R. Pryde, "Climate, Soils, Vegetation, and Wildlife," in Philip R. Pryde. ed., *San Diego: An Introduction to the Region* (Dubuque, Iowa: Kendall Hunt Publishing Company, 1984), pp. 31–49.

[8] Pryde, "Climate, Soils, Vegetation, and Wildlife," pp. 31–49.

[9] Vladimer Köppen, a climatologist and plant geographer, devised a climate classification in the early 20th century to explain boundaries he noticed in vegetation. Most modern classifications are based on his work. Harold P. Bailey, *The Climate of Southern California* (Berkeley, California: University of California Press, 1966). See, also, Edward J. Tarbuck and Frederick K. Lutgens, *Earth Science* (New York: Macmillan College Publishing Co., 1994), pp. 715–721.

[10] Pryde, "Climate, Soils, Vegetation, and Wildlife," pp. 31–49.

[11] Orme (2002), pp. 402–424.

[12] D.W. Shimwell, *The Description and Classification of Vegetation* (Seattle, Washington: University of Washington Press, 1971).

[13] A. W. Küchler, *Vegetation Mapping* (New York: Ronald Publishing, 1967).

[14] Roy Alexander and Andrew C. Milligan, eds., *Vegetation Mapping* (New York: John Wiley and Sons, 2000).

[15] Küchler, p. 30.

[16] Early vegetation mapping efforts in California are summarized in John O. Sawyer and Todd Keeler–Wolf, *A Manual of California Vegetation* (Sacramento, California: California Native Plant Society, 1995), p. 14.

[17] The use of terms to describe elements of vegetation such as *association, biome, community,* and *formation* is described thoroughly in David E. Brown, *Biotic Communities: Southwestern United States and Northwestern Mexico* (Salt Lake City, Utah: University of Utah Press, 1994), pp. 8–16 and pp. 302–306.

[18] Brown.

[19] Brown, pp. 66–69.

[20] Brown, p. 309.

[21] R. F. Holland, *Preliminary Descriptions of the Terrestrial Natural Communities of California* (Sacramento, California: The Resources Agency, Department of Fish and Game, Natural Heritage Division, 1986).

[22] Sawyer and Keeler–Wolf.

[23] Sawyer and Keeler–Wolf, p. 1.

[24] Sawyer and Keeler–Wolf, p. 6.

[25] Carl Steinitz, Michael Binford, Paul Cote, Thomas Edwards, Jr., Stephen Ervin, Richard T.T. Forman, Craig Johnson, Ross Kiester, David Mouat, Douglas Olson, Allan Shearer, Richard Toth, and Robin Wills, *Biodiversity and Landscape Planning: Alternative Futures for the Region of Camp Pendleton, California*, (Cambridge, Massachusetts: Harvard University Graduate School of Design, 1996), p. 29, figure 45.

[26] San Diego Association of Governments, *San Diego Region Generalized Vegetation Map*, (San Diego, California: San Diego Association of Governments, 1997).

[27] Nanette Pratini, The Western Riverside County vegetation map was released at our request via ftp, Dr. Thomas Scott Lab, University of California, Riverside. The vegetation map may be viewed on-line at <http://ecoregion.ucr.edu>.

[28] U.S. Geological Survey, EROS Data Center, Sioux Falls, South Dakota, Landsat Enhanced Thematic Mapper + image, Path: 040, Row: 037, imaged on November 18, 2000 (obtained January 4, 2001).

[29] U.S. Geological Survey, *7.5 minute digital elevation models*. Available on-line at <http://edcsns17.cr.usgs.gov/EarthExplorer>.

[30] Digital land use information released to research team on CDROM by AC/S Environmental Security, MCB Camp Pendleton (August 7, 2001).

[31] Digital land use information released to research team on CDROM by Environmental Management, MCAS Miramar (May 1, 2001).

[32] San Diego Association of Governments, "2000 Land Use." Available on-line at <http://www.sandag.org/resources/maps_and_gis/gis_downloads/land.asp>.

[33] Steinitz et al. (1996), p.19.

[34] Harold A. Mooney, "Southern Coastal Scrub" in Michael G. Barbour and Jack Major, eds., *Terrestrial Vegetation of California* (New York: John Wiley and Sons, 1990), pp. 471–490.

[35] Mooney, pp. 471–490.

[36] Brown, p. 89.

[37] Jon E. Keeley, "Chaparral" in Michael G. Barbour and W. Dwight Billings, eds., *North American Terrestrial Vegetation*, (Cambridge, England: Cambridge University Press, 2000), pp. 203–253.

[38] There is a large literature on the introduction of exotic species, especially annual grasses and forbs, in California. See, for example, L.T. Burcham, "Historical Backgrounds of Range Land Use in California," *Journal of Range Management* 9 (1956), pp. 81–86; L.T. Burcham, *California Range Land* (Sacramento, California: California Division of Forestry, 1957); W.W. Robbins, "Alien Plants Growing without Cultivation in California," *California Agricultural Experiment Station Bulletin*, 637 (1940).

[39] Harold F. Heady, "Valley Grassland" in Michael G. Barbour and Jack Major, eds., *Terrestrial Vegetation of California* (New York: John Wiley & Sons, 1990), p. 497.

[40] Brown, p.132.

[41] Early concepts on the nature of vegetation diverged into two camps: that of F. Clements and that of H. Gleason. Frederick E. Clements, *Plant Succession: An Analysis of the Development of Vegetation*, Carnegie Institute of Washington Publication 242 (Washington, D.C.: Carnegie Institute of Washington, 1916). Henry Gleason, "The Individualistic Concept of the Plant Association," *Bulletin of the Torrey Botanical Club* 53 (1926), pp. 1–20.

[42] Carl O. Sauer, "The Agency of Man on the Earth" in William L. Thomas, ed., *Man's Role in Changing the Face of the Earth*, (Chicago, Illinois: University of Chicago Press, 1956), pp. 49–69.

[43] James A. Young and B. Abbott Sparks, *Cattle in the Cold Desert* (Reno, Nevada: University of Nevada Press, 2002).

[44] See, for example, Rodney Hastings and Raymond Turner, *The Changing Mile* (Tucson, Arizona: The University of Arizona Press, 1965), Conrad Bahre, *A Legacy of Change*, (Tucson, Arizona: The University of Arizona Press, 1991), and Conrad Bahre, "Human Impacts on the Grasslands of Southeastern Arizona" in Mitchel P. McClaran and Thomas R. Van Devender, eds., *The Desert Grassland*, (Tucson, Arizona: The University of Arizona Press, 1995), pp. 230–264.

[45] Hastings and Turner.

[46] Bahre (1991), p. 186.

[47] Guy R. McPherson, "The Role of Fire in the Desert Grasslands" in Mitchel P. McClaran and Thomas R. Van Devender, eds., *The Desert Grassland*, (Tucson, Arizona: The University of Arizona Press, 1995), pp. 130–151.

[48] Carl Steinitz, Hector Arias, Scott Bassett, Michael Flaxman, Tomas Goode, Thomas Maddock III, David Mouat, Richard Peiser, and Allan Shearer, *Alternative Futures for Changing Landscapes: The Upper San Pedro River Basin in Arizona and Sonora*, (Washington, D.C.: Island Press, 2003).

[49] Ragan M. Callaway and Frank W. Davis, "Vegetation Dynamics, Fire, and the Physical Environment in Coastal Central California," *Ecology* 74:5 (1993), pp. 1567–1578.

[50] Keeley, p.204.

[51] Robin Wills, personal communication, March 29, 2002.

Chapter 5

Mapping Existing Conditions

David A. Mouat and Scott D. Bassett

The Geographic Information System developed for this investigation contains a variety of data and can be used to display features of the study area. However, for the purpose of providing a basic representation of the region, a single convention must be adopted. This map is referred to as Land Use/Land Cover and is shown on the next page. Shown as Map 5.1, it is derived from the same Landsat source as the Vegetation Map (Map 4.6), but emphasizes the use of the land over the characteristics of the natural environment. For example, in this study Rural Residential development is defined as 1 house per 3 to 5 acres. Even given a house, driveway, garage, and other small out buildings, a large percentage of the lot will be unbuilt. The Vegetation map emphasizes the unbuilt portion of the landscape by representing the lot based on the dominant plant community contained within it. For comparison, the Land Use/Land Cover and Vegetation Maps are shown side-by-side with Maps 10.1 and 10.2. Figures 5.1–5.12 depict conditions that are typical of the Land Use/Land Cover mapping categories.

While selected maps have been reproduced in color, most are shown in gray scale. Color .pdf format files of all the maps are available on-line for viewing on-screen or for downloading. The address of the download web site is given at the top of the map listing, which follows the table of contents. The color maps also use shaded relief in order to provide a better sense of geographical understanding. Notably, the technique can be especially useful for assessing habitat and wildlife concerns. However, it should be recognized that the scale of the maps restricts the size of mapped units that can be recognized. At a scale of 1:100,000, it is possible to distinguish each 30 meter square (or pixel) of the base Landsat image. But, printing the study area at this scale would have resulted in a map approximately 4.5 feet by 4.5 feet in size. As a result of the mapping scale used in this report, some localized impacts may not be represented on a given map. Readers should note that the tables that summarize the impacts are based on the underlying data and not on the mapped representation.

It should also be noted that the classification schemes adopted for this study are specific to the needs undertaken. The land use maps used by the municipal and regional agencies of the area have similar categories, but each jurisdiction maintains its own precise definitions based on local laws, regulations, and other needs. For example, definitions of Rural Residential development may vary from 1 house per 2.5 acres to 1 house per 5 (or more) acres. In the 1996 study of the Camp Pendleton region, over 200 different vegetation and land use classifications were described by local governments (often with some functional overlap).

All maps used in this book are projected in Universal Transverse Mercator (UTM) Zone 11 using the North American Datum of 1983 (NAD83) and the Geographic Reference System 1980 (GRS80) spheroid. Map statistics are reported in acres. Readers may note small round-off errors that result from converting from the 30 meter square pixels of the original Landsat data to English (American) units.

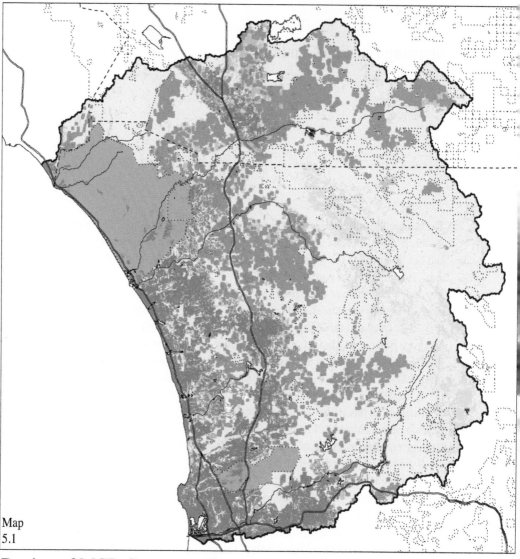

Map
5.1

Region of MCB Camp Pendleton & MCAS Miramar
Existing Conditions 2000
Land Use/Land Cover

Grassland and Shrubland
853476 acres (49.2%)

Forest and Riparian
106218 acres (6.1%)

Agriculture and Orchards
57611 acres (3.3%)

Military Maneuver and Impact
145720 acres (8.4%)

Rural and Ex-Urban
294005 acres (17%)

Urban and Suburban
260191 acres (15%)

Interstates

Rivers

Federal Lands

Counties

Figure 5.1
Beach and Barren

Figure 5.2
Grassland

While popular for recreation, some beach areas, serve as habitat. In the top left photograph a fence is used to protect native vegetation along the shore line. The center photograph shows a wash which is classified as a barren area. Areas cleared for construction, like that shown in the bottom image, have the same general appearance as barren lands in Landsat images.

The photograph at the top right shows native bunch grass on the Santa Rosa Reserve, which is managed by The Nature Conservancy. In the center, the recently burned grassland on the left can be compared with an unburned field on the right. The bottom image shows an annual grassland on MCB Camp Pendleton during peak "greenup."

Figure 5.3   Figure 5.4
Orchards and Vineyards   ## Sage and Chaparral

Top left, the Temecula Valley is an active vineyard area. Center, avocado and orange trees are common in the region. Shown is an avocado orchard on a hillside. The bottom image is a false-color infra-red photograph showing avocado orchards (in the darker tone) abutting chaparral and coastal sage vegetation.

Sage and chaparral are important components to the plant communities which support native fauna. The top right image shows coastal sage scrub which is habitat for the California Gnatcatcher, a federally listed threatened species. The center image shows disturbed coastal sage scrub. The bottom image shows a view over a large patch of chaparral.

Figure 5.5
Forest and Woodland

Figure 5.6
Riparian

The top and center left photographs show open oak woodland on the Santa Rosa Plateau east of MCB Camp Pendleton. Both Engelmann and Coast Live Oaks dominate the landscape with a perennial grass understory. The bottom photograph shows a view of the San Jacinto Mountains with a mountain meadow in the foreground and Mixed Forest dominated by ponderosa pine in the background.

Riparian habitat, shown in the top right photograph, supports a wide range of species, providing forage and nesting grounds as well as connections to the larger landscape. The center image shows disturbed riparian habitat along a corridor which passes under I-15. This path provides an imperfect, but rare east–west crossing for species such as the cougar. The bottom photograph shows a vernal pool area on MCAS Miramar.

Figure 5.7
Agricultural and Golf Courses

Figure 5.8
Ex-Urban Residential Development

The top left photograph shows a recently planted field on MCB Camp Pendleton. The center image illustrates a planting of flowers for the florist industry. The bottom photograph is a golf course.

Ex-Urban Residential Development is defined as a housing density of 1 house per 5 to 15 acres. In the top right photograph, a housing site can be seen near the bottom left corner; others are far in the distance. In the center and bottom images, houses are tucked within the natural vegetation.

Figure 5.9
Rural Residential Development

Figure 5.10
Suburban and Urban Development

All three photographs in the left column show Rural Residential Development which is defined as 1 house per 3 - 5 acres.

Suburban and Urban Development are defined as 1 house per acre and 1 house per less than one acre, respectively. The bottom right photograph shows a checkerboard or "leap frog" pattern of development which leaves some parcels undeveloped (at least, for a time).

Figure 5.11
Commercial and Industrial Development

Figure 5.12
Military Maneuver and Impact Areas

The Commercial and Industrial category includes development ranging from industrial parks (shown in the top left photograph) to retail shopping districts (center) to restaurants and hotels (bottom).

Military Maneuver and Military Impact Areas are primarily natural areas used for a wide variety of training and testing operations. In the top right photograph, a landing craft—most clearly evident by its wake—approaches the beach at MCB Camp Pendleton. In the center photograph, Marines disembark armored personnel carriers. The bottom photograph shows the placement of artillery pieces.

Chapter 6

Critical Uncertainties Which Could Influence Development in the Region

Allan W. Shearer

How might the region of MCB Camp Pendleton and MCAS Miramar change? More specifically, what broad social, political, economic, technological, and environmental changes could influence the successful outcome of local decisions? In a general sense, a scenario-based study process provides a structure to explore aspects of the possible future. More precisely, it serves to identify *assumptions* about how the future might develop. As noted in Chapter 1, the future offers no facts that can be verified or testimony that can be corroborated. Explicitly defining all of the assumptions about a given decision is not an easy task, but it is one that will help to avoid errors in judgment.

The researchers of this study sought to build upon what was generally considered a success in the San Pedro study, which was discussed in Chapter 3. In that earlier effort, a "Scenario Guide" served as a vehicle to consider sets of regional policy options that the base staff and other members of the community thought to merit consideration. The set of issues listed for consideration and the policy options were defined by the research team. The process for the current investigation was designed to further integrate the opinions of the base land managers who will be among the primary users of the research. In particular, base personnel took a more central role in defining the scope of the issues that were considered salient, identifying possible actions, and mapping possible relationships between actions. Commensurately, the role of the researchers was primarily to provide a platform for these tasks.

The framework used in this study for cataloguing and questioning these assumptions about the future is based upon the eight steps of scenario construction identified in Peter Schwartz's often-cited book, *The Art of the Long View*.[1]

Step 1: Identify Focal Issue or Decision. Like other scenario methodologies developed to examine macro-scale or contextual environments, the process begins with the identification of specific decisions faced by an individual, organization, or government agency.[2] Beginning in such a way grounds the process in aspects of the future that are relevant to expected needs. In some cases, a very precise, single action may be under consideration; in others, a set of related actions may be under review. Regardless of the relative magnitude of the decision, this first stage of the process is intended to center the investigation on the identifiable concerns. That is, questions take the form, *Should we invest in "x"?* or *Should we follow strategy "y"?*; not, *Will global event "a" occur? or Will society solve problem "b"?*

Step 2: Key Forces in the Local Environment. Given the questions identified in Step 1, what are the factors that contribute to making a successful judgment? This second step has two components. First, what types of information will the decision makers require to assess the options? Second, and equally important, ultimately how will success or failure be assessed?

Step 3: Driving Forces. Given the factors that contribute to a successful decision, what are the larger, contextual, macro-scale forces that could impact those factors? As a working aid, Schwartz distinguishes five general kinds of forces: social, political, economic, technological, and environmental. While wary of over-defining elements of the scenario process, Schwartz also makes a distinction between uncertain and predetermined forces.[3] Uncertain forces are those that are difficult to predict and that form the basis for differentiating alternative scenarios. Predetermined forces, by contrast, can be more readily anticipated and are present in every scenario. Predetermined forces were initiated in the past and cannot be easily influenced in the present or the future. The most common example of a predetermined force, cited by Schwartz and others, is demographics. For example, the baby boom generation of Americans was put "in the pipeline" in the two decades following World War II and will continue to have an impact on American society through the middle of the twenty-first century. While the exact nature of the demands that will be placed on social services by this large group might be debated, the fact of their numbers is something that will be of concern across many policy issues. Other kinds of predetermined forces include slow changing phenomena (such as birth rates, and the planning and expansion of physical infrastructure) and constrained situations (political, social, or legal limits on possible actions).[4]

Step 4: Rank by Importance and Uncertainty. The range of forces that could impact the success of a decision is broad. The next step in clarifying an understanding of the problem at hand is to rank these forces of change by the importance to the decision at hand and by the degree of uncertainty associated with anticipating future states.

Step 5: Selecting Scenario Logics. Issues that are both very important and very uncertain define the combination of situations that differentiate the alternative visions of the future. These issues frame the "scenario logics." Each of the forces ranked in Step 4 can be thought of as an "axis of uncertainty" which defines one dimension of possibilities. Often the end points are defined in terms of a range of values for some quantity (such as high financing rates - low financing rates), but they may also be defined by structural conditions (such as social stability - social unrest, coordinated action by multiple stakeholders - unilateral action by individual stakeholders, open markets - closed markets, strict regulation - little regulation). The intersection of two or more axes defines general conditions that serve to frame the internal logic of each scenario. Intersecting two axes will produce four scenarios, intersecting three will produce eight, etc. In some cases, not all scenario logics defined in this manner will be fully developed. As with all investigations, time and resources may constrain

the number of futures that can be considered in detail. Also, some pairs (or triplets, quartets, etc.) will be intuitively illogical. An example given by Schwartz uses two axes of uncertainty to define four scenarios for an automobile manufacturer. One axis is fuel prices; the other is the degree of protection given to the industry by government regulation. The four resulting scenarios are I. High fuel prices - Protectionist market; II. Low fuel prices - Protectionist market; III. Low fuel prices - Open market; IV. High fuel prices - Open Market. Intermediate scenarios might also be envisioned.

Step 6: Fleshing Out the Scenarios. After identifying futures that are composed of both important and difficult to predict situations, the next step is to examine how these might be realized. In the case for the automobile manufacturer cited above, what could bring about low fuel prices? Possibilities might include government subsidies, increased production by exporting nations, the discovery of new oil reserves, the development of new and less expensive refining techniques, engines with greater fuel efficiency, etc. In turn, one could ask how each of these more specific possibilities could come to pass.

Step 7: Implications. How does the focus issue or decision play out in each of the alternative scenarios of the future? Would proceeding with a given plan result in success in all cases? in some? or, in none? Potential failure of a plan in some scenarios might not necessarily result in its being scrapped, but the evaluation may suggest that following through with it does carry some risk. Examining the issue in light of these risks could trigger a feedback loop in which the details of the planned action are modified to create an option that is more robust against the uncertainties of the future.

Step 8: Selection of Leading Indicators and Signposts. Finally, Schwartz argues that the scenario process does not stop once the focal issue has been assessed relative to the alternative scenarios. Since the effects of the decisions will play out over a long period of time, a continual monitoring process should be put in place to follow how the future actually evolves. To aid in this process, indicators that mark steps in the potential realization of each scenario should be identified.

Implementing the Scenario Process

Beyond the individual steps used for the scenario construction process, two additional concerns must also be addressed: who participates in the scenario construction process and how do the participants interact? In order to generate a set of scenarios which would adequately reflect the particular needs of each individual base and the collective concerns at different levels of decision making and administration, the installations in the area and the office of the Regional Environmental Coordinator (REC)[5] were asked to select scenario contributors who could represent a range of land and natural resources management activities. Additionally, it was suggested that the participants not be selected from a single tier of the chain of command, but instead reflect a range of responsibilities and levels of authority. In order to maintain a manageable sized group, it was asked that no more than twelve people participate. Ultimately, ten people participated in the entire process and several others contributed to parts. Some of the participants specialize in their current positions by topic of expertise, such as wildlife management or water resources; others coordinate activities across specializations at a single installation; still others operate at the regional level across installations. All were civilians, although many had served in the armed forces at some point of their careers and some were in the Reserves. Additionally, some of the participants were on-staff at MCB Camp Pendleton in 1996 when the study, *Biodiversity and Landscape Planning: Alternative Futures for the Region of Camp Pendleton, California*[6] was completed. All of the participants knew of the existence of the previous work, but the degree of familiarity varied.

The scenario construction process for this study was designed to follow the general principles of a Delphi study. The Delphi Method was developed by Olaf Helmer and others at the RAND Corporation in the 1960s and has been employed to examine a wide range of social and technological issues.[7] At a basic level, it calls for a monitoring group or referee to distribute questionnaires to a participant group. The results are compiled by the monitor(s) and then returned to the participants for additional consideration and reevaluation in light of the group responses. Points of disagreement may prompt new questions aimed at clarification. The number of question-and-answer iterations may be limited by time or funds; however, the process may also continue until either a group consensus is reached or it becomes clear that additional convergence of opinion is unlikely. Issues which employ Delphi techniques are typified by at least one, and often several, of the following characteristics: The problem does not lend itself to precise analytical techniques, but can benefit from subjective judgments on a collective basis; the individuals needed to contribute to the examination of a broad or complex problem have no history of adequate communication and may represent diverse backgrounds with respect to experience or expertise; more individuals are needed than can effectively interact in a face-to-face exchange; time and cost make frequent group meetings infeasible; it is assumed that the efficiency of face-to-face meetings can be increased by a supplemental group communication process; disagreements among individuals are so severe or politically unpalatable that the communication process must be refereed and/or anonymously assured; the heterogeneity of the participants must be preserved to assure validity of the results, i.e., avoidance of domination by quantity or by strength of personality.[8]

An additional reason to employ a Delphi-based approach in a scenario study is elaborated by Kees Van Der Heijden. He argues that if an organization is left to its own devices, it will trend toward one of two pathologies: fragmentation, in which consensus is never reached on any topic and the organization cannot move forward, or "groupthink" (also know as the "bandwagon effect") in which consensus occurs too readily and the organization misses the ability to observe conditions outside the norm.[9] Both pathologies create impediments to a successful decision-making process. While there is no reason *a priori* to conclude that a study involving group input will suffer from either situation, it is prudent to take steps that will preempt the likelihood. Delphi-based approaches can serve to mitigate the drift into either of these unwanted situations. Foremost, the anonymity of the process ensures that ideas, good or bad, stand on their own merits. Ideas are not associated with an individual with whom others might ordinarily agree (with little thought) because of that person's rank, expertise, or social-administrative skills. Similarly, ideas are not associated with a particular individual with whom others might ordinarily disagree (with little thought) for reasons of differing professional perspective or ideological stance.[10] Also, helping to limit the threat of groupthink, the results are returned to the participants as a set, and so no identifiable person has the first word (or the last word) that might unduly direct attitudes about the topic at hand.[11]

For this study, the surveys were distributed by e-mail. This communication medium was practical in that participants were at different locations—sometimes over an hour away from each other by car—and the moderator was in Cambridge, Massachusetts. Once a week for ten weeks, a new exercise was sent to each participant. The collated results of the previous week's exercise were also sent at the same time. The process was anonymous in that while all of the participants were given a complete record of the results, information that would identify a given response with a particular individual was removed.

One limitation of the Delphi methodology implemented in this case study should be noted. Scheduling needs of the study required that the scenario construction exercises be restricted to the summer months of 2001. Because of this constraint, recycling through the question and answer process until a full consensus was reached on all issues was not possible. Instead, some exercises involved ranking elements in terms of importance. Elements of higher net scores were carried through to the next round. In order to better resolve any issues that had not reached consensus and to address any other points of confusion, a one-day face-to-face workshop was held with the participants at the end of the ten weeks. While there was general agreement on many aspects of the possible futures, some relationships required additional consideration and refinement. Because of the limitation of time, the research team made additional assumptions beyond those specified by the participants. Further, in carrying the scenario development process forward, it was assumed that all of the region's

installations would continue to operate at roughly their current levels of operation in the future. This detail qualifies as an "uncertainty stabilization."

The outline for the scenario construction process used in the case study is provided in Table 6.1. The relationship of each exercise to Schwartz's framework is noted by bold-font square brackets, [...]. In addition to the exercises listed below, other survey questions were also asked to help assess the performance of the approach. Two additional notes on the process should be mentioned. First, in order to most efficiently make use of limited time, the group of ten participants was divided into two sub-groups after the scenario logics had been identified. Sub-group "A" worked to identify how these futures might emerge; sub-group "B" worked to identify what might happen next. Second, Schwartz's Step 7 is a part of an overall and ongoing decision-making process. It is not specifically addressed by the surveys, but is instead left to the area's stakeholders to answer.

As a preface to some of the key results of the scenario development process, it should be noted that the surveys and workshop were completed before the events of September 11, 2001. Clearly, since then the region, the nation, and the world have identified new priorities. At the present time, it is not obvious how questions associated with homeland security will influence private and public life. Throughout most of history, times of war have resulted in people taking collective shelter behind city walls. The advent of artillery, in part, changed this defensive strategy and the introduction of long-range missiles and strategic bombing furthered the decentralization of people and industry. It is conceivable that the prospect of terrorist activities may prompt additional growth pressures away from urban cores, but that is only speculation.

The topic of how cities and regions in the United States might evolve in an era under the threat of terrorist attacks could be informed through a scenario-based investigation; but such an investigation was not done here. Regardless, the pressures of growth and development on natural systems have not abated and the questions, trends, and uncertainties described below are very much still germane.

Table 6.1
Scenario Construction Process

I   Identification of Issues (Round I)
- Facing the Region
- Facing the Military Installations

[Schwartz Step 1]

Identification of Significant Uncertain Future Situations (Round I)
[Schwartz Step 2]

II   Identification of Trends Impacting the Region
- Economics
- Environment
- Politics
- Society
- Technology

[Schwartz Step 3]

III   Identification of Issues (Round II)
- Facing the Region
- Facing the Institution

[Schwartz Step 1]

Identification of Significant Uncertain Future Situations (Round II)
[Schwartz Step 2]

IV   Identification of Actors in the Region
[Schwartz Step 6 and 8]

V   Assessment of Impacts of Uncertain Situations on Trends and
         Identification of Scenario Logics
[Schwartz Steps 4 and 5]

VI   Possible Events of a Surprise Free Scenario
[Schwartz Steps 6 and 8]

VII   Scenario Outlines
- Sub-Group A: What Actions Could Cause Each Scenario to
  Occur?
- Sub-Group B: What Actions Could Follow if Each Scenario
  Occurred?

[Schwartz Steps 6 and 8]

VIII   Scenario Elaborations
[Schwartz Steps 6 and 8]

IX   Scenario Review—Clarifications, Corrections
[Schwartz Steps 6 and 8]

X   Scenario Workshop for Additional Review and Development

Summary of Survey Results

Scenarios are sometimes described as flashlights that pierce the darkness of the future. What one finds is partly determined by the direction of the beam and partly by what one seeks. Tables 6.2 and 6.3 summarize questions that are facing the region and facing the installations. This two-part approach reflected the fact that, on the one hand, the bases and the surrounding communities are interrelated and share many of the same concerns; but, on the other hand, each has needs specific to its own communities. Separate lists provide a means to help identify areas of mutual interest. Given common concerns related to natural resources management, some obvious questions follow. With whom

Table 6.2

Selected Issues Facing the Region

Development and Housing

- Should a building moratorium on new development be invoked until a comprehensive regional plan is created and agreed upon by all municipalities?
- Should a building moratorium on new development be invoked until new long-term water supplies have been identified?
- Should more low- and mid-income housing be required in any new development?
- Should the region reward development for building in areas that have become rundown or have been set aside for development that do not have sensitive biological and environmental resources?
- Should governments in the region take steps to control growth?

Environmental Conservation

- Should local governments transfer some land use control to regional agencies to better support ecosystem and biodiversity conservation?
- Should the region spend more on habitat preservation and restoration than it is currently?
- Should the State of California be more aggressive with its legislation and implementation of environmental regulations even at the risk of driving business interests away?
- Should the region continue to insist upon protection of all species as a matter of principle, without regard to identifiable usefulness or to "charisma"?
- Should the region develop a grassroots effort to ensure sustainable biodiversity by sponsoring [federal, state, and local] legislation to find the funds necessary to pay for private land banks?
- Should the region establish specific limits to growth at the expense of the property owner's rights to develop/use their land?
- Should the region make regulations relating to clean air be more stringent?

Power

- Should the state and federal governments spend more or provide other incentives to spur renewable energy research and conversion to non-carbon (or reduced carbon) power sources?
- Should the region reward efforts that reduce air pollution and promote energy production and conservation by re-instituting incentives for constructing and retro-fitting homes and offices with photo-voltaic electrical generation, solar hot water systems, etc.?

Transportation

- Should taxes be increased to pay for better infrastructure?
- Should the state and federal governments spend more to develop mass transit systems?

Water

- Should the region allow the anticipated growth if it is not certain to have sufficient water to support our current lifestyle?

Table 6.3

Selected Issues Facing the Region's Military Installations

On-Base Land Management

- Should the installations prioritize the maintenance of biodiversity or the recovery of specific species?
- Should all animal and plant species, regardless of status, be protected at the expense of military training in all cases?
- Should on-base environmental services be administered by DoD or contracted to private companies?
- Should future Range Rule regulations be enforced on inactive military ranges?

Off-Base Land Management

- Should military lands be managed as part of a regional system of ecological resources? If so, what is the appropriate level of organization? Where are boundaries drawn?
- Should military bases enter into conservation management partnerships with private land holders, local governments, and/or non-governmental organizations (NGOs)?
- Should the installations be able to acquire additional lands to use for mitigating installation activities?
- Should the installations provide funds for others to perform mitigation at off-base sites?

Water

- Should DoD spend money with communities to protect watersheds?
- Should DoD develop alternative sources of potable water?

(precisely) might DoD cooperate/partner/assist? Also, if off-base land is acquired (or leased or otherwise used) for mitigation purposes, what kinds of mitigation would be most advantageous from the perspectives of 1) the individual installations, 2) the set of installations in the region, 3) the Regional Environmental Coordinator, and 4) DoD as a whole? How broadly could (should) the policy be applied? For example, would it be best to target a single specific issue, such as acquiring endangered species habitat and to use such a policy only as a stop-gap measure? Or, should it be used proactively to, say, protect riparian vegetation and thereby lessen the impact of downstream flooding? Could or should off-base land be acquired in such a way that DoD facilities could manage ecosystems?

In general, questions about potential collaborative activities are a part of what Emery and Trist refer to as the transactional environment— areas in which a single organization can influence the outcome of events, but cannot dictate them (see Chapter 1). In the past, such collaborations for natural resources conservation were limited by policy or difficult to implement; however, recent legislation—including the Readiness and Range Preservation Initiative (RRPI) which was part of the Defense Authorization Act of 2003—has made DoD's cooperation with third parties for conservation-related efforts more effective.

Table 6.4 summarizes some of the trends which have shaped the present and which could be expected to shape a Surprise Free (i.e., business-as-usual) future. On balance, trends relating to population and demographics were perceived as among the most influential. Population pressures were identified as stemming from continuing inmigration from other parts of the country and immigration from Mexico (both legal and illegal). Better medical technology allowing longer life spans was also seen as contributing to population increase. Demographic trends include an increasing split in the middle class (with some moving up and others down) and an emphasis on multiculturalism over a unified national identity. Could these trends lead to a fragmentation or Balkanization of the region or, more broadly, of the nation. If so, how would this fragmentation manifest itself?

Trends associated with development were numerous and complexly related. Several trends and policies were cited as encouraging the current ways and means of development including: actions by local government to increase revenue and federal

Table 6.4

Trends Potentially Affecting the Region (in alphabetical order)

Economics
- Energy costs are increasing.
- Land prices are increasing.
- Land use is increasingly driven by profit.
- The middle class is being eroded in favor of more "haves" and more "have-nots."
- Wealth is increasingly becoming concentrated.

Environment
- Environmental awareness and education are increasing.
- Environmental regulations are becoming increasingly inflexible.
- Environmental regulations are increasingly put in place too soon.
- Law suits to settle environmental differences are increasing.
- Public lands are increasingly seen as the solution to conservation needs.
- Water is becoming scarce.

Politics
- Elected offices are becoming weaker, due, in part, to the campaign-electoral process.
- Federal regulations are becoming more stringent.
- Greater interest is placed on local and regional politics relative to national politics.
- Opinions on environmental issues are becoming more extreme and more polarized.
- Regional planning efforts are becoming more accepted.
- Southern California is becoming increasingly Democratic.
- The influence of advocacy groups is increasing.
- The public is becoming increasingly dissatisfied with federal intervention.

Society
- Development continues at a brisk pace with little or no regard for natural resources.
- There is a growing emphasis on multiculturalism over common beliefs.
- People are generally living longer due to medical advances.
- Populace is more complacent about social ills.
- Regional demographics are changing toward a more Hispanic populace.
- Satisfaction with customer service is decreasing.
- The quality of education, particularly in the sciences, is decreasing.

Technology
- Clean and renewable energy supplies are being developed.
- Digital and communications infrastructures are improving.
- Technology is increasing, but the application to environmental concerns is slow.
- The use of cars is increasing and road infrastructure is expanding.
- There is an increasing belief that technology has all the answers.

Table 6.5

Critical Uncertainties Related to Land Use Management (in alphabetical order)

- Will conservation of biodiversity become important to the public-at-large?
- Will global warming impact the region?
- Will higher densities of development be acceptable?
- Will people be, on average, satisfied with their lifestyles?
- Will population grow slower/faster than expected?
- Will the current and proposed conservation plans be (perceived as) sufficient?
- Will there be any significant natural disasters in the region?
- Will there be mass transit alternatives?
- Will there be sufficient power for "contemporary" society?
- Will there be sufficient water for development and conservation needs?

tax breaks for home ownership. However, it was also cited that land in the region is becoming more and more expensive. This situation coupled with the changes in demographics prompts several questions: Will land prices squash demand for the single family home with a back yard? A perception of social values relating to homeownership in this country dates back to Jefferson. His thesis—adopted by others —was that with land ownership comes a greater vested interest in the community. Does homeownership still contribute to such a feeling? Will it do so in the future? A more general question of housing is will the new populations accept new (or different) patterns or styles of development? Or, will they expect the kinds of housing preferred by current residents? Several responses cited the ever increasing power of computers and noted how digital communications will allow people to work from anywhere. Reliance on automobiles and improved transportation corridors were also cited as trends. Will ubiquitous networks make cars more or less necessary? Additionally, a pro-property rights trend was cited along with increasing federal environmental regulation. How will these seemingly opposing forces co-evolve?

Political trends suggest a redirection of attention from national affairs to local concerns. Alongside this trend was a sense that the public-at-large is increasingly questioning the roles of different levels of government. How might the relationships between federal and state/local governments change in terms of support—be it political or financial—for certain initiatives? Is part of the answer a function of the types of technologies used to address environmental issues?

Is science and technology the solution to the region's problems? Aspects of several trends point toward this question. First, on the one hand, there is a trend toward a renewed or increased interest in the environment as evidenced by educational initiatives and expressions of popular culture, such as movies. On the other hand, a concern over a perceived downward trend in the quality of science education (writ large) was also cited. Could bad information cause more harm than no information? What information is used to make land use decisions and what information would be more helpful? Is this other information available now, or if not, how long can the region wait for it? Second, the actual development of clean and renewable energy sources was cited as a trend along with technological solutions to other problems; but also cited was a trend to put too much faith in expensive and unproven systems.

Assuming that the identified trends are likely to shape a Surprise Free future, what might cause one or more of them to become more significant—to exert greater influence or to cause change at a faster pace? Similarly, what might cause them to fade or to cause change at a slower pace? These questions mark the critical uncertainties of the future. Some of the critical uncertainties that were identified as salient to this study are summarized in Table 6.5.

Answering each of the questions that define the critical uncertainties with a yes or no, speculating on the means by which each answer could be achieved, and assessing the implications on the trends becomes the subject of the scenarios, which will be discussed in the next chapter.

NOTES:

[1] Peter Schwartz, *The Art of the Long View: Planning for the Future in an Uncertain World* (New York: Currency–Doubleday, 1991). The process is explained in some detail over the entire text; however, the eight steps are summarized on pp. 241 – 248. These steps are also summarized in Gill Ringland, *Scenario Planning: Managing for the Future* (New York: John Wiley & Sons, 1998) pp. 228–233.

[2] The approach used by SRI Consulting Business Intelligence also begins with the question of "What decisions need to be made?" See, "Scenario Planning," available on-line at <http://www.sric–bi.com/consulting/ScenarioPlan.shtml>.

[3] Schwartz, pp. 108ff.

[4] Schwartz, p. 111.

[5] REC regions correspond to the same administrative territories as those of the U.S. Environmental Protection Agency and other federal agencies. Region IX includes Arizona, California, Nevada, and Hawaii.

[6] Carl Steinitz, Michael Binford, Paul Cote, Thomas Edwards, Jr., Stephen Ervin, Craig Johnson, Ross Kiester, David Mouat, Douglas Olson, Allan Shearer, Richard Toth, and Robin Wills, *Biodiversity and Landscape Planning: Alternative Futures for the Region of Camp Pendleton, California* (Cambridge, Massachusetts: Harvard University Graduate School of Design, 1996).

[7] Olaf Helmer, *Analysis of the Future: The Delphi Method*, RAND Corporation Paper P–3558 (Santa Monica, California: RAND Corporation, 1967).

[8] Harlod A. Linstone and Murray Turoff, eds., *The Delphi Method: Techniques and Applications* (Reading, Massachusetts: Addison–Wesley Publishing Company, 1975), p. 4.

[9] Irving L. Janis, *Victims of Groupthink: A Psychological Study of Foreign–Policy Decisions and Fiascoes* (Boston, Massachusetts: Houghton Mifflin Company, 1972). The problem of groupthink in corporate environments is mentioned in Kees Van Der Heijden, *Scenarios: The Art of Strategic Conversation* (New York: John Wiley & Sons, 1996), ix, 48.

[10] Robert B. Cialdini, *Influence: The Psychology of Persuasion* (New York: Quill–William Morrow and Company, 1993), pp. 208–236.

[11] Philip G. Zimbardo and Michael R. Leippe, *The Psychology of Attitude Change and Social Influence* (New York: McGraw–Hill, 1991) pp. 186ff.

Chapter 7

Four Scenarios and Alternative Futures of Regional Change

Allan W. Shearer, Scott D. Bassett, and David A. Mouat

Based on the results of the scenario survey exercises and workshop, the research team prepared a set of final scenarios in narrative form and maps of corresponding alternative futures. As discussed in Chapter 1, scenarios provide a representation of possible changing conditions. However, to better understand the effects of these changes, it is necessary to isolate individual moments and consider them in closer detail. The alternative futures mark these moments.

Each scenario is rendered in two alternative futures based on progressive stages of population growth. The first stage allocates 500,000 new residents; the second stage allocates an additional 500,000 people, bringing the total number of new residents to 1,000,000. There is no assumption within the scenarios about precisely when these increases will be achieved. As many know, or could reasonably expect, the state of California provides a range of well-considered population and demographic information through the Department of Finance. Included in the published reports are population projections for the next twenty years at the county level of aggregation[1] and historic population estimates for towns, cities, and unincorporated areas.[2] This information could be used to roughly project populations for the coming years; however, there would be limitations of such an estimate given the different conditions of each of the alternative futures. The growth in population for this area (or any area) is not only a function of the number of local births and deaths, but also the number of people immigrating from other countries and inmigrating from other parts of the United States. Assuming relatively stable birth and death rates in each alternative future, the changes

in society, technology, economy, politics, and the environment that are outlined could conceivably affect the distribution of new residents within the area and even the rate at which people move to the region. Given these different assumptions, it would be inappropriate to use the published figures as an exact forecast. Therefore, for reference alone, data from the U.S. Census shows that the population of the study area was 1,441,000 in 1980; 1,702,000 in 1990; and 1,998,000 in 2000.

The allocation of houses and commercial buildings is accomplished through a GIS model. It first identifies private land which can readily be developed. Planners often call this classification of land *buildable*. For this study, private land that is on a slope greater than 25% or that is in a wetland is not considered buildable. Additionally, no lands on Native American reservations or public holdings are considered in the allocation process. The model then prioritizes building locations based on the needs of a given scenario. All existing development is left intact. That is, for the allocation of the first 500,000 people, no buildings present in the existing (built) conditions are removed; for the allocation of the second 500,000 people, no buildings present in the existing (built) conditions or added during the first allocation are removed.

The same definitions of land use which describe the existing conditions are also used to describe the future. Urban housing is 4 houses per acre, Suburban housing is 2 houses per acre, rural housing is 1 house per 5 acres, and ex-urban housing is 1 house per 5 - 15 acres. It is assumed that there are three people per household for all new development. In some cases new orchards accompany rural residential and ex-urban development. One alternative future also constructs new golf courses which are shown combined with agricultural lands— the same as extensive (i.e. row crop) agriculture.

The following pages provide narratives that describe the evolutions of possible futures, maps of new development for both stages of population growth, and maps of the resulting land cover. Table 7.1 summarizes the change in land use in each alternative. Tables 7.2-7.5 list allocations of the population by housing type and county. Each alternative future is illustrated by a digram showing the general spatial pattern, allocation maps showing only new development with 500,000 (500k) and 1,000,000 (1,000k) new residents, and comprehensive lands use land cover maps that integrate new development and unchanged conditions. Chapters 8, 9, and 10 will discuss some of the broader implications of these futures.

The translation from scenarios into alternative futures requires several important notes. While the alternative futures are based on the results of the scenario development process, the researchers exercised a significant degree of judgment in adapting specific details. Any effort to chart the course of the future is a complicated endeavor, and legitimate differences of opinion about how the world might work are bound to occur. The scenario process was designed to identify these differences (or at least some of them) and to provide a platform through which they could be compared and synthesized. In many instances, a convergence of opinion did, indeed, occur. However, given constraints of time, some assumptions about the interrelationships among possible future situations, actions, and events were left unresolved at the end of the workshop process. As necessary, the researchers made decisions to drop some ambiguous relationships from further consideration and to modify aspects of others in order to make them defensible relative to the broader logic underlying a given scenario. In some cases, these modifications required the inclusion of factors that were not considered during the development of the scenarios. As such, any faults or failings associated with the contents of the scenarios and related alternative futures rest solely with the research team and are not the responsibility of the scenario workshop participants. Among the most significant assumptions added by the researchers was the basis for precisely allocating new development within the study area. Finally, the naming of the alternative futures was done by the researchers. Throughout the development process, the scenarios were identified by number (I, II, III, and IV) and often characterized by a subset of salient features. Many possible naming conventions were considered; however, for reasons of consistency and comprehension, the alternative futures are identified by a single characteristic: the relative distribution of new development within each scenario.

The intended use of these scenarios and alternative futures that was stated in the Introduction should be repeated here. By focusing on critical uncertainties rather than common regional goals, the possible futures which are considered in this book reflect concerns about what tomorrow *might* become, not preferences for what it *should* become. In each of the scenarios, there may be what some would call winners and losers. Similarly, it is also possible that a person reading this report would very much want to live in some of the described alternative futures or, equally possible, not live in any of them. While such assessments could mark the success or failure of a comprehensive regional planning initiative, they are incidental to the scope of this work. Instead, the four alternative futures developed for analysis are best understood as different contextual models against which the risks and opportunities of local actions might be considered. For example, as asked above, can the various actions taken to manage the lands of a military base succeed given different possible changes to the larger landscape? Similar kinds of questions could be asked by other organizations, agencies, and institutions which are responsible for the conservation or preservation of natural resources. It is the hope of the researchers that the approach taken in this study and its results will also be of benefit to everyone in the region.

Table 7.1

Land Cover (in acres)

	2000	Coastal		Northern		Reg. Low-Density		Three-Centers	
		500k	1,000k	500k	1,000k	500k	1,000k	500k	1,000k
Barren and Beach	8,651	5,535	3,719	5,137	4,030	4,544	2,409	3,613	3,069
Change from 2000		-3,116	-4,932	-3,514	-4,621	-4,107	-6,242	-5,038	-5,582
% Change from 2000		**-36.0**	**-57.0**	**-40.6**	**-53.4**	**-47.5**	**-72.2**	**-58.2**	**-64.5**
Grassland	101,934	88,049	73,735	84,807	61,380	77,540	50,777	85,284	72,467
Change from 2000		-13,885	-28,199	-17,127	-40,554	-24,394	-51,157	-16,650	-29,467
% Change from 2000		**-13.6**	**-27.7**	**-16.8**	**-39.8**	**-23.9**	**-50.2**	**-16.3**	**-28.9**
Orchards and Vineyards	30,589	24,221	20,060	17,092	14,053	27,668	23,269	28,305	27,312
Change from 2000		-6,368	-10,529	-13,497	-16,536	-2,921	-7,320	-2,284	-3,277
% Change from 2000		**-20.8**	**-34.4**	**-44.1**	**-54.1**	**-9.5**	**-23.9**	**-7.5**	**-10.7**
Sage and Chaparral	751,542	705,725	644,195	666,149	554,408	638,286	497,726	690,375	614,095
Change from 2000		-45,817	-107,347	-85,393	-197,134	-113,256	-253,816	-61,167	-137,447
% Change from 2000		**-6.1**	**-14.3**	**-11.4**	**-26.2**	**-15.1**	**-33.8**	**-8.1**	**-18.3**
Forest and Woodland	60,292	59,946	57,690	59,583	55,397	57,831	57,006	59,887	58,815
Change from 2000		-346	-2,602	-709	-4,895	-2,461	-3,286	-405	-1,477
% Change from 2000		**-0.6**	**-4.3**	**-1.2**	**-8.1**	**-4.1**	**-5.5**	**-0.7**	**-2.4**
Riparian	45,924	44,594	41,977	43,707	40,050	41,985	35,461	43,971	42,761
Change from 2000		-1,330	-3,947	-2,217	-5,874	-3,939	-10,463	-1,953	-3,163
% Change from 2000		**-2.9**	**-8.6**	**-4.8**	**-12.8**	**-8.6**	**-22.8**	**-4.3**	**-6.9**
Ag. and Golf Courses	27,023	22,884	19,390	30,131	30,060	10,442	3,929	24,614	24,784
Change from 2000		-4,139	-7,633	3,108	3,037	-16,581	-23,094	-2,409	-2,239
% Change from 2000		**-15.3**	**-28.2**	**11.5**	**11.2**	**-61.4**	**-85.5**	**-8.9**	**-8.3**
Ex-Urban Residential	86,132	102,509	128,287	87,765	94,746	87,518	104,555	119,680	123,592
Change from 2000		16,377	42,155	1,633	8,614	1,386	18,423	33,548	37,460
% Change from 2000		**19.0**	**48.9**	**1.9**	**10.0**	**1.6**	**21.4**	**38.9**	**43.5**
Rural Residential	207,870	222,175	239,087	283,144	378,036	326,869	442,321	213,455	248,232
Change from 2000		14,305	31,217	75,274	170,166	118,999	234,451	5,585	40,362
% Change from 2000		**6.9**	**15.0**	**36.2**	**81.9**	**57.2**	**112.8**	**2.7**	**19.4**
Suburban and Urban	198,925	237,732	279,478	235,909	275,383	241,340	290,523	244,046	292,193
Change from 2000		38,807	80,553	36,984	76,458	42,415	91,598	45,121	93,268
% Change from 2000		**19.5**	**40.5**	**18.6**	**38.4**	**21.3**	**46.0**	**22.7**	**46.9**
Commercial and Industrial	46,489	52,005	57,755	51,950	57,831	51,347	57,391	52,143	58,056
Change from 2000		5,516	11,266	5,461	11,342	4,858	10,902	5,654	11,567
% Change from 2000		**11.9**	**24.2**	**11.7**	**24.4**	**10.4**	**23.5**	**12.2**	**24.9**
Military Maneuver Areas	130,059	130,059	130,059	130,059	130,059	130,059	130,059	130,059	130,059
Change from 2000		0	0	0	0	0	0	0	0
% Change from 2000		**0.0**	**0.0**	**0.0**	**0.0**	**0.0**	**0.0**	**0.0**	**0.0**
Military Impact Areas	15,661	15,661	15,661	15,661	15,661	15,661	15,661	15,661	15,661
Change from 2000		0	0	0	0	0	0	0	0
% Change from 2000		**0.0**	**0.0**	**0.0**	**0.0**	**0.0**	**0.0**	**0.0**	**0.0**

The Coastal Future

Feeling pulled by the electorate's growing concern for the protection of native ecosystems and pushed by agriculture's continuing need for irrigated farmland, California's elected officials initiated a series of programs to strategically manage water as the state's primary natural resource. The goal was to create a statewide water surplus by the year 2020.

Federal claims, based on the Endangered Species Act, about the environmental impact of moving water from the Sacramento-San Joaquin Delta to the central and southern parts of the state (via the Central Valley Project and State Water Project) limited the volume available for transfer. Recognizing the importance of agriculture on the state's economy and the nation's prosperity, priority was given to Central Valley irrigation at the expense of water for domestic and commercial use farther south. Additionally, the state was under a Supreme Court decision to reduce the amount of water drawn from the Colorado River, then Southern California's principal water source, from over 5 million acre feet (maf) to its lawful allotment of 4.4 maf. The subsequent voluntary reduction to 4.0 maf not only provided a comfortable reserve for emergency drought circumstances, but also put the state in the once unimaginable position of being able to sell water to Arizona (lessening the water deficit of Phoenix).

The centerpiece of the state's agenda made Southern California self-sufficient in managing its water needs. More precisely, the region became much less dependent on the Colorado River and not-at-all dependent on supplies from the northern part of the state. To accomplish this goal, a series of desalinization plants which dot the Pacific shoreline were built. Funding for construction of the first plants was financed in large part by the Army Corps of Engineers as part of a larger project designed to maintain stream flow levels, and in turn riparian habitat, throughout the western states. Additional funds were provided by the local residents through levies collected by the Metropolitan Water District of Southern California (a.k.a., "the Met"), an assembly of local water agencies. The Met also took charge of operating the plants when they came on line. Proceeds from water sales were reinvested into the

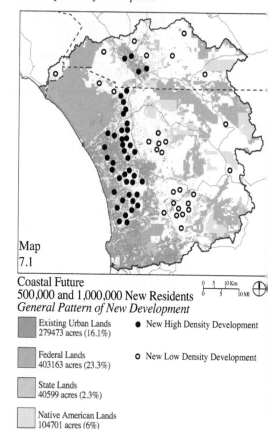

Map 7.1

Coastal Future
500,000 and 1,000,000 New Residents
General Pattern of New Development

0   5   10 Km
0   5   10 Mi

	Existing Urban Lands 279473 acres (16.1%)	● New High Density Development
	Federal Lands 403163 acres (23.3%)	○ New Low Density Development
	State Lands 40599 acres (2.3%)	
	Native American Lands 104701 acres (6%)	

construction of still more treatment facilities. It has been this continued reinvestment that has maintained the region's self sufficiency in providing water for new residential and commercial development.

Contributing to the efforts to make desalinized water affordable were much lower electricity costs. Generating and transmission facilities constructed in the years following deregulation of the electric system and long term price stability in global petroleum markets provided a basis for gradual, cost-sensitive expansion of the power grid. Perhaps more significant than the market forces at work on the conventional power supply industry were new federal incentives to decrease the nation's dependence on foreign oil—in particular, the twenty-five year subsidy that brought renewable energy sources in-line with the costs of fossil fuel plants. The first to capitalize on the subsidy was the controversial joint U.S.–Mexico venture that constructed an off-shore wind farm along four miles of the Baja California coast. Construction of the facility pitted the energy industry against the seaside resorts which feared negative impacts on tourism.

The Mexican government, while initially reticent to undermine the position of Baja California as a vacation destination, approved the project citing 1) the benefits of diversifying its energy production capabilities, 2) potential increased trade with the United States, and 3) the prospect of reducing its own Greenhouse Gas Emissions, and hence being able to sell or trade its quota to a more polluting nation, as per a United Nations accord on global warming. Making use of the North American Free Trade Act, transmission lines connected the plant to the Southern California power grid. An experimental mechanical system driven by ocean waves also showed promise, and while a large-scale facility is not yet built, additional power supplies may be on the way. The net result has been a veritable power glut in the region and some of the lowest electricity rates in the nation.

In order to keep down the costs of transporting the desalinized sea water inland (and, more significantly, up hill), development near the coast and adjacent to urban areas was encouraged. Local governments in coastal towns created new town plans and increased zoning densities to allow for three — and sometimes four — story row houses. Additional steps were taken to dissuade new rural residential and commercial construction inland. High fees tacked on to well drilling permits helped cover the costs of constructing the coastal water supply network. These costs were passed through to the buyer, significantly increasing construction costs. As a result of these actions, most development took place below 1,000 feet in elevation. Urban in-fill between the coast and Interstate-15 attracted most of the construction, with some spill over to the east of the highway. Growth in the Western Riverside County–Temecula area slowed significantly as transporting water from the coast was prohibitively expensive and long-time farmers chose to maintain their fields rather than sell their water rights.

At first, the increased density of homes and businesses along the coast brought even more automotive traffic congestion to a region which thought it had seen a peak in the late 1990s. To help alleviate this problem and to accommodate the expected continuing growth, CalTrans and the regional planning agencies prioritized improvements to major roads and light rail lines along the coast with extensive bus service running east–west to

connect coastal train stations to inland towns. The remainder of the inland road network was, at best, only maintained. The incentive to use mass transit was also helped by urban business districts where many of the region's residents worked. For example, San Diego heavily taxed car traffic.

Beyond infrastructure expenditures, the new state regulations increased property transaction costs. Would-be-sellers were required to provide a water budget for their property which specifies its water source and the volume of its allotment. Would-be-buyers had to provide a detailed plan for accommodating the water needs for any change of use, and, if appropriate, submit a plan for importing any additionally needed water.

While the decision to concentrate development along the coast produced gains in water use efficiency, it also presented a series of difficulties relating to other aspects of natural resources management. In order to protect natural habitat, the species conservation plans proposed by the municipalities and agencies of the Southern California region (circa 2000) were adopted and strictly enforced. But, given the demand for real estate in towns near the coast, all other parcels were developed; new public conservation initiatives were economically possible. Further, increasing land prices prompted some holders of private conservation land to sell their properties (that were not explicitly a part of the regional conservation plans) and re-invest the proceeds in areas where it was possible to purchase larger tracts. The lowland conservation land that was put aside at that time is all the region has now. Table 7.2 summarizes the distribution of new residents. Maps 7.1-7.2 show the spatial patterns of change.

Table 7.2
Coastal Future
Distribution of Population

Orange County	02%
Riverside County	10%
San Diego County	88%
Ex-Urban	01%
Rural	05%
Suburban	24%
Urban	70%

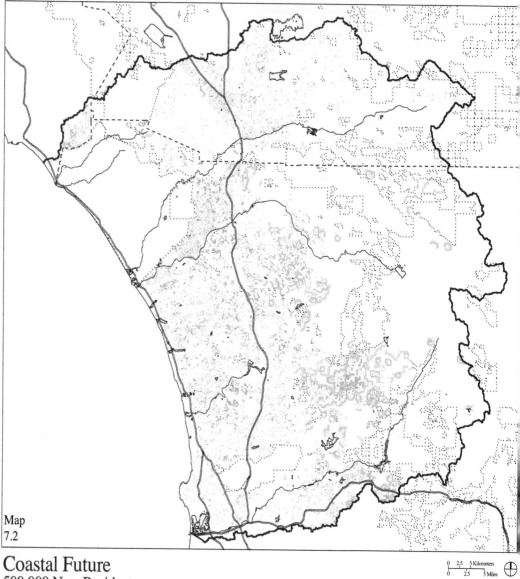

Map
7.2

Coastal Future
500,000 New Residents
New Development

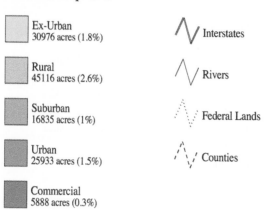

Ex-Urban
30976 acres (1.8%)

Rural
45116 acres (2.6%)

Suburban
16835 acres (1%)

Urban
25933 acres (1.5%)

Commercial
5888 acres (0.3%)

Interstates

Rivers

Federal Lands

Counties

0 2.5 5 Kilometers
0 2.5 5 Miles

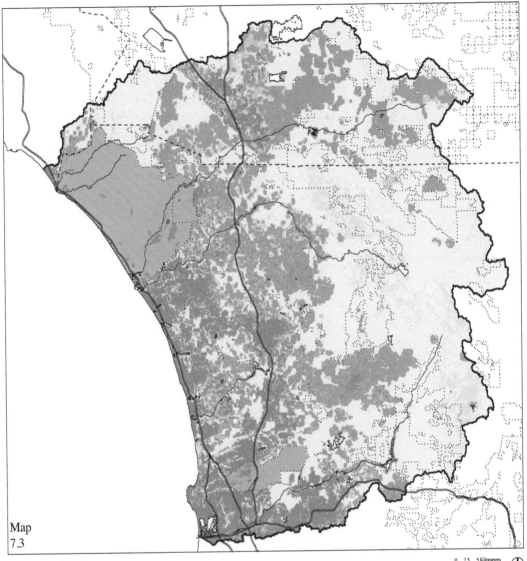

Map
7.3

Coastal Future
500,000 New Residents
Land Use/Land Cover

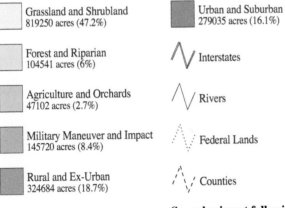

Grassland and Shrubland
819250 acres (47.2%)

Forest and Riparian
104541 acres (6%)

Agriculture and Orchards
47102 acres (2.7%)

Military Maneuver and Impact
145720 acres (8.4%)

Rural and Ex-Urban
324684 acres (18.7%)

Urban and Suburban
279035 acres (16.1%)

Interstates

Rivers

Federal Lands

Counties

— See color insert following page 204 —

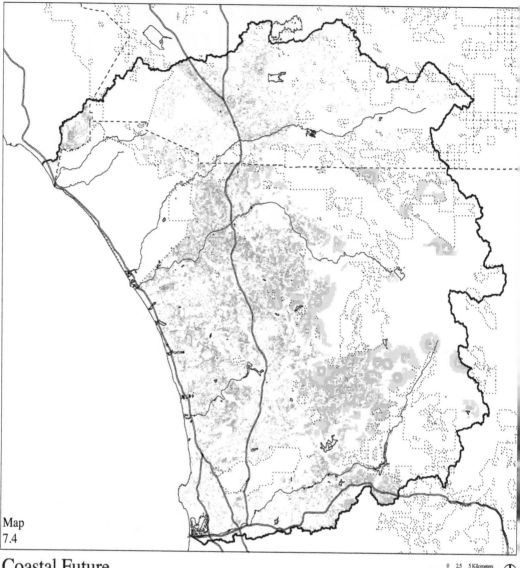

Map
7.4

Coastal Future
1,000,000 New Residents
New Development

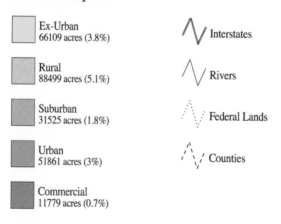

Ex-Urban 66109 acres (3.8%)	Interstates
Rural 88499 acres (5.1%)	Rivers
Suburban 31525 acres (1.8%)	Federal Lands
Urban 51861 acres (3%)	Counties
Commercial 11779 acres (0.7%)	

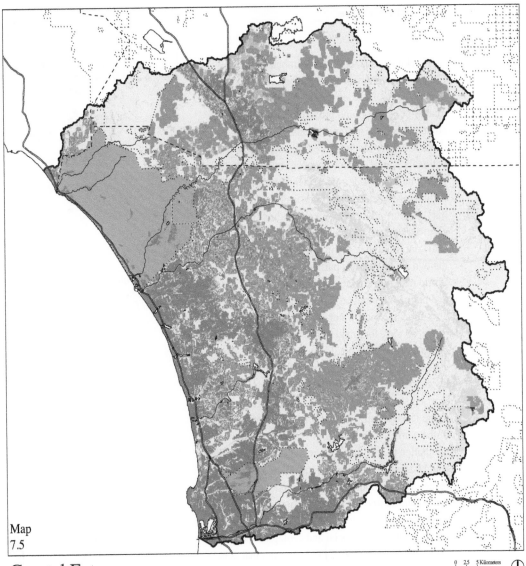

Map
7.5

Coastal Future
1,000,000 New Residents
Land Use/Land Cover

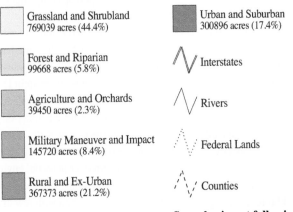

Grassland and Shrubland
769039 acres (44.4%)

Forest and Riparian
99668 acres (5.8%)

Agriculture and Orchards
39450 acres (2.3%)

Military Maneuver and Impact
145720 acres (8.4%)

Rural and Ex-Urban
367373 acres (21.2%)

Urban and Suburban
300896 acres (17.4%)

Interstates

Rivers

Federal Lands

Counties

— **See color insert following page 204** —

The Northern Future

Seeking to reduce its trade imbalance, the Canadian government agreed to sell water to the United States. The decision received far from unanimous support and proved to be the most contentious issue in Canada's history. Most—even those who very much favored the increase in national revenue—voiced concern over a potential loss of national pride. Water would be another natural resource exported to its southern neighbor only to return in a more refined product (in this case, agricultural produce). Some politicians were also concerned over the implications of binding the two nations by this particular natural resource. World markets provided multiple sources of wood and petroleum products, but no other nation was in a position to export water. Without precedent as a guide, all took for granted that future "fine tuning" of the terms-of-sale would be necessary, but some cautioned that once this process started, it could not be stopped. Those looking to the very far horizon worried about a domino effect: Would selling water to the United States prompt offers from other parts of the world? And if so, could Canada resist the temptation to cash-in on the requests? Conversely, could Canada overcome the guilt of rejecting requests from populated arid nations? And finally, could Canada find itself in a position of having an insufficient supply for its numerous trading partners and its own social and ecological needs? Nevertheless, concerns of mounting national debt and a consequent loss of the long-term standard of living proved more important, and a trade treaty was signed by both nations allowing the flow of water for at least fifty years.

Concurrent to the actions in Canada was an American initiative to tap water from the Columbia River system for use in the southwestern states. Debate on this endeavor was also heated and involved similar issues. While moving water within the country did not spark arguments of national pride, it did flame discussion about the rights of states relative to the powers of the federal government. There was also significant concern over the potential environmental impact of the transfer. Some argued on idealistic grounds that the depletion of the river system would forever change

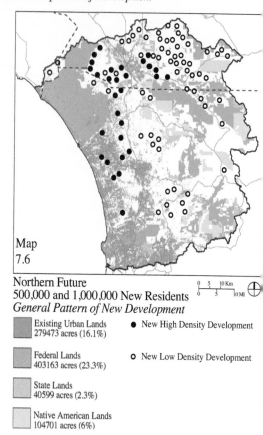

Map 7.6

Northern Future
500,000 and 1,000,000 New Residents
General Pattern of New Development

0   5   10 Km
0   5   10 Mi

	Existing Urban Lands 279473 acres (16.1%)	● New High Density Development
	Federal Lands 403163 acres (23.3%)	○ New Low Density Development
	State Lands 40599 acres (2.3%)	
	Native American Lands 104701 acres (6%)	

the ecosystem and that the government did not have that right. Others posed more pragmatic questions: If a federal project moved water out of the system, would local governments and private landowners continue to be responsible for the protection of a species which was dependent on that water? Questions such as these were not fully resolved and some savvy developers (correctly) took the "non answers" as a sign that the federal government was moving into a period when economic considerations were favored (if perhaps only slightly) over ecological concerns.

The Canadian and Columbia water projects were welcomed relief to the American southwest. Tensions in California over the transfer of water from the Sacramento-San Joaquin Delta to the South Central Valley and the urban centers farther to the south had been high for a number of years, but the promise of access to new water reserves in the north prompted agricultural expansion and urban growth.

Less concern over potential environmental impacts allowed commercial and residential development to proceed with few impediments.

The spirit of the times not only contributed to new houses, but also inspired a surge in the market for larger lots. Many families built their dream house on five acres, and others bought homes adjacent to one of the many new golf courses. Those who could not afford these kinds of homes built near federal land reserves and state parks.

Although escalating land prices tempted farmers to sell their land, most made use of the increased water supply and intensified their operations. While unable to buy new tracts of land given the local real estate market, they brought pasture and long-time fallow lands back into cultivation. Table 7.3 summarizes the distribution of new residents. Maps 7.6-7.10 show the spatial patterns of change.

Table 7.3
Northern Future
Distribution of Population

Orange County	01%
Riverside County	44%
San Diego County	55%
Ex-Urban	01%
Rural	13%
Suburban	26%
Urban	60%

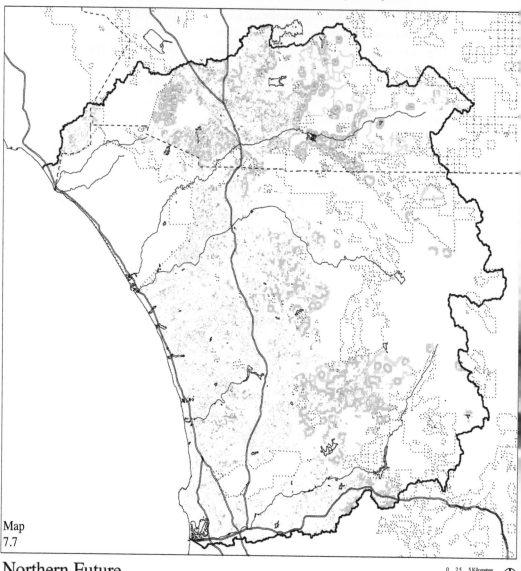

Map
7.7

Northern Future
500,000 New Residents
New Development

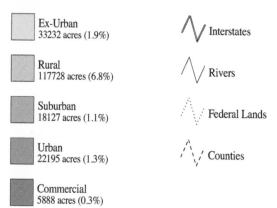

	Ex-Urban
	33232 acres (1.9%)

	Rural
	117728 acres (6.8%)

	Suburban
	18127 acres (1.1%)

	Urban
	22195 acres (1.3%)

	Commercial
	5888 acres (0.3%)

Interstates

Rivers

Federal Lands

Counties

0 2.5 5 Kilometers
0 2.5 5 Miles

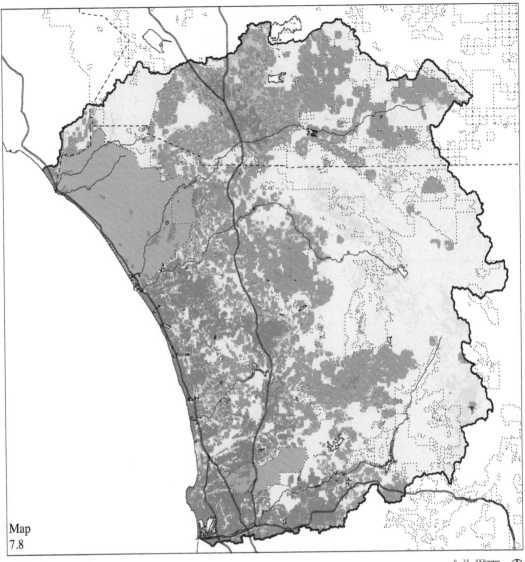

Map
7.8

Northern Future
500,000 New Residents
Land Use/Land Cover

0   2.5   5 Kilometers
0   2.5   5 Miles

Grassland and Shrubland
772902 acres (44.6%)

Urban and Suburban
280691 acres (16.2%)

Forest and Riparian
103290 acres (6%)

Interstates

Agriculture and Orchards
47223 acres (2.7%)

Rivers

Military Maneuver and Impact
145720 acres (8.4%)

Federal Lands

Rural and Ex-Urban
370907 acres (21.4%)

Counties

— See color insert following page 204 —

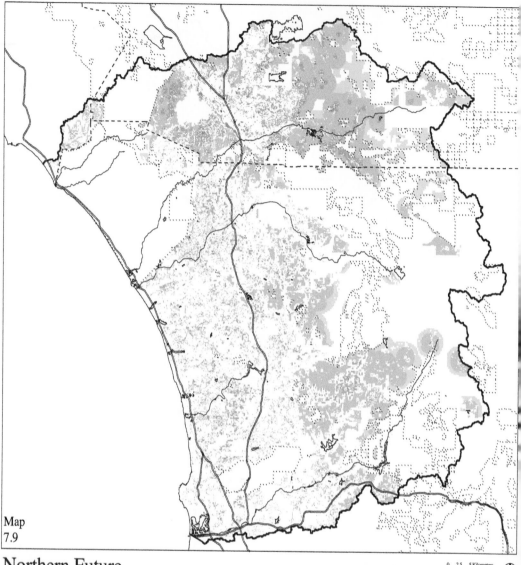

Map
7.9

Northern Future
1,000,000 New Residents
New Development

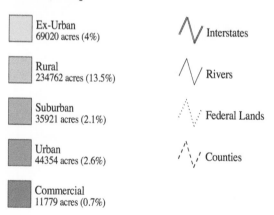

Ex-Urban 69020 acres (4%)	Interstates
Rural 234762 acres (13.5%)	Rivers
Suburban 35921 acres (2.1%)	Federal Lands
Urban 44354 acres (2.6%)	Counties
Commercial 11779 acres (0.7%)	

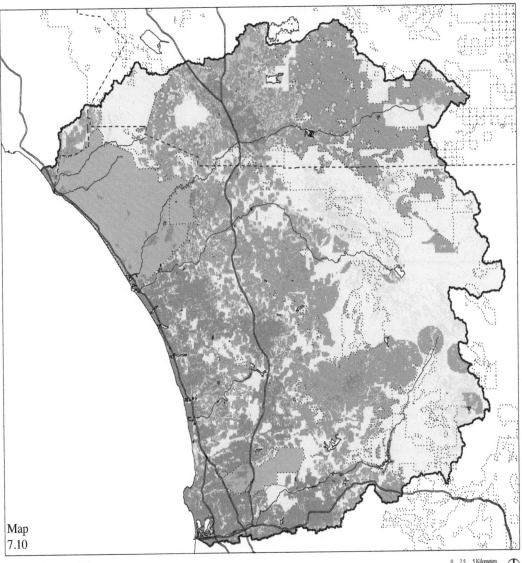

Map
7.10

Northern Future
1,000,000 New Residents
Land Use/Land Cover

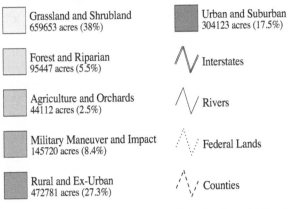

Grassland and Shrubland
659653 acres (38%)

Forest and Riparian
95447 acres (5.5%)

Agriculture and Orchards
44112 acres (2.5%)

Military Maneuver and Impact
145720 acres (8.4%)

Rural and Ex-Urban
472781 acres (27.3%)

Urban and Suburban
304123 acres (17.5%)

Interstates

Rivers

Federal Lands

Counties

— **See color insert following page 204 —**

The Regional Low-Density Future

Years of discussion, debate, and law suits had tipped the balance of power between the federal government and the states in favor of the latter. While the federal government retained its role in establishing foreign policy, maintaining national security, and managing the money supply, states took charge of most commercial, industrial, and natural resources regulation. A few states maintained a high tax rate to support a large number of social programs. Most, including California, reduced taxes and cut regulations to maintain a continuous rate of economic expansion.

While arguably all of the states enjoyed their high degree of autonomy, the reduced framework of federal regulations made some issues difficult to resolve. Perhaps the most significant of these was the relative lack of water in the American Southwest. Years of rapid, uncoordinated development with little concern for the marginal, but cumulative, environmental impacts created what economist Garrett Hardin called, "a tragedy of the commons" on an unprecedented scale. Increasing water shortages in successive years of more or less average rainfall added to the level of social anxiety.

Progressive water pricing structures became common. Under these systems the target level of 1/8 acre-foot per household per year was established and could be purchased for a nominal amount of money. However, each additional 500 gallons used above that amount became increasingly expensive. As a result of the pricing structure, water use in urban homes and apartments was greatly reduced and water savings in suburban areas was also substantial.

Water use in rural areas continued to be relatively high in part because of the difficulty of adequately monitoring well usage. More significant though, was the simple fact that the Farm Bureau had successfully lobbied within the fractious state government to exempt water for agricultural use from tight restrictions. As written, the exemption did not distinguish between types of farms or farm land, allowing so-called "hobby farms" to be included. Many homeowners with lots larger than five acres planted small orchards to qualify for farm

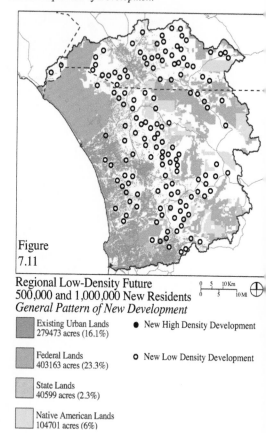

Figure 7.11

Regional Low-Density Future
500,000 and 1,000,000 New Residents
General Pattern of New Development

■ Existing Urban Lands 279473 acres (16.1%)	● New High Density Development
■ Federal Lands 403163 acres (23.3%)	○ New Low Density Development
■ State Lands 40599 acres (2.3%)	
■ Native American Lands 104701 acres (6%)	

status and thereby avoid the potentially expensive water restrictions. The same logic also resulted in a change in new home construction. Families bought larger lots even if it meant building a smaller house.

A second strategy to increase one's supply of cheap water was opening a home office to receive an additional "business" allotment. Judging by the number of applications for zoning variances, one might suspect that this period was marked by an intense entrepreneurial spirit. But, while it was true that some new businesses were created, many people used their new offices for telecommuting to a downtown firm.

A side effect of the water pricing system was a greater accounting of area residents. In order to qualify for a water allotment, each person had to provide proof of residence. Given the relatively tight allowance per person, non-residents could significantly add to a homeowner's annual water bill. Owners of apartment buildings, who had previously set lease rates by floor space, now also kept track of the number of occupants and factored semi-annual water adjustments into rent calculations.

One hopeful response to the water shortage was the construction of a desalinization facility near the city of Oceanside. City residents were offered the option of purchasing their water from the facility at a flat market rate. (That is, they were not subject to a progressive pricing scheme.) However, the price per gallon was three and one-half times the base (under 1/8 of an acre-foot) price of water from any other source. As a result few were inclined to subscribe to the new service.

While regulatory action diminished at both the state and federal levels, concern over the water shortage increased public awareness of environmental issues. Seeing few options for additional conservation lands, many environmental groups focused their attention on the management and protection of existing public lands. The incremental loss of natural habitat corridors that linked larger patches caused some to speculate that the region was possibly too fragmented to function in ecological terms. That is, while parts of the region still "looked green," the remaining areas of native vegetation were too small and too isolated to support genetically viable populations of many species. Also, many were concerned that the parks and forests of the area were being "loved to death" by high volumes of visitors and year-round recreational use. As the result of pressure generated by these groups, federal and state land managers, who were otherwise unable or unwilling to impose severe regulations, limited use of public recreation areas by adopting a schedule which stipulated when certain areas could be used, providing the natural systems with a "chance to recover."

The hands-off approach of regional and state government and the lack of federal influence resulted in sprawling development across the region. Minor roads were built on demand, but no initiatives were taken to provide an overall transportation plan. The subsequent traffic congestion further extended the creation of home offices and the use of telecommuting. Table 7.4 summarizes the distribution of new residents. Maps 7.11-.15 show the spatial patterns of change.

Table 7.4
Regional Low-Density Future
Distribution of Population

Orange County	01%
Riverside County	30%
San Diego County	69%
Ex-Urban	01%
Rural	19%
Suburban	60%
Urban	20%

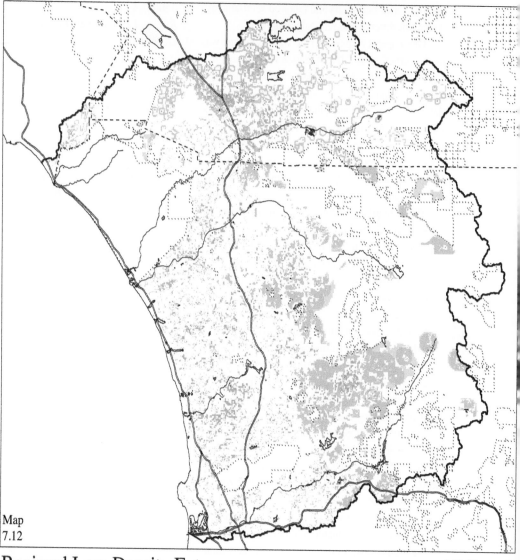

Map
7.12

Regional Low-Density Future
500,000 New Residents
New Development

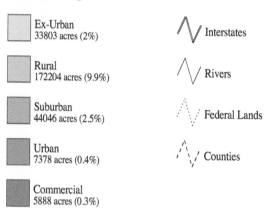

Ex-Urban 33803 acres (2%)	Interstates
Rural 172204 acres (9.9%)	Rivers
Suburban 44046 acres (2.5%)	Federal Lands
Urban 7378 acres (0.4%)	Counties
Commercial 5888 acres (0.3%)	

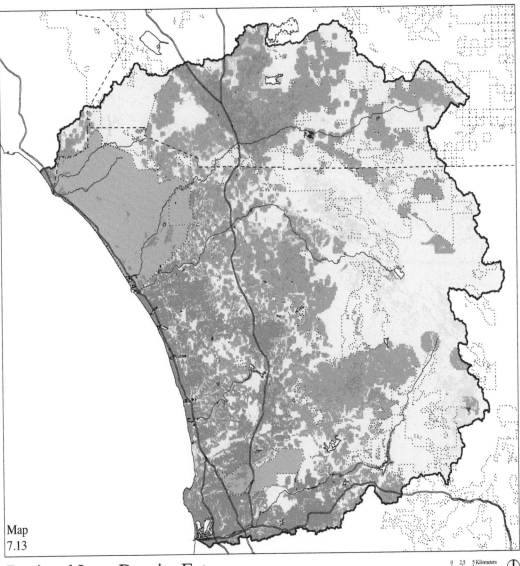

Map
7.13

Regional Low-Density Future
500,000 New Residents
Land Use/Land Cover

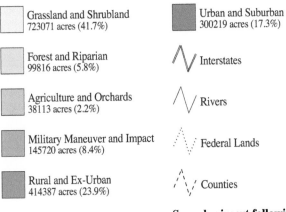

Grassland and Shrubland
723071 acres (41.7%)

Forest and Riparian
99816 acres (5.8%)

Agriculture and Orchards
38113 acres (2.2%)

Military Maneuver and Impact
145720 acres (8.4%)

Rural and Ex-Urban
414387 acres (23.9%)

Urban and Suburban
300219 acres (17.3%)

Interstates

Rivers

Federal Lands

Counties

— See color insert following page 204 —

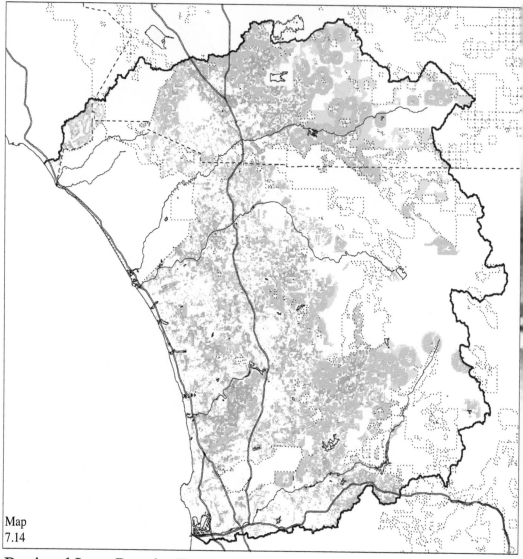

Map
7.14

Regional Low-Density Future
1,000,000 New Residents
New Development

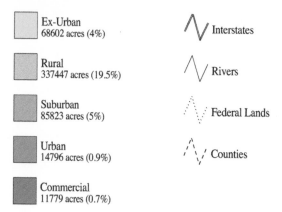

Ex-Urban 68602 acres (4%)	Interstates
Rural 337447 acres (19.5%)	Rivers
Suburban 85823 acres (5%)	Federal Lands
Urban 14796 acres (0.9%)	Counties
Commercial 11779 acres (0.7%)	

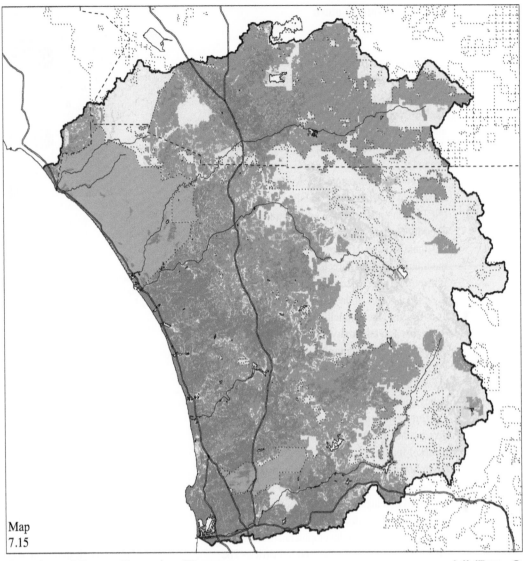

Map
7.15

Regional Low-Density Future
1,000,000 New Residents
Land Use/Land Cover

0　2.5　5 Kilometers
0　2.5　5 Miles

Grassland and Shrubland
563007 acres (32.5%)

Forest and Riparian
92467 acres (5.3%)

Agriculture and Orchards
27198 acres (1.6%)

Military Maneuver and Impact
145720 acres (8.4%)

Rural and Ex-Urban
546882 acres (31.5%)

Urban and Suburban
348184 acres (20.1%)

Interstates

Rivers

Federal Lands

Counties

— See color insert following page 204 —

The Three-Centers Future

Following through on promises made during talks on global accords for ecological protection, the United States government embarked on an aggressive agenda to lessen threats to biodiversity and climate change within its own borders, and by doing so led other post-industrialized nations by example.

Arguably the most visible action was the creation of the National Council on the Environment (NCE). Patterned after the administration's National Security Council, the group formally brought together the Secretaries of Agriculture, Energy, and the Interior, and the Director of the Environmental Protection Agency. Additionally each Department assigned a person at the Assistant-Secretary level as a liaison to the Council. While initially seen by detractors as a temporary fixture designed to appease the international community, each subsequent President has maintained the NCE as a means to elevate the debate on the country's environmental policy and coordinate its implementation.

But more immediately tangible than the debate generated by the NCE, was the continued high levels of funding provided by the Congress for environmental monitoring and the enforcement of statutory regulations. Given what had been perceived as a lag in application of technology to the environmental sciences (relative to other sectors of society), research and development contracts were established with universities and industries. The level of funding was perhaps best evidenced in the private sector by the number of defense contractors who established or built-up environmental technology divisions. Within the federal government itself, the funds allowed, for the first time in anyone's memory, for each and every natural resource-related field office to be fully staffed.

The higher profile afforded by the NCE and the financial means provided by Congress combined to enable federal regulators and resource managers with a platform for broad based action. A case in point was the management of the Santa Margarita River Basin, one of the last free-flowing rivers in Southern California. The question before the government was how to protect the habitat for

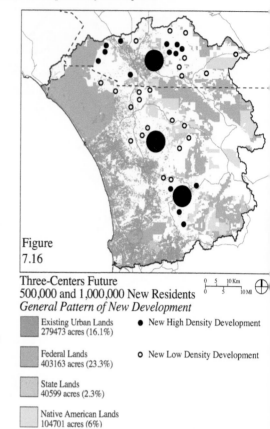

Figure
7.16

Three-Centers Future
500,000 and 1,000,000 New Residents
General Pattern of New Development

▪ Existing Urban Lands 279473 acres (16.1%)	● New High Density Development
▪ Federal Lands 403163 acres (23.3%)	○ New Low Density Development
▫ State Lands 40599 acres (2.3%)	
▫ Native American Lands 104701 acres (6%)	

the arroyo southwestern toad (*Bufo microscaphus californicus*). The species had been listed under the Endangered Species Act (ESA) since 1994 and in 2001, the U.S. Fish and Wildlife Service established more than 180,000 acres of critical habitat in other basins within Southern California.

Even before the designation of critical habitat for the arroyo southwestern toad in the Santa Margarita Basin, the government—flush with staff and monitoring equipment—had already stopped all development on land in the region which contained critical habitat for other endangered species including the California Gnatcatcher, the least Bell's vireo, the least tern, and the snowy plover. But, whereas critical habitat for these species is defined by the presence of specific vegetation, critical habitat for the toad is defined, in large part, by specific stream flow conditions. Preserving the environmental characteristics of seasonal water volume and flow rate required a more reaching solution than the standard establishment of wildlife reserves. The government's strategy was to minimize the likelihood of change in the pattern of storm water runoff and surface water flow. To accomplish this goal, it imposed strict development constraints which had the effect of zoning most of the Basin—from the headwaters to the Pacific Ocean—to low density Rural Residential Development (or one house per five acres). Property owners who had had higher local zoning allowances at the time when the policy went into effect were offered financial compensation for their development rights at a price determined by a court. Some land owners took the offer, but most held their rights in full, speculating on relaxations of the regulation over the long term.

Recognizing that limiting development in the Santa Margarita Basin could adversely impact the region's ability to accommodate its growing population, the government took steps to make other areas more attractive for development. Three areas were targeted for dense growth: Temecula (in Riverside County and in the Santa Margarita Basin) and the area to its north, Valley Center, and Ramona (both in San Diego County). Partially enabling the effort to direct development was the Transportation Department which transferred funds to the state for improvements to the local road network. Access to Temecula was improved by the construction of a new east–west toll road joining Interstate-5 near San Clemente and Oceanside. Valley Center was made more accessible by improvements to County Highway-6 which links the city of Escondido to Pala Road. Finally, State Highway-67 from Ramona to San Diego was doubled in width.

Although the efforts to protect the ecosystem of the Santa Margarita and direct development to other areas attempted to be comprehensive in scope, not all needs were sufficiently met. Most notably, nothing was done to increase the growing region's water supplies. Several proposals had been discussed at the regional, state, and federal levels of government to reapportion water rights across residential, commercial, agricultural, and conservation needs, but the complexities of water law and irreconcilable differences between special interest groups prevented any negotiated plan from moving forward. Additionally, rapid population growth throughout the southwestern states increased the already substantial draw on the Colorado River

and prompted lawsuits both within individual states and between states. International tensions also played a part as Mexico made its case to restore greater flow into the Gulf of California. With the results of multi-jurisdictional litigation pending, regional water companies were unable to make any strategic plans for local resource management, and hence became susceptible to lawsuits of their own.

Local water conservation regulations including municipal ordinances which capped annual water consumption at one-third of an acre foot of water per year per family of four (down from the end-of-the-century estimate of almost one acre foot of water per year per family) helped to mitigate some of the conflict, but were difficult to enforce in suburban and rural areas.

The relative lack of available water did slow the rate of commercial or residential development, but only slightly. Capitalizing on the economic situation, many farmers chose to sell their land rather than risk years of planting crops in such uncertainty. The farmers found ready buyers, as their former fields were accessible, flat, already cleared, and often came with modest water rights. Table 7.5 summarizes distribution of the new residents. Maps 7.16-7.20 show the spatial patterns of change.

Table 7.5
Three-Centers Future
Distribution of Population

Orange County	01%
Riverside County	30%
San Diego County	69%
Ex-Urban	01%
Rural	05%
Suburban	34%
Urban	60%

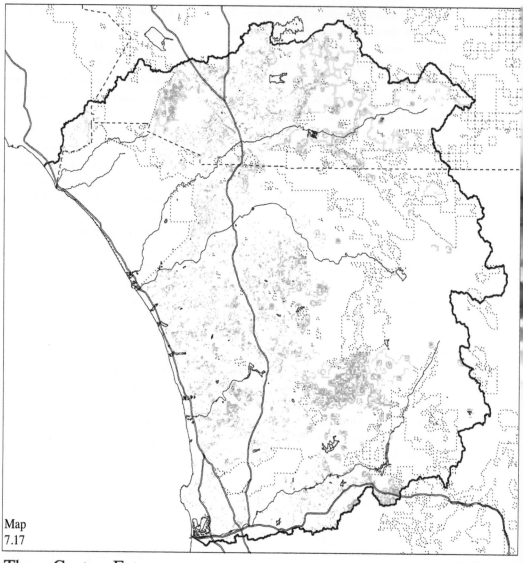

Map
7.17

Three-Centers Future
500,000 New Residents
New Development

0   2.5   5 Kilometers
0   2.5   5 Miles

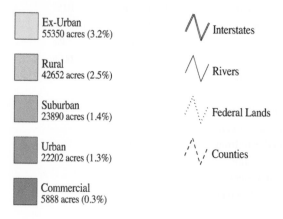

Ex-Urban 55350 acres (3.2%)	Interstates
Rural 42652 acres (2.5%)	Rivers
Suburban 23890 acres (1.4%)	Federal Lands
Urban 22202 acres (1.3%)	Counties
Commercial 5888 acres (0.3%)	

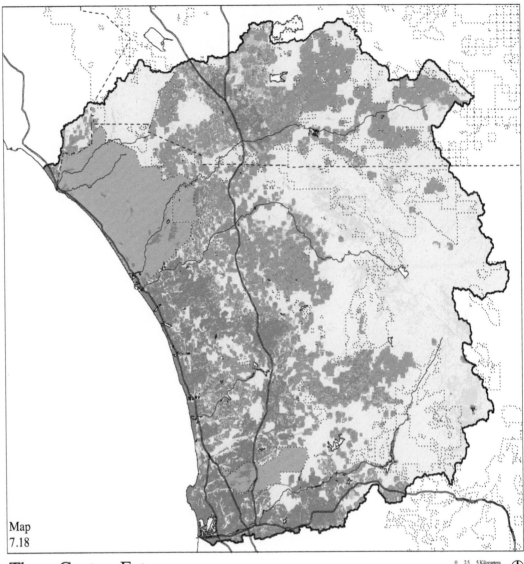

Map
7.18

Three-Centers Future
500,000 New Residents
Land Use/Land Cover

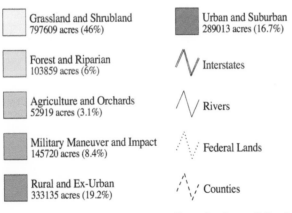

Grassland and Shrubland
797609 acres (46%)

Forest and Riparian
103859 acres (6%)

Agriculture and Orchards
52919 acres (3.1%)

Military Maneuver and Impact
145720 acres (8.4%)

Rural and Ex-Urban
333135 acres (19.2%)

Urban and Suburban
289013 acres (16.7%)

Interstates

Rivers

Federal Lands

Counties

— See color insert following page 204 —

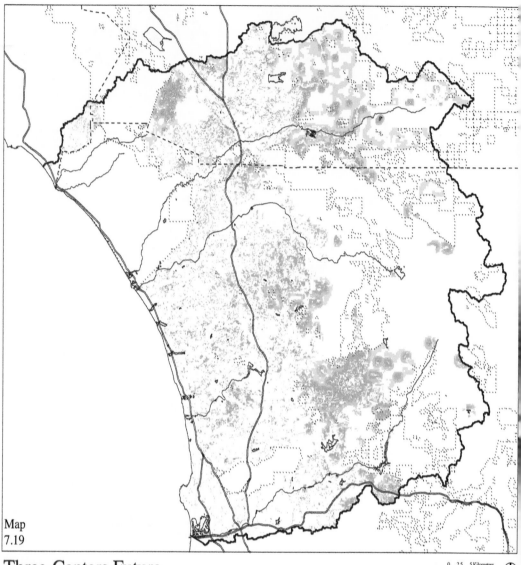

Map
7.19

Three-Centers Future
1,000,000 New Residents
New Development

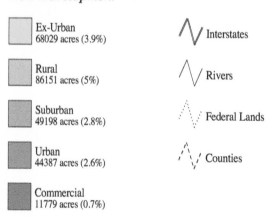

Ex-Urban
68029 acres (3.9%)

Rural
86151 acres (5%)

Suburban
49198 acres (2.8%)

Urban
44387 acres (2.6%)

Commercial
11779 acres (0.7%)

Interstates

Rivers

Federal Lands

Counties

0   2.5   5 Kilometers
0   2.5   5 Miles

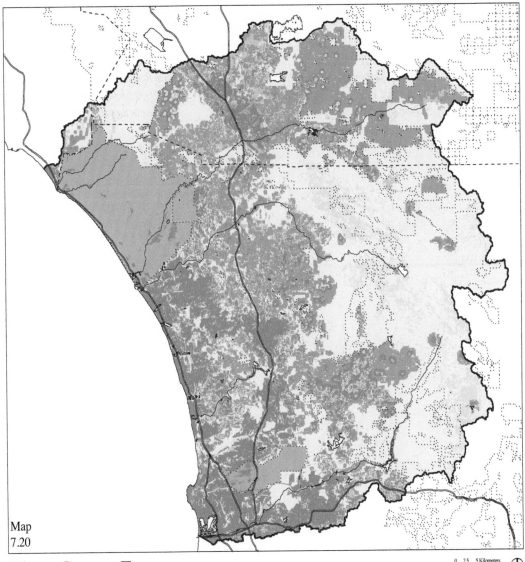

Map
7.20

Three-Centers Future
1,000,000 New Residents
Land Use/Land Cover

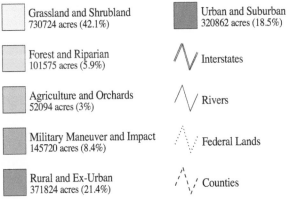

Grassland and Shrubland
730724 acres (42.1%)

Forest and Riparian
101575 acres (5.9%)

Agriculture and Orchards
52094 acres (3%)

Military Maneuver and Impact
145720 acres (8.4%)

Rural and Ex-Urban
371824 acres (21.4%)

Urban and Suburban
320862 acres (18.5%)

Interstates

Rivers

Federal Lands

Counties

0   2.5   5 Kilometers
0   2.5   5 Miles

— See color insert following page 204 —

NOTES:

[1] State of California, *Interim County Population Projections* (Sacramento, California: California Department of Finance, June 2001).

[2] State of California, *Historical Population Estimates and Components of Change, July 1, 1970–1990* (Sacramento, California: California Department of Finance, June 2002); State of California, *Revised Historical City, County, and State Population Estimates, 1991 - 2000, with 1990 and 2000 Census Counts* (Sacramento, California: California Department of Finance, March 2002).

Chapter 8

Hydrologic Consequences Associated with the Alternative Futures

Michael W. Binford and Justin A. Saarinen

It has long been known that urban development in a drainage basin will increase flood flows by rendering surfaces impervious, reducing infiltration, and increasing the rate of water flow across the landscape.[1] Consequently, floods exacerbated by these conditions can be even more catastrophic to downstream natural and human systems than floods generated in undeveloped landscapes. Civil and hydraulic engineers, hydrologists, and landscape planners use numerous mathematical models to simulate stream flow in order to better understand how different management practices address flood-related concerns. In some cases, these models are useful for forecasting development-induced changes of flood flows. All of these models simulate some subset of the processes that convert rainfall to stream flow which include interception, infiltration, evapotranspiration, sheet flow on the surface, subsurface flow through soils, shallow concentrated flow, and open-channel flow.[2] Most models aggregate the characteristics of drainage basins into single parameters. These values are usually calculated as weighted averages, although some consider the distance and velocities of various components of water flow to calculate the amount of water arriving at a given point in a given time. While the use of weighted averages is computationally efficient, the approach carries an inherent limitation because the built and unbuilt landscape is spatially heterogeneous and the disposition of environmental features can greatly influence natural processes. In order to simulate better the consequences of different urban and suburban development patterns on hydrologic regimes, the models must account not only for the characteristics of landscape elements but also for their specific locations. Few models, however, do.

The 1996 study of the Marine Corps Base (MCB) Camp Pendleton[3] region used a version of the spatially explicit CASC2D[4] model implemented in the GIS software GRASS 4.1 as the r.hydro.casc2d module, to simulate a 25-year flood event on the rivers flowing through or adjacent to the installation. This event is statistically expected to occur once in any 25-year period; it can similarly be understood to have a 4% chance of occurring in any given year. The study used the land cover that existed circa 1990 (referred to as 1990+ in the report). The characteristics of the hydrological elements of the landscape were calibrated with the known spatial and temporal distribution of a rainstorm that generated the discharge of the known 25-year flood. Possible future change was investigated through a Plans Build-Out scenario that fully implemented local land use plans. Additionally, students from the Harvard Graduate School of Design developed five alternative regional land-development designs that were intended to accommodate the 500,000 new residents expected by the year 2020 and to manage for biodiversity. Each alternative was a plausible spatial arrangement of residential, commercial, industrial, agricultural, and other land uses. These alternative plans were simulated with the calibrated CASC2D model to model the new stream flows that would result from the changed land cover. The primary product of each simulation was a hydrograph that graphically displayed the volume of discharge over time.

The results of the earlier study forecast significant increases in the 25-year flood discharge for nearly all the scenarios in all the drainage basins. Of particular concern was, and still is, the potential impact to the Santa Margarita Basin. Of all the drainages which originate within, or which pass through Camp Pendleton, the Santa Margarita River hydrologic regime has the greatest effect on its activities, has the most sensitive response to land-cover changes, and is likely to be the location of a significant amount of future development. Simulated peak discharge for the Santa Margarita River nearly doubled under the Plans Build-out future. Additionally, the total volume of water discharged into the Pacific Ocean increased by approximately 50%. A consequence of the increase is a reduction in the amount of water available to recharge groundwater aquifers. Peak and total discharges for each of the five alternative futures developed by the students varied considerably, indicating that the spatial distribution of land cover can greatly influence hydrological processes.

The current study updates the 1996 results by using a newer version of the CASC2D program which incorporates channel-routing methods and allows for the explicit simulation of the changes in imperviousness in urban areas.[5] The new version is more realistic because it allows channel flow to disperse over only one dimension (down the channel) instead of two (across the landscape) and it is limited to defined cross sections and volumes. Overbank flow can be simulated when the flow in the channel exceeds the dimensions of the cross section in high flows. The addition of explicit imperviousness allows direct simulation of zero infiltration, where the older version of the program required some infiltration to occur even if impossible. In addition to the 25-year (4% probability) flood event, the 100-year (1% probability) event is also modeled in this project for the Santa Margarita River. Possible change is modeled in terms of the four alternative futures that have been described in the preceding chapter of this book. A point of principal interest along the river is the U.S. Geological Survey's (USGS) gaging station named "Santa Margarita River at Ysidora" (USGS Gage #11046000), which is near the lowest point of the river without tidal influence. The historical record gathered there allows real and simulated flood events to be directly compared. All of the simulations and analyses in this report are focused on this point along the river. The catastrophic flood of January 1993 was approximately the 100-year flood, and examination of what might happen if similar precipitation and soil-moisture conditions

were to re-occur in the drainage basin with different land cover could help the Marine Corps plan for possible flooding problems in the future.

The CASC2D Model

CASC2D is a two-dimensional, deterministic model for simulating hydrologic processes that propagate precipitation to stream flow.[6] CASC2D is distinguished from most other operational hydrological models in that it is spatially explicit: all the input and output parameters of CASC2D are distributed across two-dimensional representations of the landscape. These representations are derived from a set of cartographic models for the study area. All of the information for these models, including elevation, watershed boundaries, stream channels, and outlet points are extracted from a Digital Elevation Model (DEM).[7] The data are structured in the form of a 2 by 2 array of cells or grid of rasters. Each cell contains a value for a defined parameter and each X,Y pair of coordinates designates a cell with a finite, user-defined, dimension. Although the parameters of the model are distributed over two-dimensions, the model itself simulates three dimensions by linking adjacent cells with a finite-difference solution to water depth in each cell.

Rainfall data taken from meteorological stations are interpolated into a series of maps that distribute the amount of precipitation over the entire space of the region. Each single map denotes the amount of precipitation at a single moment of time. The series, as a whole, represents the changing conditions over the rainfall event. An Inverse Distance Weighted (IDW) method is used for interpolating the point-based station data to area-wide representations.

The simulation starts when each cell receives the initial volume of rainfall and converts it to a uniform depth over the area of the grid cell. The CASC2D algorithm assumes that the relationship between rainfall and stream flow is Hortonian. Hortonian flow (after Robert E. Horton) occurs when rainfall intensity (precipitation rate, LT^{-1}, e.g., millimeters hour^{-1} or millimeters per hour) exceeds the infiltration capacity of the soil.[8] Infiltration of rainfall into the soil occurs at a rate dependent on the properties of the soil and the nature of the land cover. For example, the infiltration capacity of dense clay soil (which is around 0–1 millimeters per hour) is lower than a loose, deep, sandy soil (which can exceed 12 millimeters of water per hour). If the intensity of precipitation is greater than infiltration capacity over the course of a storm, the overabundance of water flows directly over the soil surface. In some storm events, the soil can become 100% saturated and all precipitated water runs off the landscape through increasingly larger channels, eventually passing through the watershed outlet.

Infiltration rate is calculated with the Green–Ampt model, which uses grids of physical properties represented by three parameters and two variables.[9] The Green–Ampt model is:

$$\text{Eq. 8.1} \quad f = K\,I\left[1 + \frac{(\phi - \theta_i)\,S_f}{F}\right]$$

where f = infiltration capacity (LT^{-1}); K = effective hydraulic conductivity, or rate of water movement through the soil (LT^{-1}); I = percent impervious surface in the cell (dimensionless); ϕ = soil porosity, proportion of void space (dimensionless); S_f = effective suction at the wetting front (L); θ = initial water content as a proportion of volume (dimensionless); and F is accumulated infiltration (L).[10] This approach assumes that the soil is infinitely deep and has uniform properties.

Water flow over land from cell to cell is simulated by CASC2D with a two–dimensional finite-difference method. The depth of overland flow in any given cell is a function of the equilibration of water levels with its neighboring cells. Discharge at any point and any time depends primarily on flow direction, velocity, and friction. The routing calculation is very sensitive to the elevation of the landscape. For example, the simulation algorithm can calculate negative water depths when the surface slopes upward more than the change in height of water. Flow velocity is calculated by the empirical relationship between slope, surface roughness, and velocity as described by Manning in 1891.[11] Surface roughness is measured by a value known as Manning's n, which is a friction index, and is inversely related to velocity. The formula for flow velocity is:

$$\text{Eq. 8.2} \quad V = \frac{1}{N}\sqrt{S}\;R^{2/3}$$

where V = flow velocity (LT^{-1}); N = Manning's n (T^{-1}); S = slope (dimensionless); and R = hydraulic radius or flow depth (L). Hence, the lower the n value, the greater the velocity. Smoother land covers (such as short grasses and bare soil) have lower Manning's n values than rougher covers (such as dense forests and tall grasses).

Within the CASC2D program, the movement of rainwater over each grid cell to the ultimate outlet point is modeled over time. For the Santa Margarita River, the Pacific Ocean is the outlet point, but for these simulations the Ysidora gage is the outlet point. The rates of these physical processes, rainfall, infiltration and flow velocity, are calculated simultaneously during one time step to determine the water depth in each cell. The current version of CASC2D simulates the movement of water over cells through channels. The volume of discharge at the outlet point is reported for each time step to produce a hydrograph. In the general case, the simulation finishes when all the rain-generated flow exits through the outlet point. The simulations in this study were ended after 5,000 minutes, when nearly all the flood flow had passed and additional run time would add nothing to the comparisons among the futures.

Methods

The program ArcINFO GRID[12] was used to prepare the DEM, and the Watershed Modeling System[13] was used to prepare channel routing data. The Harvard Graduate School of Design (GSD) made available data used in the 1996 study of the Camp Pendleton Region. These data included a DEM with a 30-meter grid cell size, which was compiled from publicly available USGS files. Two transformations were made to the DEM for modeling purposes. First, it was re-sampled to 90-meter grid cells to more appropriately reflect the scale of the simulated processes. Testing (not reported here) showed that the larger grid cell generated results very similar to those from 30-meter cells. Further, the resampling vastly reduced the likelihood that the program would crash. The 90-meter cell size loses (to downhill cells) or gains (from uphill cells) a smaller proportion of water during each time step both in the real landscape and in the simulation. For the simulation, this change made it less likely to have a negative water-level elevation, or to have an anomalous uphill water slope. These conditions crashed the computer program. The second transformation smoothed the DEM in order to accommodate CASC2D's rules for simulated water movement. The program requires that cell-to-cell transitions occur across a change of elevation (i.e., water moves downhill) and that simulated flow can only occur in the four cardinal directions (i.e., not along diagonals). The smoothed surface closely approximates the original DEM, but allows flow across the study area through the removal of localized depressions and level areas. The watershed boundary, channel network, and outlet points were extracted from the corrected DEM.

The Watershed Modeling System (WMS) has modules that format data specifically for CASC2D input. After importing the corrected DEM into WMS, files for links, nodes, and channels were created. Links, nodes, and channels are representations of channels with attributes of cross-sectional area and shape, and channel roughness. A link represents a reach of the river; and each reach is assigned a unique link number. The individual grid cells that comprise each link are assigned a node number. This scheme allows any grid cell to be located in the network given the link and node pair. The channel input file includes the information about the dimensions of the channel including width, depth, and side slope. Thalweg elevation (the thalweg is the line down the channel along its deepest path, which is also the region of greatest flow velocity), and Manning's n for channel roughness are also required. Thalweg elevation for the channel at each cell was less than or equal to DEM elevation but smoothed so that water was continuously running downhill through the network.

The parameters of the Green–Ampt Infiltration (GAI) model are based on soil texture, or particle-size distribution (e.g., percent sand, silt, and clay). The GSD data set included soil maps for the region, which had been digitized from U.S. Soil Conservation Service (now Natural Resources

Conservation Service) county surveys. Soil textures were re-coded into GAI parameters based on tables published by Rawls et al.[14]

Each alternative future is simulated at two levels of population growth: 500,000 and 1,000,000 people. An appropriate value of Manning's n for each land cover class was determined by reference to tables in Chow and photographs in Barnes.[15] The input parameter map was generated by re-coding the given land cover values.

Precipitation data for the 25–year storm were also included in the GSD data as a grid of total accumulations for the event. These data were converted into a contemporary rain gage file by systematically selecting fifteen points and dividing their total accumulation values by 28 hours, then weighting the hourly accumulations to the same temporal distribution as the storm event of February 23 and 24, 1998, which lasted 28 hours and resulted in a flood that was very close in peak discharge to the statistically determined 25–year flood.

The amount of moisture in the soil before a rain event is called the antecedent moisture. The soil may have no moisture at all, or may have some as the result of previous rains. For example, the largest-of-record flood of January 1993 resulted from nearly ten days of rainfall accumulation in the soil, which was probably close to saturation when rain fell on January 14, 15, and 16. Storm simulations for this study assume soil moisture content to be 90% of saturation. This value limits the amount of space into which water can infiltrate and thus overland flow occurs relatively early in the precipitation event.

The magnitude of the 100-year flood on the Santa Margarita River was estimated by WEST Consulting to be approximately 46,000 cubic feet per second (cfs) at the Ysidora Gage.[16] The estimated discharge of the January 1993 flood was 44,000 cfs, but that figure is only an estimate because the Ysidora gaging station was washed away when the flood discharge was 33,000 cfs. While evidence that would allow a definitive comparison is lacking, the approximated 100-year storm and the January 1993 storm are similar. For this study, the 100-year flood was modeled by using precipitation events equivalent to those in January 1993 and assuming 90% soil saturation. Thirty-minute precipitation accumulations were recorded at 25 gaging stations

Figure 8.1
25- and 100-Year Storm Event Hydrographs for the Santa Margarita River

25-Year Storm Event

Coastal Future

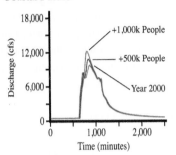

100-Year Storm Event

Coastal Future

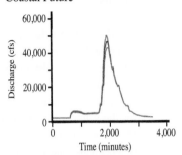

Northern Future

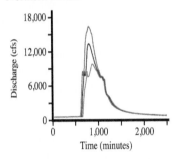

Northern Future

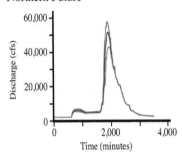

Regional Low-Density Future

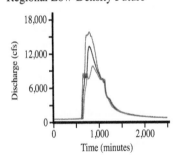

Regional Low-Density Future

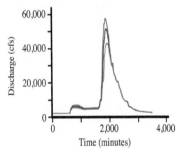

Three-Centers Future

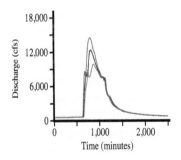

Three-Centers Future

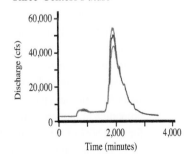

within or near the Santa Margarita Basin. Half of these gages are located within 500 meters of another and recorded similar data, so only 12 unique gages were used to create the precipitation file.

Results

Maps 8.1 - 8.9 display the spatial pattern of change in Manning's *n* values for each alternative future relative to existing conditions; Maps 8.10 - 8.18 show change in impervious surfaces. Greater impervious area results in more runoff because smaller volumes of water infiltrate the soil. Higher Manning's *n* values result in slower overland and channel flows, and a more attenuated hydrograph. That is, the hydrograph will have a lower peak given equal total discharge. As all of the alternative futures add development to existing conditions, they are all characterized by an increase in impervious surface and a lower Manning's *n*. Consequently, all should have larger peak and total discharges, and lower infiltration volumes. The Northern Future generates the largest change of hydrological input variables in the Santa Margarita Basin and should have the largest peak flood and total discharge of storm water. Most of the development in the Coastal Future occurs outside the Basin and so has the least amount of change in terms of hydrologic variables. The Three-Centers and Regional Low-Density Futures will have an intermediate, and similar (to one another), impact, but both will cause an increase of flood discharge over existing conditions.

Hydrographs for each of the futures, compared with existing conditions, are shown in Figure 8.1. Summary hydrograph statistics, including change of peak discharge (Q), stage, discharge volume, and infiltration volume for each of the scenarios are also presented in Tables 8.1 and 8.2. As predicted by the analysis of drainage basin characteristics (specifically, land cover values Manning's *n* and amount impervious surface), all of the alternative futures result in an increase in peak flood and total discharge volume, and a decrease in the amount of water that infiltrates into the soil. Note that a major proportion of infiltrated water (which can account for up to 80% of the amount precipitated) is used by vegetation and is lost from the basin through evapotranspiration; the remaining infiltrated water

is the only source of recharge for aquifers of the drainage basin—with, of course, the exception of the recharge basins in the Santa Margarita River floodplain on Camp Pendleton. Furthermore, reduced infiltration volume during the flood implies that infiltration is reduced during lower flows in other times of the year, and that the different development patterns would probably all result in less recharge of groundwater. This inference is a basis for additional research on the change of groundwater recharge as a function of basin development.

Peak Flows

The Northern Future scenario caused the greatest changes in peak flood and the discharge volume of the 25-year flood to increase by 38% for the 500,000 person allocation and 71% for the 1,000,000 person allocation. The Coastal Future scenario had the lowest increase of 12% and 26%. Even the lower increase of peak discharge would have a significant impact. Note that the 500,000-person Northern Future impact was even greater than the 1,000,000-person Coastal Future, indicating the effects of different spatial patterns of development. Doubling the population has less effect than the distribution of that population. The 100-year responses were different, with the Northern Future causing the greatest increase in peak Q (25% and 42% increases for the 500,000 and 1,000,000 allocations, respectively), the Coastal Future the least (9% and 17%), and the other two intermediate. The other two futures were, as expected, intermediate in increasing the flood peak.

Flood Stage

Figure 8.2 shows a statistically derived rating curve of the historical relationship between peak flood flow and the stage of the river at the Ysidora Gaging Station. River stage is defined as the elevation of the water surface at a given point. Although the relationship is noisy, the regression equation is somewhat useful for estimating the stage of future floods. The reader is cautioned to note that the larger model floods that resulted from the different scenarios extend beyond the range of the historical data used to calculate the regression

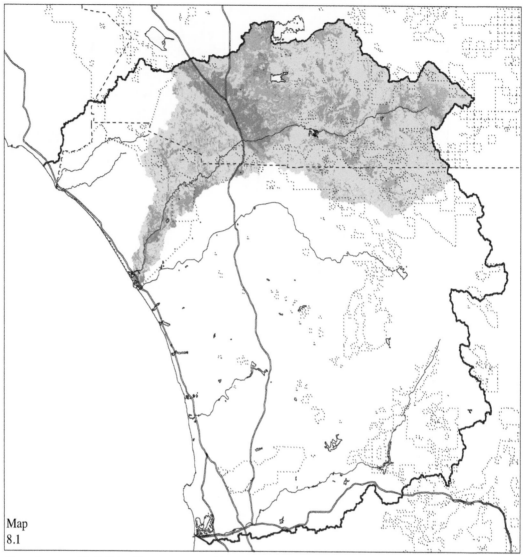

Map
8.1

Region of MCB Camp Pendleton & MCAS Miramar
Existing Conditions 2000
Manning's n for the Santa Margarita River Basin

0 2.5 5 Kilometers
0 2.5 5 Miles

0 to 0.10
48387 acres (2.8%)

0.11 to 0.20
57851 acres (3.3%)

0.21 to 0.30
46477 acres (2.7%)

0.31 to 0.40
27898 acres (1.6%)

0.41 to 0.70
301452 acres (17.4%)

Interstates

Rivers

Federal Lands

Counties

— See color insert following page 204 —

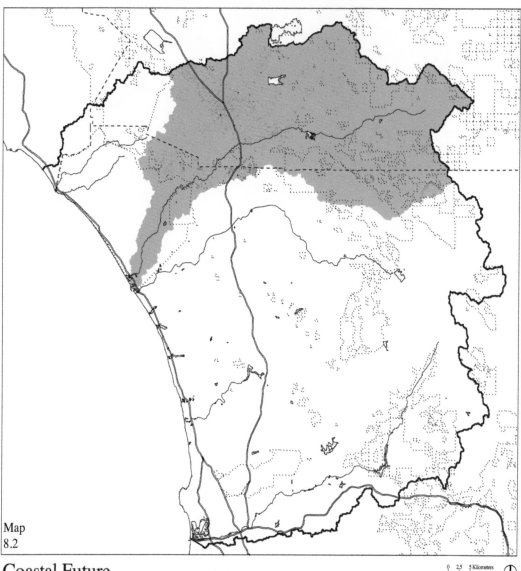

Map
8.2

Coastal Future
500,000 New Residents
Manning's n Change for the Santa Margarita River Basin

0 2.5 5 Kilometers
0 2.5 5 Miles

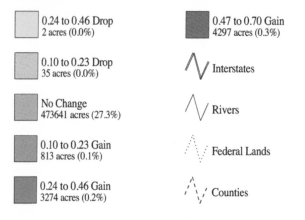

	0.24 to 0.46 Drop 2 acres (0.0%)		0.47 to 0.70 Gain 4297 acres (0.3%)
	0.10 to 0.23 Drop 35 acres (0.0%)		Interstates
	No Change 473641 acres (27.3%)		Rivers
	0.10 to 0.23 Gain 813 acres (0.1%)		Federal Lands
	0.24 to 0.46 Gain 3274 acres (0.2%)		Counties

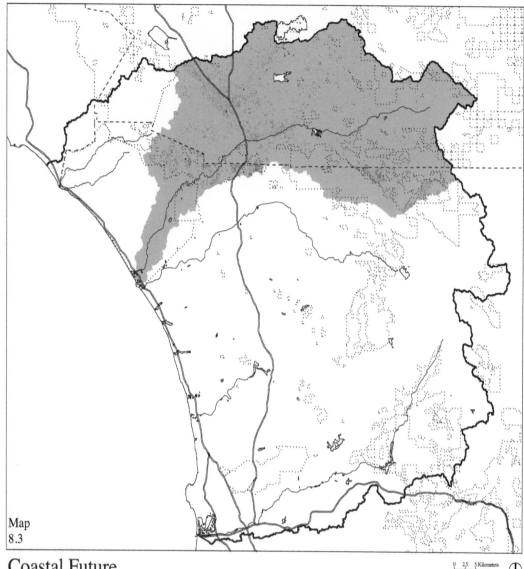

Map
8.3

Coastal Future
1,000,000 New Residents
Manning's n Change for the Santa Margarita River Basin

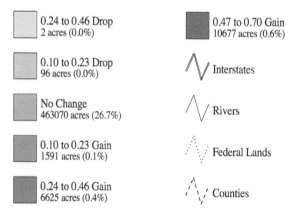

0.24 to 0.46 Drop 2 acres (0.0%)	0.47 to 0.70 Gain 10677 acres (0.6%)
0.10 to 0.23 Drop 96 acres (0.0%)	Interstates
No Change 463070 acres (26.7%)	Rivers
0.10 to 0.23 Gain 1591 acres (0.1%)	Federal Lands
0.24 to 0.46 Gain 6625 acres (0.4%)	Counties

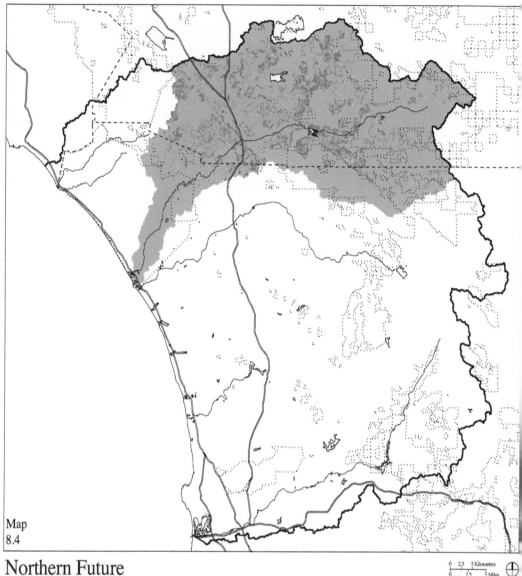

Map
8.4

Northern Future
500,000 New Residents
Manning's n Change for the Santa Margarita River Basin

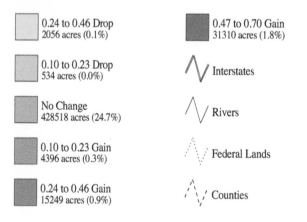

0.24 to 0.46 Drop
2056 acres (0.1%)

0.10 to 0.23 Drop
534 acres (0.0%)

No Change
428518 acres (24.7%)

0.10 to 0.23 Gain
4396 acres (0.3%)

0.24 to 0.46 Gain
15249 acres (0.9%)

0.47 to 0.70 Gain
31310 acres (1.8%)

Interstates

Rivers

Federal Lands

Counties

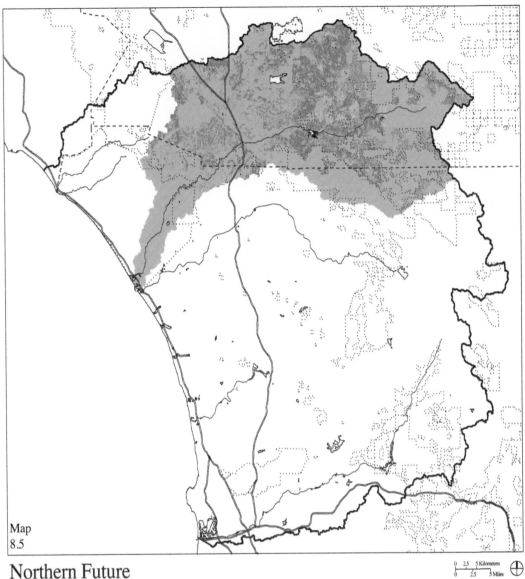

Map
8.5

Northern Future
1,000,000 New Residents
Manning's n Change for the Santa Margarita River Basin

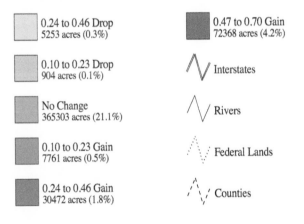

| | 0.24 to 0.46 Drop
5253 acres (0.3%) | | 0.47 to 0.70 Gain
72368 acres (4.2%) |

0.10 to 0.23 Drop
904 acres (0.1%)

No Change
365303 acres (21.1%)

0.10 to 0.23 Gain
7761 acres (0.5%)

0.24 to 0.46 Gain
30472 acres (1.8%)

Interstates

Rivers

Federal Lands

Counties

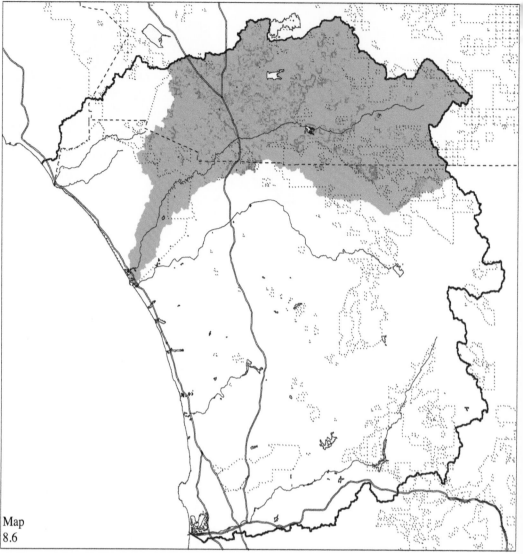

Map
8.6

Regional Low-Density Future
500,000 New Residents
Manning's n Change for the Santa Margarita River Basin

0 2.5 5 Kilometers
0 2.5 5 Miles

0.24 to 0.46 Drop
0 acres (0.0%)

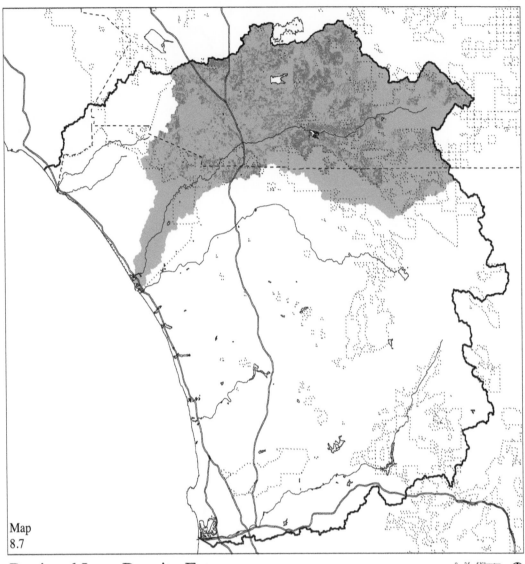

Map
8.7

Regional Low-Density Future
1,000,000 New Residents
Manning's n Change for the Santa Margarita River Basin

0   2.5   5 Kilometers
0   2.5   5 Miles

0.24 to 0.46 Drop
0 acres (0.0%)

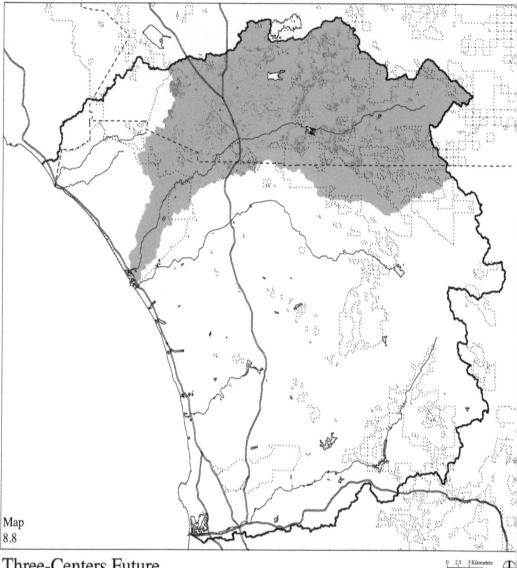

Map
8.8

Three-Centers Future
500,000 New Residents
Manning's n Change for the Santa Margarita River Basin

0.24 to 0.46 Drop
0 acres (0.0%)

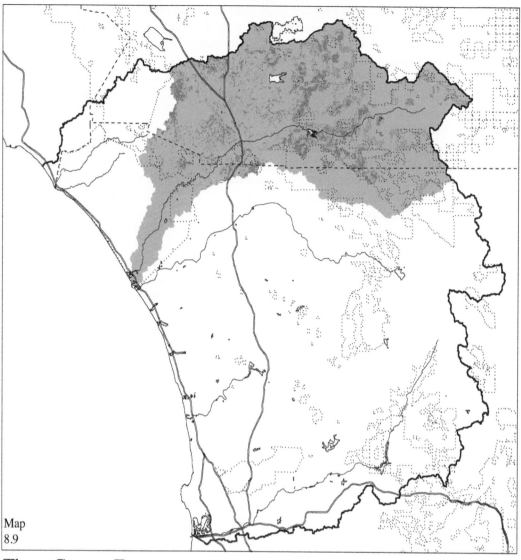

Map
8.9

Three-Centers Future
1,000,000 New Residents
Manning's n Change for the Santa Margarita River Basin

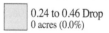

0.24 to 0.46 Drop
0 acres (0.0%)

equation, so the values are at best approximations of potential future flood stages. The morphology of the valley within which the flood occurs affects flood stage as flood flow and stage increase. If the valley broadens, as is the case in the Santa Margarita River basin, then increased discharge results in smaller stage increases as the extra water has a larger area over which to spread. The regression model makes no provision for valley morphology. Nonetheless, estimates of flood stage are useful for planning flood-control structures and other infrastructure improvements as long as the uncertainty of extrapolation beyond the data is recognized.

Calculated river stages above both the Ysidora gage datum (given as 75 feet in the official USGS description of the gage)[17] and above mean sea level for existing conditions and each future are presented in Tables 8.1 and 8.2. The measured stage of the January 1993 flood was 20.47 feet,[18] which is very close to the simulated 100-year flood stage of 20.8 feet, and therefore the simulated floods seem to produce reasonably accurate stage estimates. The change in stage for both the 25- and 100-year floods mirrored the changes in peak discharge, with Northern Future (1.2 feet increase in peak stage for the 500,000 person allocation, 2.2 feet for the 1,000,000 allocation), Three-Centers (1.2 feet and 2.3 feet), and Regional Low-Density (1.2 feet and 2.2 feet) had nearly the same effect on the 25-year flood stage, while the Coastal Future (0.4 feet and 0.9 feet) caused only a small increase. For the 100-year flood, the Northern Future scenario has the greatest impact, with stage 3.9 feet and 6.5 feet higher (500,000 and 1,000,000 populations, respectively). The Regional Low-Density Future causes almost as much change as the Northern Future. These changes may not seem very large, but must be evaluated with consideration of the height of flood control structures such as the levee around the Air Station at Camp Pendleton, and the elevation of buildings, roads, and other structures.

Discharge and Infiltration Volumes

Discharge volume is the total amount of water that flows out of the basin during the flood, and is calculated by integrating the hydrograph over the time of simulation. All these hydrographs were simulated with the same amount of precipitation, so

if runoff and consequent flood volume increases, the volume of water that infiltrates into the soil must be reduced. Sharp-eyed readers will note that the sums of discharge and infiltration volumes shown in Table 8.1 are not all equal. The difference is the volume of water that remains in the stream channels at the end of each simulation; this volume is not reported in the table.

As the Northern Future had the greatest amount of impervious surface, it usually causes the greatest discharge volume and lowest infiltration volumes. The development pattern of the Northern Future may cause as much as 3,200 (for the 25-year flood) or 10,550 acre-feet (100-year flood) of water to be prevented from infiltrating. (Note: an acre-foot is the volume of water that covers one acre to a depth of one foot, or 43,560 cubic feet.) So, if we assume that 80% of infiltrated water is later evaporated or transpired back to the atmosphere,[19] then 640 acre-feet and 2,110 acre-feet, respectively, are lost by the increased flood and not available for percolation to groundwater. The other futures have lesser impacts, but all cases result in loss of groundwater. To put this in perspective, an average U.S. family in a three-bedroom house is estimated to use approximately 1 acre-foot of water in a year.[20]

Discussion

All scenarios result in increases in flood peak flow, stage, volume flow, and decreases in infiltration. The Northern Future has the greatest impact, while the Coastal Future has the least. These results are expected, first because the Northern Future resulted in the largest area of impervious surface and second, because it has the greatest decrease in surface roughness, modeled with Manning's n parameter. More water flowed over the landscape faster to contribute to the flood.

Higher discharges during floods necessarily result in less recharge to groundwater aquifers. The amounts of water lost to the ocean are very high (467 acre-feet for the Coastal Future with 500,000 people during the 25-year flood to 11,635 acre-feet for the Three-Centers Future with 1,000,000 people during the 100-year storm). These volumes are during a single flood event, but can be extrapolated to some degree over the entire rainy season as

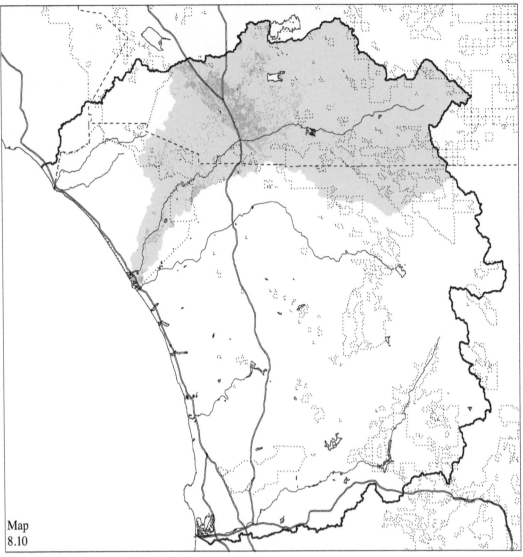

Map
8.10

Region of MCB Camp Pendleton & MCAS Miramar
Existing Conditions 2000
Impervious Surface for the Santa Margarita River Basin

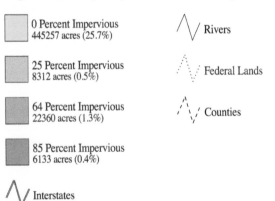

0 Percent Impervious
445257 acres (25.7%)

25 Percent Impervious
8312 acres (0.5%)

64 Percent Impervious
22360 acres (1.3%)

85 Percent Impervious
6133 acres (0.4%)

Rivers

Federal Lands

Counties

Interstates

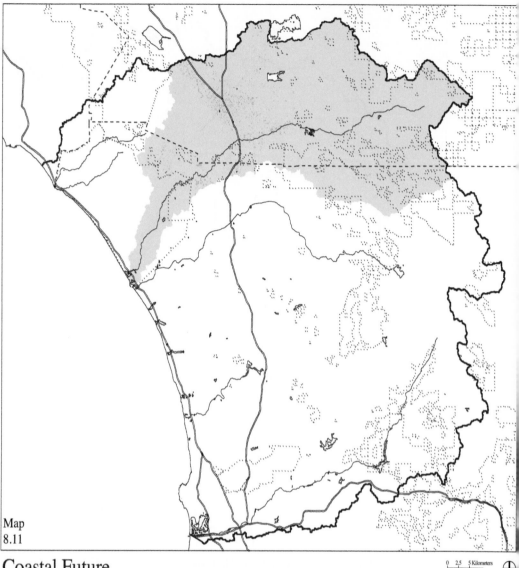

Map
8.11

Coastal Future
500,000 New Residents
Impervious Surface Change for the Santa Margarita River Basin

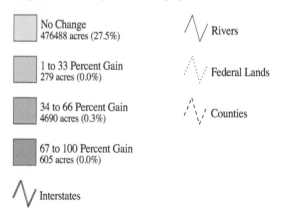

No Change
476488 acres (27.5%)

1 to 33 Percent Gain
279 acres (0.0%)

34 to 66 Percent Gain
4690 acres (0.3%)

67 to 100 Percent Gain
605 acres (0.0%)

Interstates

Rivers

Federal Lands

Counties

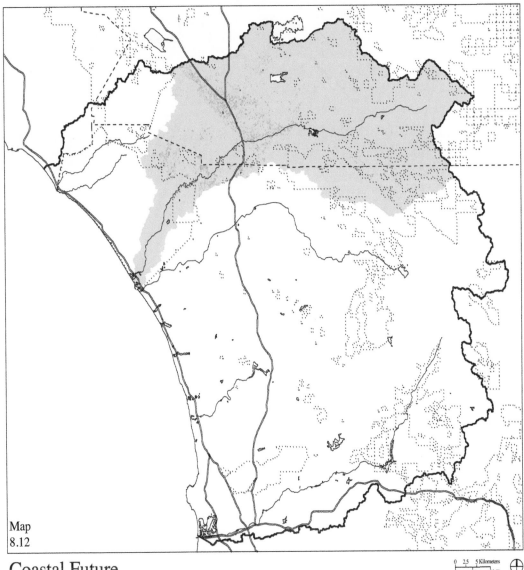

Map
8.12

Coastal Future
1,000,000 New Residents
Impervious Surface Change for the Santa Margarita River Basin

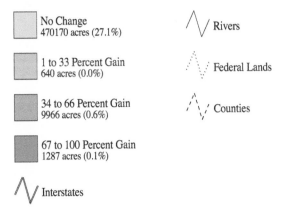

No Change 470170 acres (27.1%)	Rivers
1 to 33 Percent Gain 640 acres (0.0%)	Federal Lands
34 to 66 Percent Gain 9966 acres (0.6%)	Counties
67 to 100 Percent Gain 1287 acres (0.1%)	
Interstates	

0   2.5   5 Kilometers
0   2.5   5 Miles

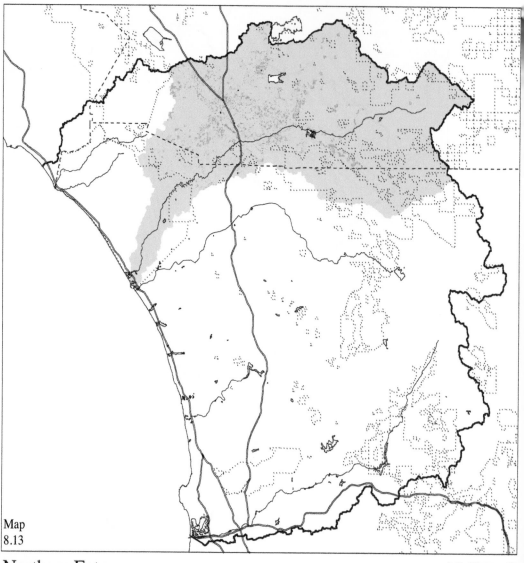

Map
8.13

Northern Future
500,000 New Residents
Impervious Surface Change for the Santa Margarita River Basin

0 2.5 5 Kilometers
0 2.5 5 Miles

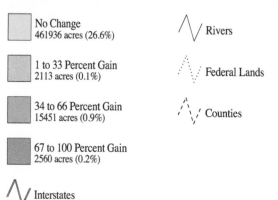

No Change
461936 acres (26.6%)

Rivers

1 to 33 Percent Gain
2113 acres (0.1%)

Federal Lands

34 to 66 Percent Gain
15451 acres (0.9%)

Counties

67 to 100 Percent Gain
2560 acres (0.2%)

Interstates

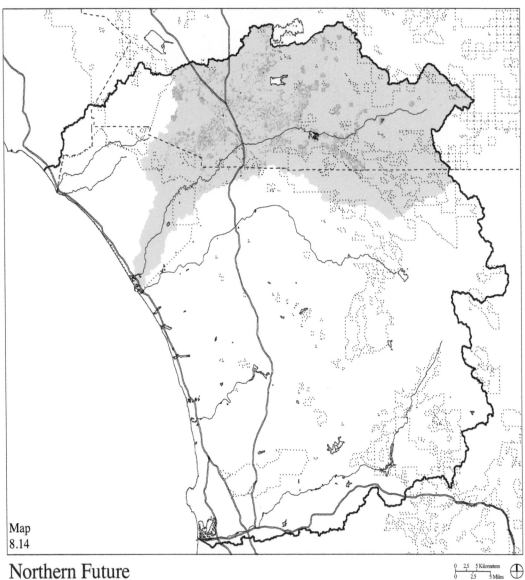

Map
8.14

Northern Future
1,000,000 New Residents
Impervious Surface Change for the Santa Margarita River Basin

0   2.5   5 Kilometers
0   2.5   5 Miles

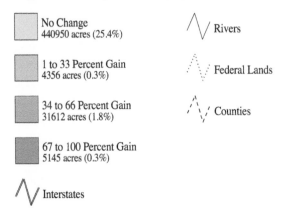

No Change
440950 acres (25.4%)

1 to 33 Percent Gain
4356 acres (0.3%)

34 to 66 Percent Gain
31612 acres (1.8%)

67 to 100 Percent Gain
5145 acres (0.3%)

Rivers

Federal Lands

Counties

Interstates

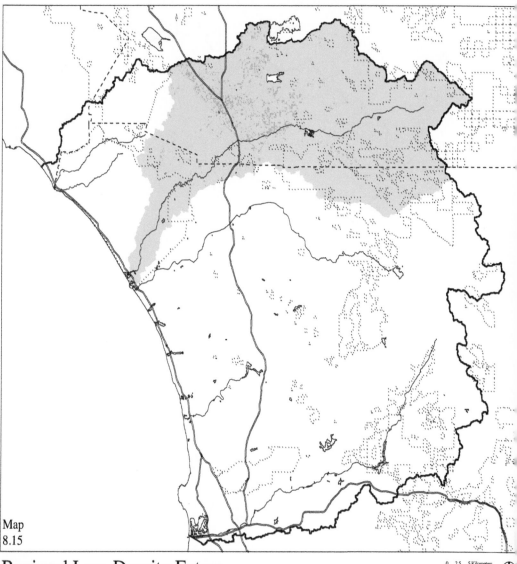

Map
8.15

Regional Low-Density Future
500,000 New Residents
Impervious Surface Change for the Santa Margarita River Basin

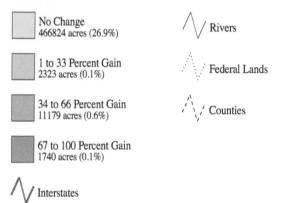

No Change
466824 acres (26.9%)

1 to 33 Percent Gain
2323 acres (0.1%)

34 to 66 Percent Gain
11179 acres (0.6%)

67 to 100 Percent Gain
1740 acres (0.1%)

Interstates

Rivers

Federal Lands

Counties

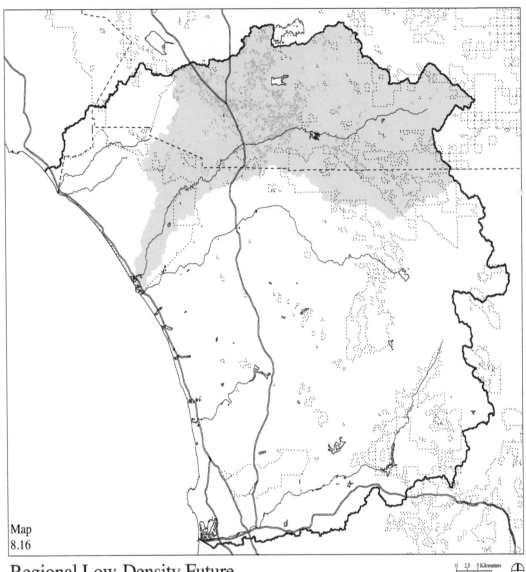

Map
8.16

Regional Low-Density Future
1,000,000 New Residents
Impervious Surface Change for the Santa Margarita River Basin

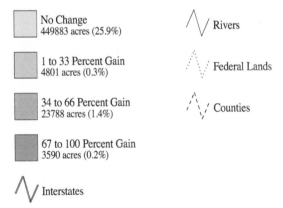

No Change
449883 acres (25.9%)

1 to 33 Percent Gain
4801 acres (0.3%)

34 to 66 Percent Gain
23788 acres (1.4%)

67 to 100 Percent Gain
3590 acres (0.2%)

Rivers

Federal Lands

Counties

Interstates

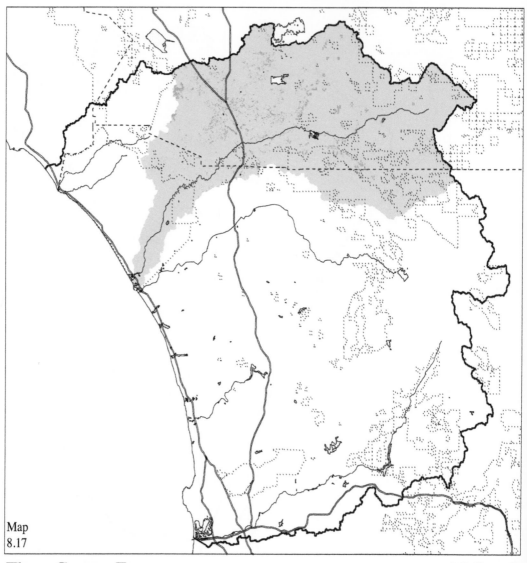

Map
8.17

Three-Centers Future
500,000 New Residents
Impervious Surface Change for the Santa Margarita River Basin

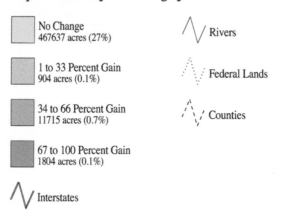

No Change
467637 acres (27%)

1 to 33 Percent Gain
904 acres (0.1%)

34 to 66 Percent Gain
11715 acres (0.7%)

67 to 100 Percent Gain
1804 acres (0.1%)

Interstates

Rivers

Federal Lands

Counties

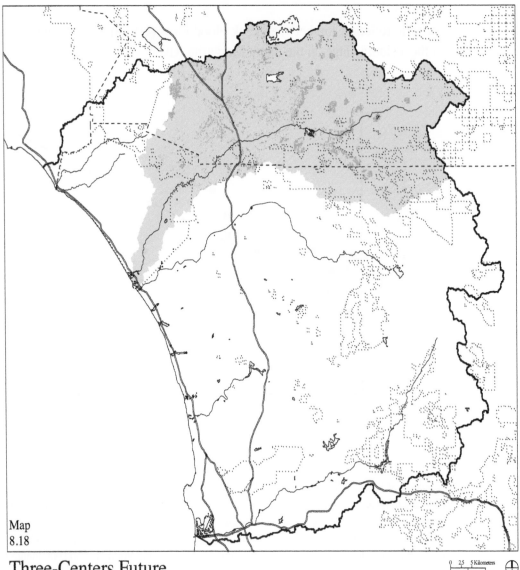

Map
8.18

Three-Centers Future
1,000,000 New Residents
Impervious Surface Change for the Santa Margarita River Basin

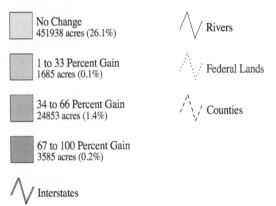

No Change
451938 acres (26.1%)

Rivers

1 to 33 Percent Gain
1685 acres (0.1%)

Federal Lands

34 to 66 Percent Gain
24853 acres (1.4%)

Counties

67 to 100 Percent Gain
3585 acres (0.2%)

Interstates

Figure 8.2

Historic Relationship between Peak Flood Flow (Q) and River Stage at the Ysidora Gage on the Santa Margarita River

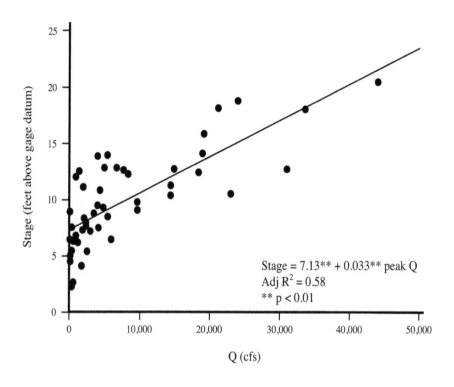

Stage = 7.13** + 0.033** peak Q
Adj R^2 = 0.58
** p < 0.01

indicators of how much water would no longer be available to recharge the aquifer.

Larger floods have greater sediment erosion and transport power, so riparian habitats will be more susceptible to washout and re-colonization by opportunistic exotic species. The January 1993 flood resulted in major amounts of sediment scouring followed by colonization by *Arundo donax*. Eight years of eradication efforts have reduced the extent of *Arundo*, but future floods that occur may cause the same pattern to be repeated.

The pattern of development has a major influence on the change of flood characteristics. In both the 25– and 100–year floods, the Coastal Future with 1,000,000 people has less impact than all the other 500,000 person futures except Three-Centers peak discharge and Three-Centers and Regional Low-Density discharge volumes. This observation is not surprising, as the Coastal Future alters the Santa Margarita River drainage basin the least. The other three futures had similar impacts

on the 25–year flood, but showed large variance among the simulated results for the 100-year flood. For example, the Three-Centers 1,000,000 Future increased the 25–year peak discharge by 50%, while the Northern 1,000,000 Future increased it 71%. The same contrast for change of stage at the Ysidora gage for the 100–year flood was 5.1 feet (Three-Centers) and 6.5 feet (Northern) for a 1.4 feet difference, and the difference of change in infiltration volume amounted to 1,235 acre-feet (10,550 fewer acre-feet infiltrated during the flood of the Northern Future relative to present conditions, and 11,785 fewer acre-feet infiltrated as a result of the Three-Centers Future). The Three-Centers Future clusters development, results in a higher rate of runoff across impervious surfaces and more rapid delivery to stream channels, but also leaves larger areas for infiltration.

The floods examined in this study are statistically expected to return in intervals of 25 and 100 years. This empirical understanding links

Table 8.1

25-Year Storm Event Flood Statistics for the Santa Margarita River

	2000	Coastal		Northern		Reg. Low-Density		Three-Centers	
		500k	1,000k	500k	1,000k	500k	1,000k	500k	1,000k
Peak Discharge in cfs	9,432	10,530	11,920	13,050	16,140	12,930	15,520	11,970	14,160
Change from 2000		1,098	2,488	3,618	6,708	3,498	6,088	2,528	4,728
% Change from 2000		**11.6**	**26.4**	**38.4**	**71.1**	**37.1**	**64.5**	**26.9**	**50.1**
Discharge Volume in ac-ft	6,944	7,427	8,029	8,409	10,040	8,495	9,971	7,963	9,009
Change from 2000		483	1,085	1,465	3,096	1,551	3,027	1,019	2,065
% Change from 2000		**7.0**	**15.6**	**21.1**	**44.6**	**22.3**	**43.6**	**14.7**	**29.7**
Infiltration Volume in ac-ft	203,411	202,944	202,351	201,885	200,242	202,431	200,381	202,350	201,251
Change from 2000		-467	-1,060	-1,526	-3,169	-980	-3,030	-1,061	-2,160
% Change from 2000		**-0.2**	**-0.5**	**-0.8**	**-1.6**	**-0.5**	**-1.5**	**-0.5**	**-1.1**
Stage above Ysidora in ft	10.3	10.7	11.2	11.5	12.5	11.5	12.5	11.5	12.6
Change from 2000		0.4	0.9	1.2	2.2	1.2	2.2	1.2	2.3
Stage above Sea Level in ft	85.3	85.7	86.2	86.5	87.5	86.5	87.5	86.5	87.6
Change since 2000		0.4	0.9	1.2	2.2	1.2	2.2	1.2	2.3

Table 8.2

100-Year Storm Event Flood Statistics for the Santa Margarita River

	2000	Coastal		Northern		Reg. Low-Density		Three-Centers	
		500k	1,000k	500k	1,000k	500k	1,000k	500k	1,000k
Peak Discharge in cfs	45,670	49,650	53,590	56,910	64,780	55,330	62,960	53,950	60,520
Change from 2000		3,980	7,920	11,240	19,110	9,660	17,290	8,280	14,850
% Change from 2000		**8.7**	**17.3**	**24.6**	**41.8**	**21.5**	**37.9**	**18.1**	**32.5**
Discharge Volume in ac-ft	37,880	40,195	42,990	43,335	48,220	42,890	47,200	41,490	49,515
Change from 2000		2,315	5,110	5,455	10,340	5,010	9,320	3,610	11,635
% Change from 2000		**6.1**	**13.5**	**14.4**	**27.3**	**13.2**	**24.6**	**9.5**	**30.7**
Infiltration Volume in ac-ft	514,060	511,720	508,910	508,470	503,510	509,060	504,720	510,320	502,275
Change from 2000		-2,340	-5,150	-5,590	-10,550	-5,000	-9,340	-3,740	-11,785
% Change from 2000		**-0.5**	**-1.0**	**-1.1**	**-2.1**	**-1.0**	**-1.8**	**-0.7**	**-2.3**
Stage above Ysidora in ft	22.0	23.5	24.8	25.9	28.5	25.4	27.9	24.9	27.1
Change from 2000		1.5	2.8	3.9	6.5	3.4	5.9	2.9	5.1
Stage above Sea Level in ft	97.2	98.5	99.8	100.9	103.5	100.4	102.9	99.9	102.1
Change since 2000		1.3	2.6	3.7	6.3	3.2	5.7	2.7	4.9

them to all other flood events which all have similar expected recurrence rates. But, as demonstrated by this analysis, managing for the impacts of a given storm (or a season of storms) is not only a matter of how much precipitation may fall at a given time, but is also a function of the conditions on which that precipitation falls and over which it moves. Deceasing the area available for infiltration into the soil and increasing the velocity of water movement will amplify the impacts of any storm. That is, the 2.5-year flood (one with a 40% chance of occurring in a given year) discharge will increase some amount, as will the 10-year (10% probability), the 50-year (2% probability), and all other floods. This link between the volume of water falling from the sky and the situation on the ground presents significant implications for land use. It means that assuming a continuation of the historic pattern of rainfall—an assumption that ignores threats associated with long-term climate change—new urban and suburban development will exacerbate and accelerate flood events. That is, the current 25-year flood will become the future flood of a shorter interval (such as, perhaps the 20-year flood) as natural ground covers are replaced by impervious surfaces. In short, greater floods are more likely to occur more often. More specifically, this situation means that the entire hydrological regime (long-term hydrographs) is altered by new development to one of higher peak flows and greater flow volume more often, lower infiltration and less aquifer recharge, and more disturbance of riparian habitat.

The simulations performed in this study do not include the modeling of any engineering works that can mitigate flood flows. Standard flood-control structures such as strategically placed dams, retention and detention basins, and infiltration basins can be used to reduce flow in the channels. Maintenance of existing wetlands, and perhaps restoration of degraded wetlands, in the basin can also provide extra water storage and reduce runoff. The issue here is whether the cost of flood-control engineering can be borne by the communities whose population will increase in each of the futures.

In summary, all the futures considered will cause significant changes in the flows of all floods. Careful landscape design and planning may be able to mitigate some of the effects. Based on the principles of the hydrological model, several options can be suggested. First, areas of high infiltration capacity and areas with high Manning's n should be left undeveloped. Second, instead of channelizing the streams which speeds the flow of water to the ocean, a strategy of maintaining or restoring meandering channels could be followed to slow flood flows.

NOTES:

[1] I.R. Calder, "Hydrologic Effects of Land Use Change," Ch. 13, in D.R. Maidment, ed., *Handbook of Hydrology* (New York: McGraw–Hill, 1993).

[2] Baxter E. Vieux, *Distributed Hydrologic Modeling Using GIS*, Water Science and Technology Library, Volume 38 (Dordrecht, Netherlands: Kluwer Academic Publishers, 2001).

[3] Carl Steinitz, Michael Binford, Paul Cote, Thomas Edwards, Jr., Stephen Ervin, Craig Johnson, Ross Kiester, David Mouat, Douglas Olson, Allan Shearer, Richard Toth, Robin Wills, *Biodiversity and Landscape Planning: Alternative Futures for the Region of Camp Pendleton, California* (Cambridge, Massachusetts: Harvard University Graduate School of Design, 1996).

[4] Citing this software is problematic. There never has been a published reference to GRASS itself, and all the manuals have been electronic. The CASC2D model is first reported in the Julien et al. 1995 paper (for this reference) and the Ogden and Julien 2002, both are cited below.

[5] F.L. Ogden, *CASC2D Reference Manual* (Storrs, Connecticut: University of Connecticut, Department of Civil and Environmental Engineering, 1997). NB: The source code of version 1.19 was used, but modified to accommodate certain minor ways of handling data input.

[6] P.Y. Julien, B. Dahafian, and F.L. Ogden, "Raster–Based Hydrologic Modeling of Spatially–Varied Surface Runoff," *Water Resources Bulletin* 31(1995), pp. 523–536; F.L. Ogden and P.Y. Julien, "CASC2D: A Two–Dimensional, Physically–Based, Hortonian, Hydrologic Model," in V.J. Singh and D. Freverts, eds., *Mathematical Models of Small Watershed Hydrology and Applications* (Littleton, Colorado: Water Resources Publications, 2002)

[7] R.D. Dodson, "Advances in Hydrologic Computation" in David R. Maidment, ed., *Handbook of Hydrology*, (New York: McGraw–Hill, 1993); Also, Vieux.

[8] T. Dunne and L.B. Leopold, *Water in Environmental Planning* (San Francisco: W.H. Freeman, 1978).

[9] W.J. Rawls, D.L. Brakensiek, and N. Miller, "Green–Ampt Infiltration Parameters from Soils Data," *Transactions of the American Society of Agricultural Engineers* 26 (1983), pp. 62–70.

[10] W.J. Rawls, et al.

[11] Rawls, et al.

[12] Environmental Systems Research Institute, Redlands, California. See also on-line <http://www.esri.com>.

[13] Environmental Modeling Systems, Inc, *The Watershed Modeling System* (South Jordan, Utah: 2002). Also see on-line <http://www.environmental–center.com/software/ems–I/watershed.htm>.

[14] Rawls et al.

[15] Ven Te Chow, ed. *Handbook of Applied Hydrology: A Compendium of Water–Resources Technology* (New York: McGraw–Hill, 1964); Harry H. Barnes, Jr., United States Geological Survey, *Roughness Characteristics of Natural Channels*, US Geological Survey Water Supply Paper 1849 (Washington, D.C: U.S. Government Printing Office, 1967), available on –line <http://www.rcamnl.wr.usgs.gov/sws/fieldmethods/Indirects/nvalues/>.

[16] WEST Consultants, *Final Report: Santa Margarita River Hydrology, Hydraulics, and Sedimentation Study* (San Diego, California: WEST Consultants, Inc., 2000).

[17] U.S. Geological Survey, *California Hydrologic Data Report— 11046000 Santa Margarita River at Ysidora, CA*. See on-line <http://ca.water.usgs.gov/archive/waterdata/99/11046000.html>. Note that this link is data for the year 1999.

[18] U.S. Geological Survey, *Peak Streamflow for California*. See on-line <http://www.usgs.gov/ca/nwis/peak>, search for 11046000 (the site number for the Ysidora Gage on the Santa Margarita River).

[19] Dunne and Leopold state on page 126: "The annual amount of water leaving a drainage basin as runoff varies from less than 10 percent of the yearly precipitation in hot deserts to more than 90 percent in the Cascade Mountains of Washington. The difference between rainfall and runoff is largely explained by evapotranspiration." The Santa Margarita River drainage basin is located in a semi–arid and Mediterranean climate, the assumption of about 20% of the water flowing out is reasonable for this example, which is meant to be illustrative and not quantitatively accurate.

[20] Dunne and Leopold, pp. 458–459 (from calculations).

Chapter 9

Air Quality Consequences Associated with the Alternative Futures

Alan W. Gertler and Jülide Kahyaoğlu-Koračin

Air pollution is often associated with urbanization and industrialization, and starting from the early 20th century, governments have imposed regulations intended to reduce air pollutant emissions from industrial sources. In measurable ways, these efforts have been successful in improving the quality of the built environment. But while laws such as the U.S. Clean Air Act of 1970 have contributed to significant improvements in air quality, the relationships between urbanization and the atmosphere continue to be important issues at both local and regional scales.

Of increasing concern is the structure and design of urban developments, which can contribute to adverse ecological effects.[1] Uncoordinated or poorly planned urban development spreads the locations of homes and businesses over a large extent and as a result, there is an increased need to use motor vehicles to get to and from work, school, the grocery store, etc. In part, the longer the trips one has to take, the more pollutants are emitted into the atmosphere. Further, it must also be recognized that uncoordinated development—that is, sprawl—can make the use of mass transit operations difficult, if not impossible, to implement. Therefore, not only do people travel far on each trip, most travel alone, adding to the number of vehicles on the road.

As part of this study, the spatially allocated alternative futures were assessed with an atmospheric chemistry modeling system. To implement this analysis, the land use/land cover information (described earlier) was augmented with transportation, meteorological, emissions, and photochemical modeling components. The methods and modularity of the coupled land–atmospheric system allow its application to a broad region of interest.

In order to provide metrics to assess air quality, standards have been defined by both federal and state agencies. These standards are defined in terms of concentration levels for specific periods and pollutants that can be easily and routinely measured. They are developed based on effects data and are primarily designed to protect human health and welfare. Currently, there are seven criteria pollutants covered by the federally mandated National Ambient Air Quality Standards (NAAQS) (Table 9.1). These include: ozone (O_3), respirable particulate matter (PM_{10}, particulate matter with aerodynamic diameter less than 10 μm), fine particulate matter ($PM_{2.5}$, particulate matter with aerodynamic diameter less than 2.5 μm), carbon monoxide (CO), nitrogen dioxide (NO_2), sulfur dioxide (SO_2), and lead (Pb). Federal primary standards provide a margin of safety to protect human health. Secondary standards are designed to protect health and welfare from known and anticipated impacts. Currently only SO_2 has a secondary standard that differs from the Federal primary standard.

In addition to the federal standards, states can promulgate standards that are more stringent than those set in Washington, D.C. This ability can include regulating additional pollutants. The California standards are also listed in Table 9.1. Notably, these standards are more stringent than federal standards and include four additional pollutants (visibility, sulfates, hydrogen sulfide (H_2S), and vinyl chloride (CH_2CHCl)). Thus, an area can be in attainment for the federal standard but in non-attainment for the California standard. While the air quality in San Diego County is significantly better than in other areas of California (e.g., Los Angeles County), it still violates a number of federal and California air quality standards for

Table 9.1
Ambient Air Quality Standards

Pollutant	Averaging Time	California Standards[1] Concentration[3]	Method	Federal Standards[2] Primary[3,5]	Secondary[3,6]	Method[7]
Ozone (O_3)[8]	1 Hour	180 µg/m³	Ultraviolet Photometry	—	Same as Primary Standard	Ultraviolet Photometry
	8 Hour	137 µg/m³		147 µg/m³		
Respirable Particulate Matter (PM_{10})	24 Hour	50 µg/m³	Gravimetric or Beta Attenuation	150 µg/m³	Same as Primary Standard	Inertial Separation and Gravimetric Analysis
	Annual Arithmetic Mean	20 µg/m³		—		
Fine Particulate Matter ($PM_{2.5}$)	24 Hour	No Separate State Standard		35 µg/m³ *	Same as Primary Standard	Inertial Separation and Gravimetric Analysis
	Annual Mean	12 µg/m³	Gravimetric or Beta Attenuation	15 µg/m³		
Carbon Monoxide (CO)	8 Hour	10 µg/m³	Non-Dispersive Infrared Photometry (NDIR)	10 µg/m³	None	Non-Dispersive Infrared Photometry (NDIR)
	1 Hour	23 µg/m³		40 µg/m³		
	8 Hour @ Lake Tahoe	7 µg/m³		—	—	
Nitrogen Dioxide (NO_2)	Annual Mean	57 µg/m³	Gas Phase Chemilumin-escence	100 µg/m³	Same as Primary Standard	Gas Phase Chemilumin-escence
	1 Hour	339 µg/m³		—		
Sulfur Dioxide (SO_2)	Annual Mean	—	Ultraviolet Fluorescence	80 µg/m³	—	Spectro-photometry (Pararosaniline Method)
	24 Hour	105 µg/m³		365 µg/m³	—	
	3 Hour	—		—	1300 µg/m³	
	1 Hour	655 µg/m³		—	—	
Lead (Pb)	30 Day Avg.	1.5 µg/m³	Atomic Absorption	—	—	—
	Calendar Qtr.	—		1.5 µg/m³	Same as Primary Standard	Atomic Absorption
Visability Reducing Particles	8 Hour	Vis. of 10 ml (30 @ Tahoe) due to particles w/ 70% relative humidity	Beta Attenuation & Transmittence through Filter Tape	No Federal Standards	—	—
Sulfates	24 Hour	25 µg/m³	Method	No Federal Standards	—	—
Hydrogen Sulfide	1 Hour	42 µg/m³	Ultraviolet Fluorescence	No Federal Standards	—	—
Vinyl Chloride[9]	24 Hour	26 µg/m³	Gas Chromotography	No Federal Standards	—	—

From the California Air Resources Board <<http://www.arb.ca.gov/research/aaqs/aaqs2.pdf>> (June 26, 2008)

Notes to Table 9.1:

1. California standards for ozone, carbon monoxide (except Lake Tahoe), sulfur dioxide (1- and 24-hour), nitrogen dioxide, suspended particulate matter—PM_{10}, $PM_{2.5}$, and visibility reducing particles, are values that are not to be exceeded. All others are not to be equaled or exceeded. California ambient air quality standards are listed in the Table of Standards in Section 70200 of Title 17 of the California Code of Regulations.

2. National standards (other than ozone, particulate matter, and those based on annual averages or annual arithmetic mean) are not to be exceeded more than once a year. The ozone standard is attained when the fourth highest eight–hour concentration in a year, averaged over three years, is equal to or less than the standard. For PM_{10}, the 24-hour standard is attained when the expected number of days per calendar year with a 24-hour average concentration above $150 \, \mu g/m3$ is equal to or less than one. For $PM_{2.5}$, the 24-hour standard is attained when 98 percent of the daily concentrations, averaged over three years, are equal to or less than the standard. Contact U.S. EPA for further clarification and current federal policies.

3. Concentration expressed first in units in which it was promulgated. Equivalent units given in parentheses are based upon a reference temperature of 25°C and a reference pressure of 760 torr. Most measurements of air quality are to be corrected to a reference temperature of 25°C and a reference pressure of 760 torr; ppm in this table refers to ppm by volume, or micromoles of pollutant per mole of gas.

4. Any equivalent procedure which can be shown to the satisfaction of the ARB to give equivalent results at or near the level of the air quality standard may be used.

5. National Primary Standards: The levels of air quality necessary, with an adequate margin of safety to protect the public health.

6. National Secondary Standards: The levels of air quality necessary to protect the public welfare from any known or anticipated adverse effects of a pollutant.

7. Reference method as described by the EPA. An "equivalent method" of measurement may be used but must have a "consistent relationship to the reference method" and must be approved by the EPA.

8. New federal 8-hour ozone and fine particulate matter standards were promulgated by U.S. EPA on July 18,1997. Contact U.S. EPA for further clarification and current federal policies.

9. The ARB has identified lead and vinyl chloride as 'toxic air contaminants' with no threshold level of exposure for adverse health effects determined. These actions allow for the implementation of control measures at levels below the ambient concentrations specified for these pollutants.

*   While the latest federal $PM_{2.5}$ standard is $35 \, \mu g/m^3$, it was $65 \, \mu g/m^3$ at the time of this study.

O_3, PM_{10}, and $PM_{2.5}$. Table 9.2 summarizes the non-attainment status for the pollutants that exceed either the federal or California standards. The San Diego area has yet to be designated with respect to the state $PM_{2.5}$ standard; however, since it was recently designated as non-attainment for the federal annual $PM_{2.5}$ standard, it should receive the same classification for the annual state standard, which is more stringent than the federal standard.

Sources of Pollutants

In order to develop effective strategies to improve ambient air quality, it is necessary to identify the sources of pollutants and their precursors. Some pollutants, such as CO, are directly emitted from sources. These are referred to as primary pollutants. Controlling emissions at the source will directly influence the observed ambient concentration of these pollutants. Others, such as O_3, are formed by chemical reactions in the atmosphere and are referred to as secondary pollutants. Since these are not directly emitted, it is necessary to understand the chemical and physical processes leading to their formation and to control the appropriate pollutant precursors. $PM_{2.5}$ is unusual in that it is both a primary (e.g., directly emitted from combustion sources such as diesel engines) and a secondary pollutant (formed by chemical reactions). PM_{10} tends to be dominated by primary emissions; although some PM_{10} is formed as a secondary pollutant. Prior to the development of our current understanding of O_3 formation, the mechanism producing O_3 in the troposphere was believed to consist of only three reactions:

Table 9.2

Current Non-Attainment Designations for San Diego County

Pollutant	Average Time	Federal Standard	State Standard
O_3	1-Hour	Maintenance Area	Non-Attainment
O_3	8-Hour	Non-Attainment	Not Applicable
PM_{10}	Annual	Attainment	Non-Attainment
$PM_{2.5}$	Annual	Non-Attainment	Not Yet Designated

Eq. 9.1 $NO_2 + h\upsilon \rightarrow NO + O$

Eq. 9.2 $O + O_2 \rightarrow O_3$

Eq. 9.3 $O_3 + NO \rightarrow NO_2 + O_2$

In the first reaction (Equation 9.1), sunlight ($h\upsilon$) causes NO_2 to photodissociate ($\lambda < 420$ nm) to yield NO and atomic oxygen, which, in turn, reacts (Equation 9.2) with molecular oxygen to yield O_3. In the final reaction, O_3 is destroyed. Equation 9.2 is the only reaction that forms O_3 in the troposphere. Based on this mechanism, the amount of O_3 should be proportional to the NO_2/NO ratio; however, the amount observed is often greater than predicted by this simple mechanism. Additional studies highlighted the importance of other reactions. Key to this better understanding is the reaction of hydrocarbons (volatile organic compounds, also known as VOCs) with OH to eventually form additional NO_2 (Figure 9.1). This results in an enhanced NO_2/NO ratio and increased levels of O_3. Thus, in order to control O_3, one needs to control the sources of VOCs or NO_x (the sum of NO_2 and NO). It is important to note that the cycle that forms O_3 is dependent on the VOC/NO_x ratio.

Reducing both VOCs and NO_x proportionately will not yield any change in the amount of O_3 formed. In addition, depending on whether an area is NO_x or VOC limited, reducing one of these species may not reduce O_3 and, under some circumstances, can actually increase the amount of O_3. For this reason it is critical to employ an appropriate atmospheric chemistry model prior to implementing any control strategies.

As mentioned above, a significant fraction of $PM_{2.5}$ is a secondary pollutant. There are many reactions that can lead to fine particle formation in the atmosphere. For example, one of the products in the O_3 cycle is gaseous nitric acid (HNO_3). If ammonia (NH_3) is present, it will react with HNO_3 to form the particle ammonium nitrate (NH_4NO_3). This type of process is often referred to as gas-to-particle conversion. Similarly, SO_2 can react to form ammonium sulfate ($(NH_4)_2SO_4$). Secondary organic aerosols (SOA) can also be formed by the oxidation of VOCs.

To develop an understanding of the impact of future growth and development on air quality, one must take into account changing land use patterns, emissions, and chemistry. This is especially true for the San Diego region, where O_3 is solely formed by chemical reactions in the atmosphere and a fraction of the $PM_{2.5}$ is formed by secondary processes.

Figure 9.1
Photochemical Cycle Leading to Ozone Production

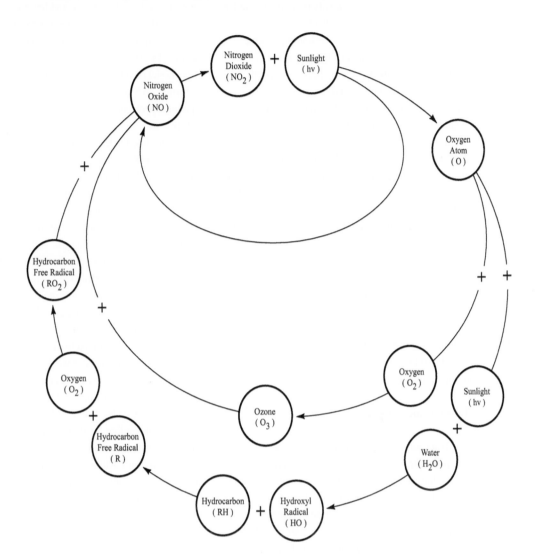

Description of Modeling System

Future emission assessments require knowledge on how an area may change. Factors include changes in population, use of technologies, and adoption of laws. Generally, the processes of change are described in the scenarios, mapped in the alternative futures, and assessed through evaluation models. To develop, test, and apply a modeling system to assess the impact of future scenarios on regional air quality in the San Diego area, we coupled a Geographic Information System that includes land cover and infrastructure data with pollutant emissions and air chemistry models. The modeling framework encompassed a number of discrete models involving future land use, emissions, air quality, and their subsequent linkages. These may be envisioned as a series of loosely coupled models with outputs and inputs being shared among models (Figure 9.2). In execution, the models are linked as follows: the model, the population, and land use change (the scenarios and alternative futures) is an input to the transportation model which is an input to the emissions model which is an input to the air quality model.

Transportation Model

Transportation models attempt to describe the flow of traffic between locations to allow for forecasting and analyzing future passenger and/ or freight movement.[2] A goal of a transportation model may be to produce an optimal route between a starting and stopping position or alternatively a goal of the transportation model may be to assess the consequences of certain traffic volumes over a given route.[3] Regardless of the specific goal, certain basic conditions must be understood before execution of any transportation model. The basic conditions are:

1. Potential beginning and ending locations must be known.

2. An overall transportation network (i.e., road map) must be present.

3. An estimate of overall potential traffic volume must be calculated.

The transportation model developed and described below accommodates all the basic conditions described above and has a low level of complexity. The model's intent is to route all the potential future passenger cars in the region from a home location to work along the quickest path. The model developed consists of a Geographic Information System (GIS) to coordinate all the spatial aspects and has linkages with aspatial information.

To initiate the transportation modeling process, the potential beginning and ending locations were identified. The beginning locations were based on regional census data that were adjusted to the logic and spatial distributions underlying each scenario. The study identified each potential new house location as a 30 m by 30 m grid cell within a GIS layer. House locations not only identified the house itself, but included added area for landscaping. Although this added area was indicative of the house lot plus the house, the added area is negligible when factored into a model for traffic. The ending locations were determined by identifying thirteen major commuting points or work centers (Map 9.1) within the study area or as locations where commuters would exit the study area heading mainly north to Los Angeles or to Northern Riverside County. The ending locations represented areas of major commercial–industrial centers within the study region. Although substantially more work centers (more than thirteen) exist within the study area, many of these are clumped into these commercial–industrial centers. Further, in identifying the number of centers it was necessary both to provide for distribution across the region and to accommodate computational constraints. It was judged by the researchers that thirteen centers accommodated these needs.

The transportation model incorporated the nationally available Tiger line transportation files and the associated attribute information as a foundation for determining the vehicular miles traveled between home and work.[4] At the time of the modeling, the 2002 Tiger line files represented the best readily available road network inclusive of the entire study region; however, it was recognized that critical connection errors existed within the Tiger line files. Attempts were made to correct errors resulting in the lack of connectivity

Figure 9.2
Air Quality Modeling System

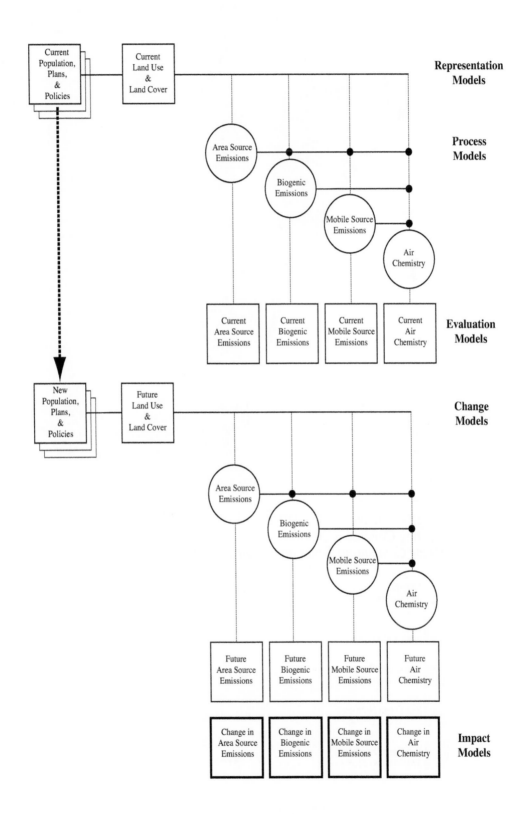

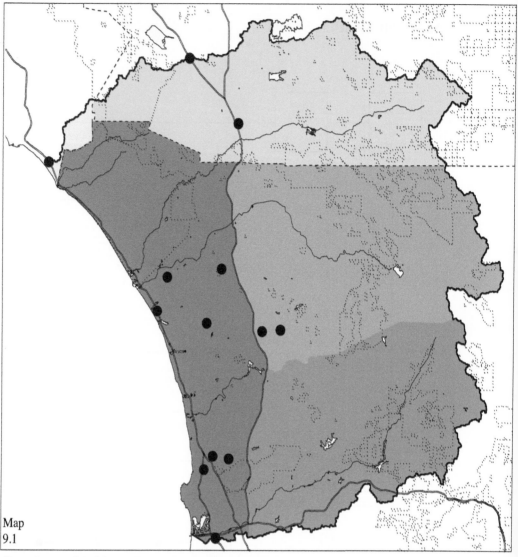

Map
9.1

Region of MCB Camp Pendleton & MCAS Miramar
Existing Conditions 2000
Major Commuting Zones and Associated Work Centers

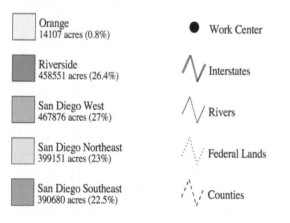

Orange
14107 acres (0.8%)

Riverside
458551 acres (26.4%)

San Diego West
467876 acres (27%)

San Diego Northeast
399151 acres (23%)

San Diego Southeast
390680 acres (22.5%)

● Work Center

Interstates

Rivers

Federal Lands

Counties

along known major road routes before execution of the transportation model. With over 196,000 connections or intersections contained within the 2002 Tiger line file, many were missed. As such, the data errors inherent within the Tiger line files have the potential to produce future travel routes that are excessively long.

Homes were randomly assigned a path to a commercial–industrial work center as a percentage derived from U.S. Census county commuter information.[5] The work center for each house was a function of the county of residence and the percentage of residents commuting from the county of residence to a particular county (i.e., if 100 new homes are added to Riverside County and 3% of Riverside residents commute to San Diego County for work, then 3 of the new Riverside homes would be randomly assigned to work centers in San Diego County for the calculation of travel routes to work). For the purpose of this analysis, San Diego County was divided into three zones while the other two counties, Orange and Riverside, consisted of single zones. The percentage of commuters between and within these zones is shown in Table 9.3.

Once homes were randomly assigned to a work place as a percentage of commuters, then the optimal travel route was determined. The optimal travel route was calculated as the road path which took the shortest time to travel from home to work. Calculations for determining the route were a function of impedance values as calculated using road segment speed limits and their lengths. The total path along a series of road segments with the lowest impedance value was the optimal path. Each road segment impedance value was determined according to the equation:

$$\text{Eq. 9.4} \quad \text{impedance} = \frac{60 * \text{road length (mi.)}}{\text{speed limit (mi/hr)}}$$

Once a path was determined, the total length of each commuting path by road speed was summarized for use by the mobile emissions model.

Emissions Model

The emissions model incorporated biogenic, area wide, and mobile source components. Future changes in stationary source emissions were not considered since the growth and development scenarios did not project the addition of new producers. This assumption is based on trends that indicate a reticence to add additional stationary sources that would increase emissions in the region. Existing emissions estimates from this and other categories served as a base for estimating future emissions.

Table 9.3

Percentage of Commuters Traveling from Home (rows) to Work Centers (columns)

	Orange Co. and Los Angeles Co.	Riverside Co.	San Diego Co. (West)	San Diego Co. (North East)	San Diego Co. (South East)
Orange Co. and Los Angeles Co.	97	2	1	0	0
Riverside Co.	25	72	2	1	0
San Diego Co. (West)	1	1	97	1	0
San Diego Co. (North East)	1	1	86	12	0
San Diego Co. (South East)	1	1	93	4	1

Biogenic Emissions

Spatially and temporally variable biogenic emissions were estimated for areas altered by future growth. Estimated pollutants consisted of biogenic volatile organic carbons (BVOC) such as isoprene, monoterpenes, and methylbutenol. BEIGIS, the biogenic emissions model from the California Air Resources Board (CARB), was used as the basis for this component of the larger model. BEIGIS is based on biomass and emissions studies performed in Southern California.[6] Default emission rates given in BEIGIS come mostly from Horie et al.[7] and Benjamin et al.[8] Urban and natural isoprene and monoterpene emission factors were based on Benjamin et al., which developed biogenic emission rates as a function of land use.[9]

BEIGIS was coupled within the GIS system at 30 m resolution. After determining future residential and commercial land cover, emission rates were given at standard conditions (30° C and 1000 μmol m^{-2} s^{-1}). Values were corrected using the isoprene and monoterpene algorithms from Guenther et al.[10] and the methylbutenol algorithm from Harley et al.[11] Temperature and photosynthetically active radiation (PAR) fields were generated by the meteorological forecast model. For low density housing, emission rates for isoprene and monoterpene were in the range for open set houses given in Benjamin et al.,[12] while for the high density development the rates given for close-set houses were used. Similarly, the emission rates given for commercial-industrial land use types were used for future commercial-industrial areas. To estimate the total future biogenic emissions, new emission factor layers were created in the GIS platform. Computations were then performed using the emission factor layers and the temperature and PAR fields provided by the meteorological model.

Area Wide Emissions

Area wide emissions were estimated based on the use of consumer products, residential natural gas consumption, dry cleaning, and residential and commercial lawn maintenance. If a source was not added or updated in the future scenarios (e.g., commercial solvent use), base emissions were used to represent the future emissions. Emissions were computed for all six criteria pollutants.

Methodologies used for area sources were based on the CARB's Emissions Inventory Procedure Manual.[13] Total emission rates were allocated spatially and temporally for each future and superimposed with the current emission layer.

Mobile Source Emissions

To estimate future on-road mobile source emissions, the GIS based travel simulation algorithm described in the transportation model section was utilized in conjunction with EMFAC2002, the on-road emissions model specific to California.[14] This model calculates emission factors and/or emission rates for the vehicle fleet in California as categorized in thirteen vehicle classes and accounts for six criteria pollutant types. Emission calculation processes include running, idle, and starting exhaust, diurnal HC, resting and running loss, hot soak as well as tire and brake wear. The model outputs total emission rates per geographic region in a format that shows the distribution of total emissions with respect to vehicle class and model year, speed, and technology class as well as different modes of operation (i.e., cold-start, hot-stabilized, etc.). Total vehicle miles traveled (VMT) served as the major input for EMFAC2002 as calculated by the transportation model. For the future scenarios we assumed that the growth in the number of vehicles consisted of light-duty personal vehicles. Every household unit was assigned 1.5 vehicles, which totaled, on average, 249,789 cars for the scenarios with 500k new residents and 498,270 cars for the scenarios with 1,000k new residents. Although the U.S. 2000 Census denotes that the average number of cars per household is 1.74, a value of 1.5 was used as a modifier to account for possibilities of carpooling and other means for arriving at work. Based on these parameters, mobile source emission factors were calculated for specified temperature and relative humidity values and for nine speed bins from 5 to 70 mi/hr. Emissions were then distributed spatially to the corresponding roadway links within the GIS framework. CARB weight fractions were used to temporally allocate the mobile emissions throughout the day. Commercial land use was assumed to be composed of mostly retail shops, and so sources associated with this type of activity were included in the future predictions.

Examples of Emissions Model Output

To investigate the possible maximum impact, all future emissions simulations were generated on daily basis for an episode from July 7 through July 12, 2003. During this period, San Diego experienced high levels of air pollution, and so it provided an opportunity to investigate a scenario where violations of air quality were likely to occur. Given that the issue is violation of the NAAQS and California AQS, this period is particularly relevant from an air quality management perspective. The final product of emissions modeling was a 5x5 km gridded hourly emissions inventory. Tables 9.4a - 9.4e present the area, mobile, stationary, and biogenic emissions estimates for the base case and eight alternative futures (four scenarios and two populations) for CO, NO_x, SO_2, TOG (VOCs), and $PM_{2.5}$. Also included in the tables are the incremental differences between the base case and alternative futures. Overall, the population increase drives the future emissions and there are few differences among futures having the same population. The data are somewhat misleading in that total study area results do not reflect the differences in the spatial distribution of the emissions. As seen in the modeling results section,

the variation in spatial distribution will lead to differences in the peak concentration and location of secondary pollutants (e.g., O_3).

As stated above, most of the change in emissions is driven by population and is due to the increase in VMT and differences in average vehicle speed. Scenarios with the maximum VMT or greatest speed variability will have the highest mobile source emissions. This result is apparent in the observed changes in CO, NO_x, TOG (VOCs), and PM emissions. For a number of scenarios, emissions in these categories increased by more than 10%, with the overwhelming fraction of this increase due to changes in mobile source emissions. Since mobile sources are the major source of CO, NO_x, and SO_2, along with being a significant source of TOG (VOCs), scenarios that maximize VMT will tend to have the greatest impact on future emissions. A number of other categories remain unchanged from the base scenario (i.e., stationary source emissions). Since the alternative futures did not include an increase in the number of stationary sources, it was assumed emissions from this category would remain constant. Other source categories often had zero associated emissions. For example, biogenic sources do not emit CO, NO_x, SO_2, and PM. Hence emissions from this source were always zero.

Table 9.4a

Emission Estimates—CO (tons/day)

	2003	Coastal		Northern		Reg. Low-Density		Three-Centers	
		500k	1,000k	500k	1,000k	500k	1,000k	500k	1,000k
Area	6.9	7.3	7.5	7.3	7.5	7.3	7.5	7.3	7.5
Change from 2003		0.4	0.6	0.4	0.6	0.4	0.6	0.4	0.6
% Change from 2003		5.8	8.7	5.8	8.7	5.8	8.7	5.8	8.7
Mobile	1,817.1	1,968.8	2,152.5	1,941.8	2,165.2	1,941.8	2,145.3	1,958.7	2,144.1
Change from 2003		151.7	335.3	124.7	348.1	124.7	328.2	141.6	327.0
% Change from 2003		8.3	18.5	6.9	19.2	6.9	18.1	7.8	18.0
Stationary	17.3	17.3	17.3	17.3	17.3	17.3	17.3	17.3	17.3
Change from 2003		0	0	0	0	0	0	0	0
% Change from 2003		0	0	0	0	0	0	0	0
Biogenic	0	0	0	0	0	0	0	0	0
Change from 2003		0	0	0	0	0	0	0	0
% Change from 2003		0	0	0	0	0	0	0	0

Table 9.4b

Emission Estimates—NO$_x$ (tons/day)

	2003	Coastal		Northern		Reg. Low-Density		Three-Centers	
		500k	1,000k	500k	1,000k	500k	1,000k	500k	1,000k
Area	2.6	3.3	3.9	3.3	3.9	3.3	3.9	3.3	3.9
Change from 2003		0.7	1.3	0.7	1.3	0.7	1.3	0.7	1.3
% Change from 2003		26.9	50.0	26.9	50.0	26.9	50.0	26.9	50.0
Mobile	220.2	233.0	247.3	230.6	249.2	230.8	247.3	228.6	247.0
Change from 2003		12.8	27.1	10.4	29.0	10.6	27.1	8.4	26.8
% Change from 2003		5.8	12.3	4.7	13.2	4.8	12.3	3.8	12.2
Stationary	18.5	18.5	18.5	18.5	18.5	18.5	18.5	18.5	18.5
Change from 2003		0	0	0	0	0	0	0	0
% Change from 2003		0	0	0	0	0	0	0	0
Biogenic	0	0	0	0	0	0	0	0	0
Change from 2003		0	0	0	0	0	0	0	0
% Change from 2003		0	0	0	0	0	0	0	0

Table 9.4c

Emission Estimates—SO$_2$ (tons/day)

	2003	Coastal		Northern		Reg. Low-Density		Three-Centers	
		500k	1,000k	500k	1,000k	500k	1,000k	500k	1,000k
Area	0.029	0.033	0.037	0.033	0.037	0.033	0.037	0.033	0.037
Change from 2003		0.004	0.008	0.004	0.008	0.004	0.008	0.004	0.008
% Change from 2003		13.8	27.6	13.8	27.6	13.8	27.6	13.8	27.6
Mobile	12.1	12.2	12.3	12.1	12.3	12.2	12.3	12.2	12.3
Change from 2003		0.1	0.2	0.1	0.2	0.1	0.2	0.1	0.2
% Change from 2003		0.8	1.7	0.8	1.7	0.8	1.7	0.8	1.7
Stationary	1.4	1.4	1.4	1.4	1.4	1.4	1.4	1.4	1.4
Change from 2003		0	0	0	0	0	0	0	0
% Change from 2003		0	0	0	0	0	0	0	0
Biogenic	0	0	0	0	0	0	0	0	0
Change from 2003		0	0	0	0	0	0	0	0
% Change from 2003		0	0	0	0	0	0	0	0

Examples of the spatial distribution of selected emissions, emissions by source category, and spatial variation in emissions for alternative futures are presented in Maps 9.2 to 9.19. Total NO$_x$ base case emissions are shown in Map 9.2 and emissions by source category are contained in Maps 9.3 to 9.6. NO$_x$ is dominated by mobile source emissions (Map 9.4) and most of these emissions occur along the coast and highways. As seen in this figure, biogenic NO$_x$ emissions are zero and area emissions are minimal. Maps 9.7 to 9.10 show the changes in emissions between the base case and the four scenarios with 1,000k new residents. There is a clear spatial difference in the pattern of emissions due to changes in the location of new residences.

Similar sets of data is presented for TOGs or VOCs in Maps 9.11 to 9.19, and Total Particulate Matter in Maps 9.20 to 9.24. Area, mobile, stationary and biogenic sources are all significant hydrocarbon emitters. Again, the major changes in future emissions are due to increased mobile sources. Stationary source emissions are assumed constant. As before, changes in the spatial distribution are evident among the four futures.

Table 9.4d

Emission Estimates—Total Organic Gas (Volatile Organic Compounds) (tons/day)

	2003	Coastal		Northern		Reg. Low-Density		Three-Centers	
		500k	1,000k	500k	1,000k	500k	1,000k	500k	1,000k
Area	82.5	87.9	93.2	87.8	93.2	87.9	93.2	87.9	93.2
Change from 2003		5.4	10.7	5.4	10.7	5.4	10.7	5.4	10.7
% Change from 2003		6.5	13.0	6.5	13.0	6.5	13.0	6.5	13.0
Mobile	193.8	207.6	220.3	207.6	222.4	208.2	223.9	207.9	221.4
Change from 2003		13.8	26.5	13.8	28.6	14.4	30.1	14.1	27.6
% Change from 2003		7.1	13.7	7.1	14.6	7.4	15.5	7.3	14.2
Stationary	345.7	345.7	345.7	345.7	345.7	345.7	345.7	345.7	345.7
Change from 2003		0	0	0	0	0	0	0	0
% Change from 2003		0	0	0	0	0	0	0	0
Biogenic	157.3	160.1	162.6	158.7	160.0	157.1	157.8	160.8	163.3
Change from 2003		2.8	5.3	1.4	2.7	-0.2	0.5	3.5	6.0
% Change from 2003		1.8	3.7	0.9	1.7	-0.1	0.3	2.2	3.8

Table 9.4e

Emission Estimates—Total Particulate Matter (tons/day)

	2003	Coastal		Northern		Reg. Low-Density		Three-Centers	
		500k	1,000k	500k	1,000k	500k	1,000k	500k	1,000k
Area	287.1	287.2	287.2	287.2	287.2	287.2	287.2	287.2	287.2
Change from 2003		0.1	0.1	0.1	0.1	0.1	0.1	0.1	0.1
% Change from 2003		>0.1	>0.1	>0.1	>0.1	>0.1	>0.1	>0.1	>0.1
Mobile	6.1	7.0	8.0	7.0	8.3	7.0	8.4	7.0	8.1
Change from 2003		0.9	1.9	0.9	2.2	0.9	2.3	0.9	2.0
% Change from 2003		14.8	31.1	14.8	36.1	14.8	37.7	14.8	32.8
Stationary	20.2	20.2	20.2	20.2	20.2	20.2	20.2	20.2	20.2
Change from 2003		0	0	0	0	0	0	0	0
% Change from 2003		0	0	0	0	0	0	0	0
Biogenic	0	0	0	0	0	0	0	0	0
Change from 2003		0	0	0	0	0	0	0	0
% Change from 2003		0	0	0	0	0	0	0	0

Air Quality Modeling

To evaluate the impact of the future scenarios on the formation of secondary pollutants, the spatially and temporally resolved output from the emissions model was coupled with an air quality modeling system that can operate over multiple domains and included grid sizes from 1 to 36 km. Modeling was completed in two steps: a meteorological analysis and subsequent air chemistry analysis.

Meteorological Analysis

The Fifth Generation Penn State/NCAR Mesoscale Model (MM5) was used to generate all required field variables and the parameters for the air quality model.[15] MM5 is a well-known mesoscale, nonhydrostatic, terrain-following sigma coordinate model that is used in predictions of mesoscale air circulation. The model is supported with a number of physics options including six planetary boundary layer (PBL) schemes and eight moisture schemes. Nesting capabilities of the model suit its use for both

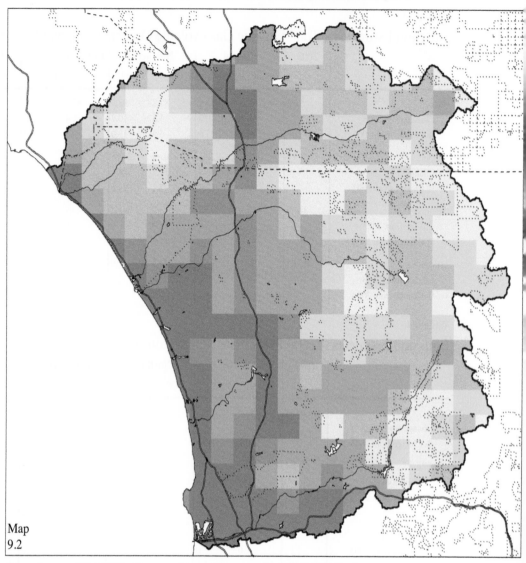

Map
9.2

Region of MCB Camp Pendleton & MCAS Miramar
Existing Conditions 2000
NO$_x$ Emissions

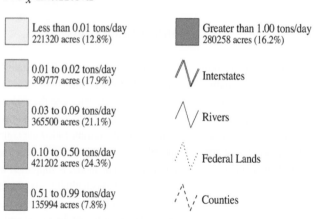

Less than 0.01 tons/day
221320 acres (12.8%)

0.01 to 0.02 tons/day
309777 acres (17.9%)

0.03 to 0.09 tons/day
365500 acres (21.1%)

0.10 to 0.50 tons/day
421202 acres (24.3%)

0.51 to 0.99 tons/day
135994 acres (7.8%)

Greater than 1.00 tons/day
280258 acres (16.2%)

Interstates

Rivers

Federal Lands

Counties

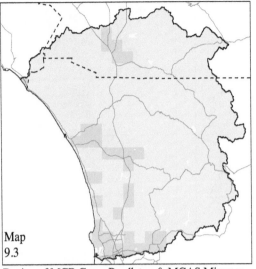

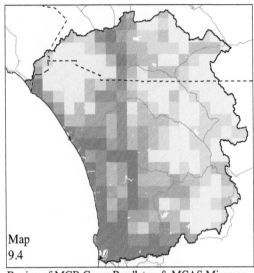

Map
9.3

Region of MCB Camp Pendleton & MCAS Miramar
Existing Conditions 2000
Area Wide Source NO$_x$ Emissions

- Less than 0.01 tons/day
  1526371 acres (88%)

- 0.01 to 0.02 tons/day
  206343 acres (11.9%)

- 0.03 to 0.09 tons/day
  1334 acres (0.1%)

Map
9.4

Region of MCB Camp Pendleton & MCAS Miramar
Existing Conditions 2000
Mobile Source NO$_x$ Emissions

- Less than 0.01 tons/day
  527835 acres (30.4%)

- 0.10 to 0.50 tons/day
  210490 acres (12.1%)

- 0.01 to 0.02 tons/day
  280355 acres (16.2%)

- 0.51 to 0.99 tons/day
  158188 acres (9.1%)

- 0.03 to 0.09 tons/day
  343706 acres (19.8%)

- Greater than 1.00 tons/day
  213477 acres (12.3%)

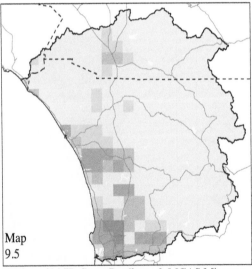

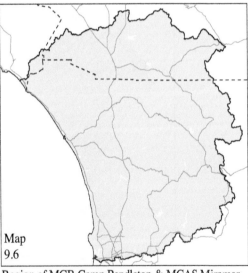

Map
9.5

Region of MCB Camp Pendleton & MCAS Miramar
Existing Conditions 2000
Stationary Source NO$_x$ Emissions

- Less than 0.01 tons/day
  1363822 acres (78.7%)

- 0.10 to 0.50 tons/day
  111076 acres (6.4%)

- 0.01 to 0.02 tons/day
  130237 acres (7.5%)

- 0.51 to 0.99 tons/day
  18495 acres (1.1%)

- 0.03 to 0.09 tons/day
  110236 acres (6.4%)

- Greater than 1.00 tons/day
  183 acres (0.0%)

Map
9.6

Region of MCB Camp Pendleton & MCAS Miramar
Existing Conditions 2000
Biogenic Source NO$_x$ Emissions

- Less than 0.01 tons/day
  1734049 acres (100%)

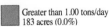

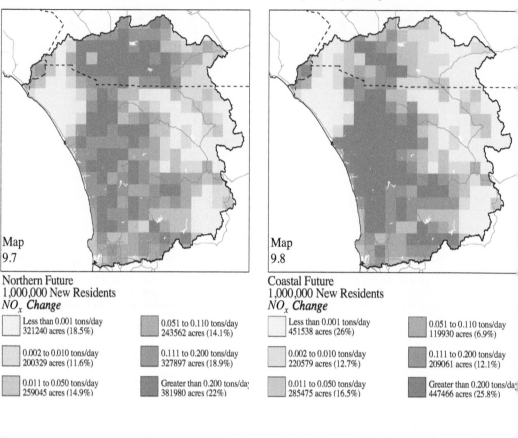

Map
9.7

Northern Future
1,000,000 New Residents
NO$_x$ Change

	Less than 0.001 tons/day 321240 acres (18.5%)		0.051 to 0.110 tons/day 243562 acres (14.1%)
	0.002 to 0.010 tons/day 200329 acres (11.6%)		0.111 to 0.200 tons/day 327897 acres (18.9%)
	0.011 to 0.050 tons/day 259045 acres (14.9%)		Greater than 0.200 tons/day 381980 acres (22%)

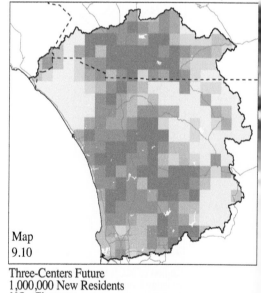

Map
9.8

Coastal Future
1,000,000 New Residents
NO$_x$ Change

	Less than 0.001 tons/day 451538 acres (26%)		0.051 to 0.110 tons/day 119930 acres (6.9%)
	0.002 to 0.010 tons/day 220579 acres (12.7%)		0.111 to 0.200 tons/day 209061 acres (12.1%)
	0.011 to 0.050 tons/day 285475 acres (16.5%)		Greater than 0.200 tons/day 447466 acres (25.8%)

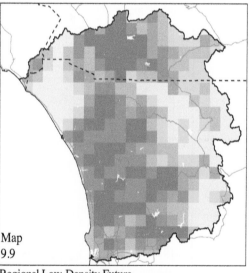

Map
9.9

Regional Low-Density Future
1,000,000 New Residents
NO$_x$ Change

	Less than 0.001 tons/day 336679 acres (19.4%)		0.051 to 0.110 tons/day 206267 acres (11.9%)
	0.002 to 0.010 tons/day 194300 acres (11.2%)		0.111 to 0.200 tons/day 327613 acres (18.9%)
	0.011 to 0.050 tons/day 280323 acres (16.2%)		Greater than 0.200 tons/day 388866 acres (22.4%)

Map
9.10

Three-Centers Future
1,000,000 New Residents
NO$_x$ Change

	Less than 0.001 tons/day 418397 acres (24.1%)		0.051 to 0.110 tons/day 210962 acres (12.2%)
	0.002 to 0.010 tons/day 154744 acres (8.9%)		0.111 to 0.200 tons/day 345206 acres (19.9%)
	0.011 to 0.050 tons/day 230799 acres (13.3%)		Greater than 0.200 tons/day 373941 acres (21.6%)

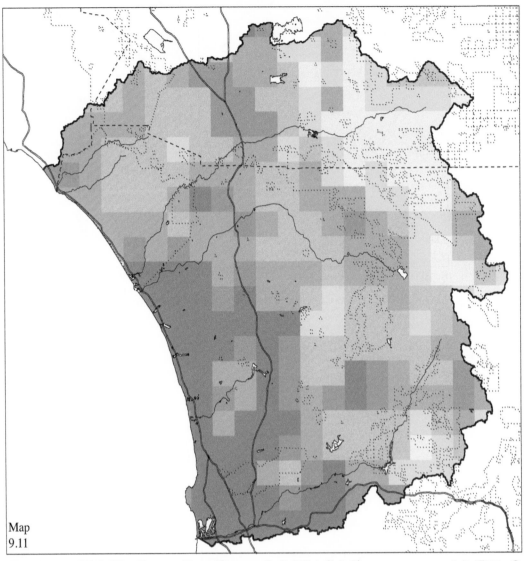

Map
9.11

Region of MCB Camp Pendleton & MCAS Miramar
Existing Conditions 2000
VOCs Emissions

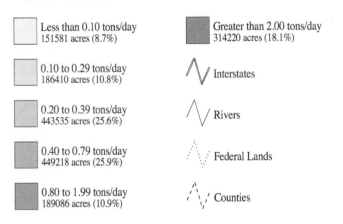

Less than 0.10 tons/day
151581 acres (8.7%)

0.10 to 0.29 tons/day
186410 acres (10.8%)

0.20 to 0.39 tons/day
443535 acres (25.6%)

0.40 to 0.79 tons/day
449218 acres (25.9%)

0.80 to 1.99 tons/day
189086 acres (10.9%)

Greater than 2.00 tons/day
314220 acres (18.1%)

Interstates

Rivers

Federal Lands

Counties

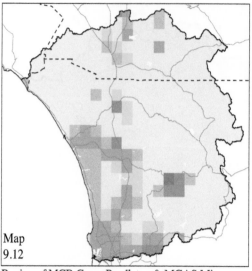

Map
9.12

Region of MCB Camp Pendleton & MCAS Miramar
Existing Conditions 2000
Area Wide VOCs Emissions

Less than 0.10 tons/day 1265987 acres (73%)		0.40 to 0.79 tons/day 158361 acres (9.1%)	
0.10 to 0.29 tons/day 102220 acres (5.9%)		0.80 to 1.99 tons/day 42538 acres (2.5%)	
0.20 to 0.39 tons/day 158777 acres (9.2%)		Greater than 2.00 tons/day 6165 acres (0.4%)	

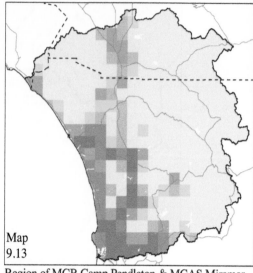

Map
9.13

Region of MCB Camp Pendleton & MCAS Miramar
Existing Conditions 2000
Mobile Source VOCs Emissions

Less than 0.10 tons/day 1172218 acres (67.6%)		0.40 to 0.79 tons/day 118687 acres (6.8%)	
0.10 to 0.29 tons/day 100612 acres (5.8%)		0.80 to 1.99 tons/day 114971 acres (6.6%)	
0.20 to 0.39 tons/day 117081 acres (6.8%)		Greater than 2.00 tons/day 110481 acres (6.4%)	

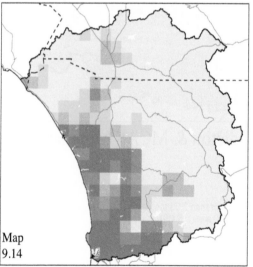

Map
9.14

Region of MCB Camp Pendleton & MCAS Miramar
Existing Conditions 2000
Stationary Source VOCs Emissions

Less than 0.10 tons/day 1083551 acres (62.5%)		0.40 to 0.79 tons/day 45321 acres (2.6%)	
0.10 to 0.29 tons/day 162490 acres (9.4%)		0.80 to 1.99 tons/day 117019 acres (6.8%)	
0.20 to 0.39 tons/day 113542 acres (6.6%)		Greater than 2.00 tons/day 212125 acres (12.2%)	

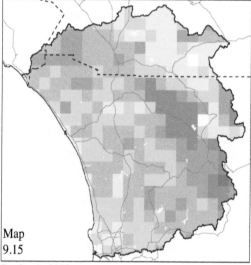

Map
9.15

Region in 2000
VOCs Emissions
Biogenic

Less than 0.10 tons/day 272230 acres (15.7%)		0.40 to 0.79 tons/day 248788 acres (14.4%)	
0.10 to 0.29 tons/day 476765 acres (27.5%)		0.80 to 1.99 tons/day 138008 acres (8%)	
0.20 to 0.39 tons/day 579153 acres (33.4%)		Greater than 2.00 tons/day 5903 acres (0.3%)	

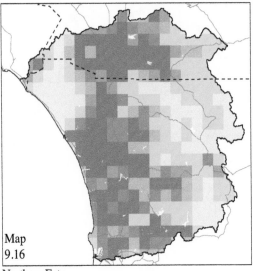

Northern Future
1,000,000 New Residents
VOCs Change

Less than 0.001 tons/day	0.201 to 0.300 tons/day
307649 acres (17.7%)	109604 acres (6.3%)
0.002 to 0.100 tons/day	0.301 to 0.500 tons/day
465452 acres (26.8%)	180027 acres (10.4%)
0.0101 to 0.200 tons/day	Greater than 0.500 tons/day
160175 acres (9.2%)	511144 acres (29.5%)

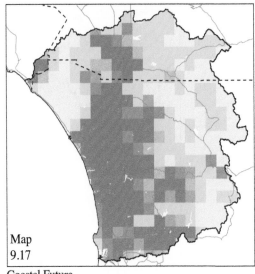

Coastal Future
1,000,000 New Residents
VOCs Change

Less than 0.001 tons/day	0.201 to 0.300 tons/day
415921 acres (24%)	65185 acres (3.8%)
0.002 to 0.100 tons/day	0.301 to 0.500 tons/day
509300 acres (29.4%)	117422 acres (6.8%)
0.0101 to 0.200 tons/day	Greater than 0.500 tons/day
84985 acres (4.9%)	541235 acres (31.2%)

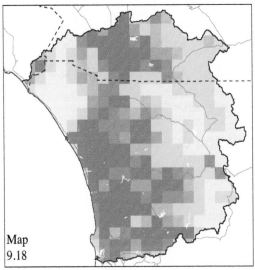

Regional Low-Density Future
1,000,000 New Residents
VOCs Change

Less than 0.001 tons/day	0.201 to 0.300 tons/day
329923 acres (19%)	115591 acres (6.7%)
0.002 to 0.100 tons/day	0.301 to 0.500 tons/day
461118 acres (26.6%)	192605 acres (11.1%)
0.0101 to 0.200 tons/day	Greater than 0.500 tons/day
140234 acres (8.1%)	494581 acres (28.5%)

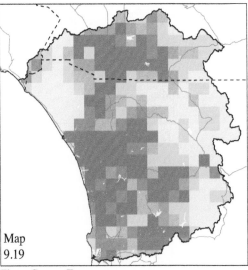

Three-Centers Future
1,000,000 New Residents
VOCs Change

Less than 0.001 tons/day	0.201 to 0.300 tons/day
407752 acres (23.5%)	94713 acres (5.5%)
0.002 to 0.100 tons/day	0.301 to 0.500 tons/day
338841 acres (19.5%)	166753 acres (9.6%)
0.0101 to 0.200 tons/day	Greater than 0.500 tons/day
196978 acres (11.4%)	529011 acres (30.5%)

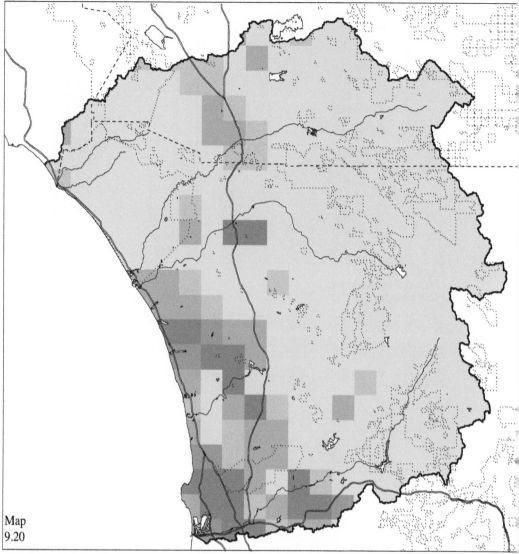

Map
9.20

Region of MCB Camp Pendleton & MCAS Miramar
Existing Conditions 2000
PM Emissions

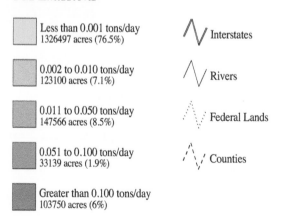

Less than 0.001 tons/day 1326497 acres (76.5%)		Interstates
0.002 to 0.010 tons/day 123100 acres (7.1%)		Rivers
0.011 to 0.050 tons/day 147566 acres (8.5%)		Federal Lands
0.051 to 0.100 tons/day 33139 acres (1.9%)		Counties
Greater than 0.100 tons/day 103750 acres (6%)		

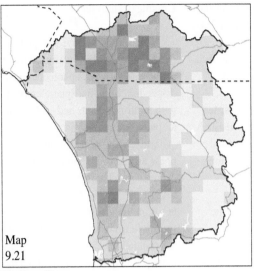

Map
9.21

Northern Future
1,000,000 New Residents
PM Change

Less than 0.00001 tons/day 396079 acres (22.8%)	0.00101 to 0.00150 tons/da 165747 acres (9.6%)
0.00002 to 0.00050 tons/day 679406 acres (39.2%)	0.00151 to 0.00200 tons/da 86571 acres (5%)
0.00051 to 0.00100 tons/day 332211 acres (19.2%)	Greater than 0.00200 tons/(74034 acres (4.3%)

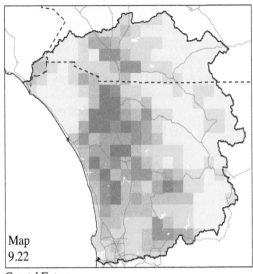

Map
9.22

Coastal Future
1,000,000 New Residents
PM Change

Less than 0.00001 tons/day 602892 acres (34.8%)	0.00101 to 0.00150 tons/da 128475 acres (7.4%)
0.00002 to 0.00050 tons/day 491601 acres (28.4%)	0.00151 to 0.00200 tons/da 123525 acres (7.1%)
0.00051 to 0.00100 tons/day 288615 acres (16.6%)	Greater than 0.00200 tons/(98939 acres (5.7%)

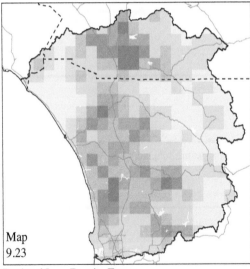

Map
9.23

Regional Low-Density Future
1,000,000 New Residents
PM Change

Less than 0.00001 tons/day 416050 acres (24%)	0.00101 to 0.00150 tons/da 185046 acres (10.7%)
0.00002 to 0.00050 tons/day 638121 acres (36.8%)	0.00151 to 0.00200 tons/da 105030 acres (6.1%)
0.00051 to 0.00100 tons/day 334168 acres (19.3%)	Greater than 0.00200 tons/(55635 acres (3.2%)

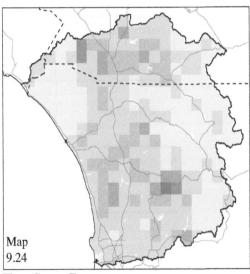

Map
9.24

Three-Centers Future
500,000 New Residents
PM Change

Less than 0.00001 tons/day 597552 acres (34.5%)	0.00101 to 0.00150 tons/da 67883 acres (3.9%)
0.00002 to 0.00050 tons/day 790310 acres (45.6%)	0.00151 to 0.00200 tons/da 18473 acres (1.1%)
0.00051 to 0.00100 tons/day 253628 acres (14.6%)	Greater than 0.00200 tons/(6202 acres (0.4%)

local and large area applications. MM5 provides the air quality model with 3-dimensional field variables such as wind and other parameters, which are translated by the air chemistry component (described below) for use in the atmospheric transport and dispersion calculations.

The meteorological modeling domain consisted of two nested grids (Map 9.25). The grids consisted of 75x75x35 and 61x55x35 points with horizontal cell sizes of 15 km and 5 km for the outer and inner grids, respectively. Both grids had 35 vertical layers extending up to the 100 mb level. The outer domain transfers synoptic processes into the inner domain, which was centered at 33.40N and 118.40W. Physical options used in meteorological simulations include Eta PBL scheme, Grell cumulus parameterization, and simple ice moisture scheme (Table 9.5).

Chemical Analysis

The Comprehensive Air quality Model with extensions (CAMx)[16] was used for chemical analysis. CAMx is a photochemical Eulerian dispersion "one-atmosphere" modeling system with multi-pollutants and scaling that can predict all phases of air chemistry including air toxics, visibility, and acid deposition. In addition to gas phase and aerosol chemistry, it provides two types of chemical mechanisms: Carbon Bond IV (CBIV) and SAPRC99. CAMx has a number of advantages in terms of applications and analyses. Its flexi-nesting feature (introducing or removing new nests during the simulation) makes high-resolution applications more practical. It also provides a tool to track the contribution of every individual source to ozone formation at each grid point. Another advanced

Table 9.5

Air Quality Mesoscape Model (MM5) Simulation Specifications

Domain	Two nested domains with a horizontal resolution of 15x15 kilometers for outer domain and 5x5 kilometers for the inner domain. Vertical resolution extends up to 100 mb in 35 sigma levels.
Planetary Boundary Layer (PBL) Scheme	Eta PBL scheme.
Cumulus Parameterization	Grell cumulus parameterization.
Moisture Scheme	Mixed phase explicit moisture scheme.
Soil Temperature	Multi-layer soil temperature model.
Radiation Scheme	Cloud top cooling.
Version	MM5 V3.6.2.

feature embedded in the model is a plume-in-grid module which simulates chemical transformation and dispersion from large point NO_x sources using a Lagrangian approach in a sub grid scale until the plume evolves to grid size levels. A parallel version of the model reduces the execution cost and allows long range and high-resolution applications to be more efficient.

CAMx simulations were performed using MM5 meteorological fields. As noted above, MM5 provided the necessary meteorological inputs such as 3-dimensional wind, temperature, pressure, water vapor, and vertical diffusivity. Another major input to the model is the emissions data generated by the modeling system. The geographical domain used by CAMx was the innermost domain of the meteorological model MM5 shown in Map 9.25 with 5x5 km horizontal grid cells and 11 vertical layers. Some of the technical options used in the simulations are given in Table 9.6. A total of nine separate simulations were performed including: base case (existing conditions) and four different scenarios, each with 500k and 1,000k additional population.

Results for Selected Pollutants as a Function of Alternative Futures

Rather than presenting the results for all pollutants, only those that are most important from an air quality management perspective and are difficult to control (i.e., the pollutants that exceed air quality standards and are secondary in nature formed by chemical reactions in the atmosphere) are reported. As shown in Table 9.2, the San Diego

Table 9.6

Comprehensive Air Quality Model (CAMx) Simulation Specifications

Domain	Area covered 294x264 kilometers; horizontal grid cell 5x5 kilometers.
Vertical Layers	33. 66. 133. 338, 617, 978, 1202, 1698, 2239, 2823, 3773, m AGL
Chemistry Solver	The CMC fast chemistry solver.
Advection Solver	The Bott advection solver.
Chemical Mechanism	Chemical mechanism-4 with aerosols; 34 gas species, 15 aerosol species.
Emissions	Gridded (5x5 kilometer) and point source emissions with hourly rates.
Version	CAMx version 4.03.

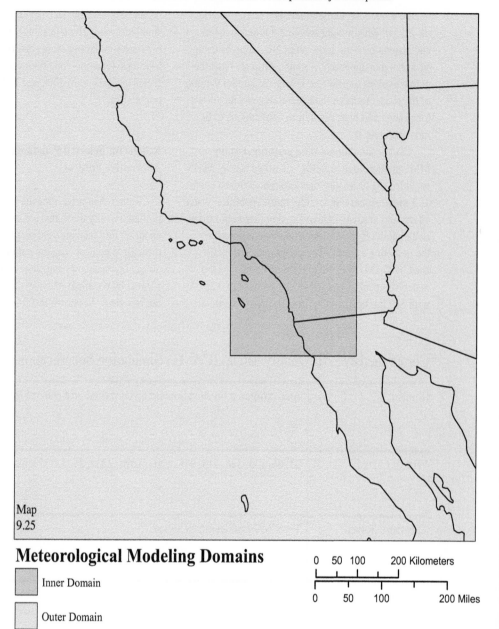

Map
9.25

Meteorological Modeling Domains

Inner Domain

Outer Domain

0   50 100   200 Kilometers

0   50   100   200 Miles

Table 9.7

Predicted Maximum Ozone (parts per billion)

	2003	Coastal		Northern		Reg. Low-Density		Three-Centers	
		500k	1,000k	500k	1,000k	500k	1,000k	500k	1,000k
July 7	115	124	128	123	138	120	150	121	139
Change from 2000		9	13	8	23	5	35	6	24
% Change from 2000		7.8	11.3	7.0	20.0	4.3	30.4	5.2	20.9
July 8	132	133	134	133	134	133	134	133	134
Change from 2000		1	2	1	2	1	2	1	2
% Change from 2000		0.8	1.5	0.8	1.5	0.8	1.5	0.8	1.5
July 9	129	132	136	131	142	131	142	131	142
Change from 2000		3	7	2	13	2	13	2	13
% Change from 2000		2.3	5.4	1.6	10.1	1.6	10.1	1.6	10.1
July 10	135	139	143	138	146	139	146	138	146
Change from 2000		4	8	3	11	4	11	3	11
% Change from 2000		3.0	5.9	2.2	8.1	3.0	8.1	2.2	8.1
July 11	121	125	128	122	133	124	137	124	134
Change from 2000		4	7	1	12	3	16	3	13
% Change from 2000		3.3	5.8	0.8	9.9	2.5	13.2	2.5	10.7

area currently violates the 1-hr California and 8-hr federal O_3 standard, annual California PM_{10} standard, and annual federal $PM_{2.5}$ standard. In terms of developing control strategies, it is more difficult to control secondary pollutants (those formed by chemical reactions in the atmosphere) than primary pollutants (those emitted directly from sources). Of the three species whose levels exceed the standards, O_3 is solely secondary, PM_{10} is dominated by primary emissions, and $PM_{2.5}$ is both primary and secondary. Thus controlling PM_{10} requires a reduction in source emissions, and detailed chemical modeling is generally not required in developing a control strategy. On the other hand, a comprehensive modeling system, such as that developed in this study, is key to controlling O_3 and $PM_{2.5}$.

Results for peak O_3 are presented in Table 9.7 — base case maximum O_3 for July 10. The highest values are outside the study area (over Los Angeles). The base case modeling shows exceedances of the federal 1-hr O_3 standard (120 ppb) on four of the five days. (These results are within 10% of the values we observed during this period as part of a preceding study.[17]) Future scenarios also show elevated levels of O_3 and violations of the 1-hr federal standard on all modeled days. Based on the data shown in Table 9.7, the Regional Low Density future leads to the highest O_3 maximum when there are 1,000k additional residents and the Coastal future leads to the smallest increase. Differences are not as pronounced with 500k new residents, but all scenarios lead to increased O_3. Thus, it appears that San Diego will continue to violate the 1-hr California standard (90 ppb) and, depending on the location of the O_3 peak and the surface monitoring sites, may be in violation of the federal 1-hr standard. It is important to note in Table 9.7 that the O_3 peak in the base case may not have the same spatial location as the peak in the scenarios and the differences listed in the table represent the difference between the base case peak and scenario peak. Thus the differences may exceed the values in the table due to a shift in the O_3 maximum.

The O_3 distribution in the future scenarios is similar; however, there are differences related to changes in emissions coupled with transport and transformation in the atmosphere. For this reason the location of the maximum change in concentration does not always coincide where there is a maximum change in emissions. In addition, large increases in NO_x lead to a reduction in O_3 near the source and increase downwind. This is due to the titration of O_3 by NO (Equation 9.3).

Since $PM_{2.5}$ is both primary and secondary in nature, its distribution and the difference between the base case maximum and an alternative future maximum is closely related to the change in the location of the emissions. There is some shift in the pattern due to dispersion and some secondary aerosol formation. Considering the four cases, the differences tend to reflect the population distribution and primary emissions in the future scenarios. From a regulatory viewpoint, the results suggest that $PM_{2.5}$ control should rely on primary emissions as opposed to controlling the precursors that form secondary PM.

Summary and Conclusions

Adding 500k or 1,000k new residents to the study area appears to have a greater overall impact on air quality than does their distribution. Nevertheless, their distribution and the resultant differences in human activity significantly impacts air quality. Other points include:

- San Diego County is currently in violation of the 1-hr California and 8-hr federal O_3 standard, annual California PM_{10} standard, and annual federal $PM_{2.5}$ standard.

- As part of this study we developed a modeling system to predict future emissions and their air quality impacts based on changes in land use patterns resulting from development and growth.

- The modeling system coupled land use predictions with transportation, emissions, meteorology, dispersion, and transformation models.

- This tool can allow potential users to evaluate and assess other "least likely impacts" stemming from the impacts of future growth.

- The spatial distribution in changes in emissions largely followed the future population distribution.

- The distribution of O_3 was more complicated and reflected both changes in emissions and the chemistry of O_3 formation.

- Given the increase in O_3-forming precursors, San Diego County will continue to violate the 1-hr state standard and may violate the federal 1-hr standard.

- Unless an effort is made to reduce primary $PM_{2.5}$, the area will continue to be in violation of the annual federal $PM_{2.5}$ standard.

NOTES:

[1] M.T. Southerland, "Environmental Impacts of Dispersed Development from Federal Infrastructure Projects," *Environmental Monitoring and Assessment* 94 (2004), pp. 163–178.

[2] Edward Beimborn and Rob Kennedy, and William Schaefer, *Inside the Blackbox: Making Transportation Models Work for Livable Communities* (New York: Citizens for a Better Environment and the Environmental Defense Fund, 1996).

[3] ibid.

[4] U.S. Census Bureau, 2002 Tiger/Line Files (2002). Available online at <http://www.census.gov/geo/www/tiger/tiger2002/tgr2002.html>.

[5] U.S. Census Bureau, Census 2000, Summary File 3, Area Name: California (2002). Available online at <www.census.gov>.

[6] Y. Horie, S. Sidawi, and R. Ellefsen, *Inventory of Leaf Biomass and Emission Factors for Vegetation in the South Coast Air Basin*, Technical Report, Contract No. 90163 (Diamond Bar, California: South Coast Air Quality Management District, 1991); M.T. Benjamin, M. Sudol, L. Bloch, and A.M. Winer, "Low-Emitting Urban Forests: A Taxonomic Methodology for Assigning Isoprene and Monoterpene Emission States," *Atmospheric Environment* 30 (1996), pp. 1437–1452.

[7] Horie et al., 1991.

[8] Benjamin et al., 1996; M.T. Benjamin, M. Sudol, D. Vorsatz, and A.M. Winer, "A Spatially and Temporally Resolved Biogenic Hydrocarbon Emissions Inventory for the California South Coast Air Basin," *Atmospheric Environment* 31 (1997), pp. 3087–3100.

[9] Benjamin et al., 1997.

[10] A.B. Guenther, P.R. Zimmermann, P.C. Harley, "Isoprene and Monoterpene Emission Rate Variability: Model Evaluations and Sensitivity Analyses," *JGR Atmospheres* 98 (D7), pp. 12609–12617.

[11] P.C. Harley, V. Fridd-Stroud, J. Greenberg, A. Guenther, P. Vasconcellos, "Emission of 2-Methyl-3-Buten-2-ol by Pines: A Potentially Large Natural Source of Reactive Carbon to the Atmosphere," *JGR Atmospheres* 103 (1998), pp. 25479–25486.

[12] Benjamin et al., 1997.

[13] California Air Resources Board, *Emission Inventory Procedural Manual, Vol. III: Methods for Assessing Area Source Emissions* (Sacramento, California: California Air Resources Board, 1997).

[14] California Air Resources Board, *User's Guide for EMFAC2002* (Sacramento, California: California Air Resources Board, 2002). Software and documentation available online at <http://www.arb.ca.gov/msei/onroad/ latest_version.htm>.

[15] G.A. Grell, J. Dudhia, D.R. Stauffer, *A Description of the Fifth-Generation Penn State/ NCAR Mesoscale Model (MM5)*, NCAR/TN-398+STR (Boulder, Colorado: National Center for Atmospheric Research, 1995), 122 pp.

[16] Environ International Corporation, *User's Guide for CAMx* (Arlington, Virginia: Environ International Corporation, 2003). Software and documentation are available online at <http://www.camx.com>.

[17] Alan W. Gertler, Julide Kahyaoglu, Darko Koracin, Menachem Luria, William Stockwell, and Erez Weinroth, *Development and Validation of a Predictive Model to Assess the Impact of Coastal Operations on Urban Scale Air Quality*, Strategic Environmental Research and Development Program (SERDP) Project No. CP-1253 (Arlington, Virginia: SERDP, July 31, 2006).

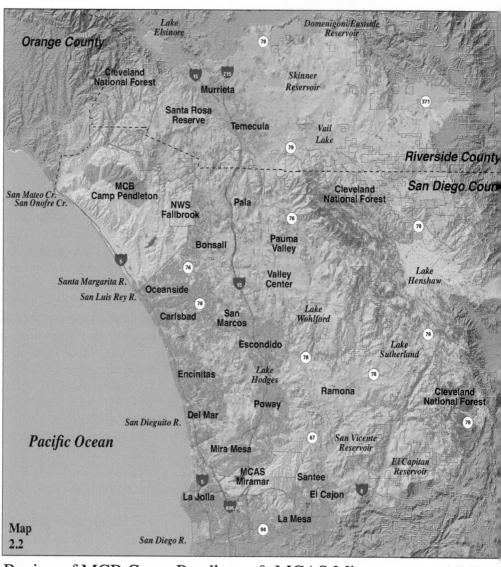

Map
2.2

Region of MCB Camp Pendleton & MCAS Miramar
Existing Conditions 2000
Study Area Locations

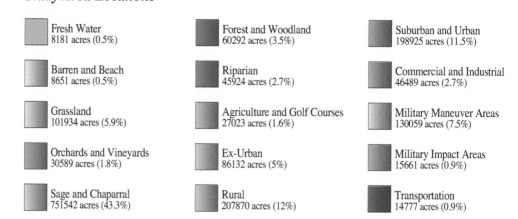

Fresh Water 8181 acres (0.5%)	**Forest and Woodland** 60292 acres (3.5%)	**Suburban and Urban** 198925 acres (11.5%)
Barren and Beach 8651 acres (0.5%)	**Riparian** 45924 acres (2.7%)	**Commercial and Industrial** 46489 acres (2.7%)
Grassland 101934 acres (5.9%)	**Agriculture and Golf Courses** 27023 acres (1.6%)	**Military Maneuver Areas** 130059 acres (7.5%)
Orchards and Vineyards 30589 acres (1.8%)	**Ex-Urban** 86132 acres (5%)	**Military Impact Areas** 15661 acres (0.9%)
Sage and Chaparral 751542 acres (43.3%)	**Rural** 207870 acres (12%)	**Transportation** 14777 acres (0.9%)

Region of MCB Camp Pendleton & MCAS Miramar
Satellite Image from November 18, 2000
Composite of Landsat ETM Bands 7, 4, and 2

**Map
4.6**

Region of MCB Camp Pendleton & MCAS Miramar
Existing Conditions 2000
Vegetation

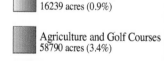

Fresh Water
8510 acres (0.5%)

Wetland
1216 acres (0.1%)

Riparian
61271 acres (3.5%)

Meadow
2002 acres (0.1%)

Mixed Forest
30087 acres (1.7%)

Oak Woodland
31822 acres (1.8%)

Orchards and Vineyards
57871 acres (3.3%)

Coastal Sage Scrub
285069 acres (16.4%)

Coastal Sage Scrub-Chaparral Mix
26516 acres (1.5%)

Chaparral
696029 acres (40.1%)

Grassland
179155 acres (10.3%)

Barren and Beach
16239 acres (0.9%)

Agriculture and Golf Courses
58790 acres (3.4%)

Urban
279473 acres (16.1%)

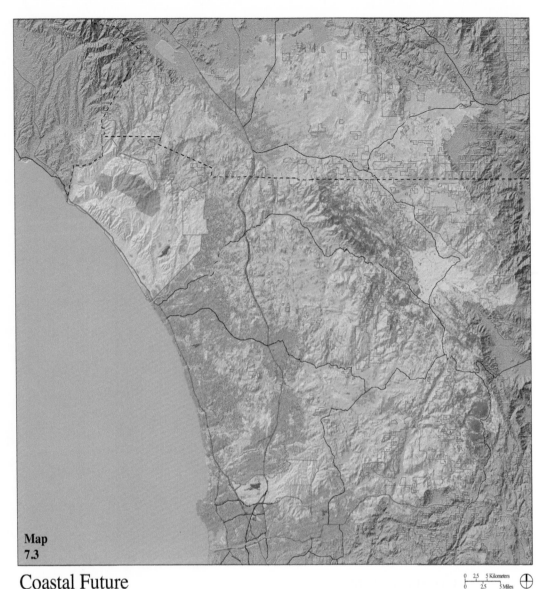

**Map
7.3**

0 2.5 5 Kilometers
0 2.5 5 Miles

Coastal Future
500,000 New Residents
Land Use/Land Cover

Fresh Water
8181 acres (0.5%)

Barren and Beach
5535 acres (0.3%)

Grassland
88049 acres (5.1%)

Orchards and Vineyards
24221 acres (1.4%)

Sage and Chaparral
705725 acres (40.7%)

Forest and Woodland
59946 acres (3.5%)

Riparian
44594 acres (2.6%)

Agriculture and Golf Courses
22884 acres (1.3%)

Ex-Urban
102509 acres (5.9%)

Rural
222175 acres (12.8%)

Suburban and Urban
237732 acres (13.7%)

Commercial and Industrial
52005 acres (3%)

Military Maneuver Areas
130059 acres (7.5%)

Military Impact Areas
15661 acres (0.9%)

Transportation
14777 acres (0.9%)

**Map
7.5**

Coastal Future
1,000,000 New Residents
Land Use/Land Cover

0   2.5   5 Kilometers
0   2.5   5 Miles

 Fresh Water
8181 acres (0.5%)

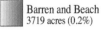 Barren and Beach
3719 acres (0.2%)

Grassland
73735 acres (4.3%)

 Orchards and Vineyards
20060 acres (1.2%)

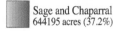 Sage and Chaparral
644195 acres (37.2%)

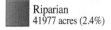 Forest and Woodland
57690 acres (3.3%)

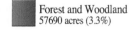 Riparian
41977 acres (2.4%)

Agriculture and Golf Courses
19390 acres (1.1%)

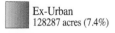 Ex-Urban
128287 acres (7.4%)

 Rural
239087 acres (13.8%)

Suburban and Urban
279478 acres (16.1%)

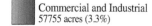 Commercial and Industrial
57755 acres (3.3%)

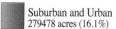 Military Maneuver Areas
130059 acres (7.5%)

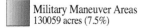 Military Impact Areas
15661 acres (0.9%)

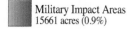 Transportation
14777 acres (0.9%)

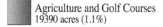 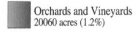 

**Map
7.8**

Northern Future
500,000 New Residents
Land Use/Land Cover

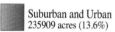 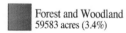 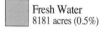 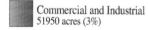 

0   2.5   5 Kilometers
0   2.5   5 Miles

| Fresh Water | Forest and Woodland | Suburban and Urban |
| 8181 acres (0.5%) | 59583 acres (3.4%) | 235909 acres (13.6%) |

| Barren and Beach | Riparian | Commercial and Industrial |
| 5137 acres (0.3%) | 43707 acres (2.5%) | 51950 acres (3%) |

| Grassland | Agriculture and Golf Courses | Military Maneuver Areas |
| 84807 acres (4.9%) | 30131 acres (1.7%) | 130059 acres (7.5%) |

| Orchards and Vineyards | Ex-Urban | Military Impact Areas |
| 17092 acres (1%) | 87765 acres (5.1%) | 15661 acres (0.9%) |

| Sage and Chaparral | Rural | Transportation |
| 666149 acres (38.4%) | 283144 acres (16.3%) | 14777 acres (0.9%) |

Map
7.10

Northern Future
1,000,000 New Residents
Land Use/Land Cover

0 2.5 5 Kilometers
0 2.5 5 Miles

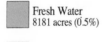 Fresh Water
8181 acres (0.5%)

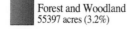 Forest and Woodland
55397 acres (3.2%)

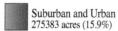 Suburban and Urban
275383 acres (15.9%)

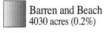 Barren and Beach
4030 acres (0.2%)

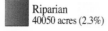 Riparian
40050 acres (2.3%)

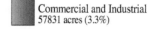 Commercial and Industrial
57831 acres (3.3%)

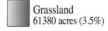 Grassland
61380 acres (3.5%)

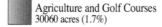 Agriculture and Golf Courses
30060 acres (1.7%)

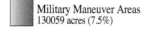 Military Maneuver Areas
130059 acres (7.5%)

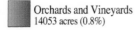 Orchards and Vineyards
14053 acres (0.8%)

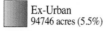 Ex-Urban
94746 acres (5.5%)

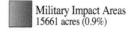 Military Impact Areas
15661 acres (0.9%)

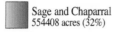 Sage and Chaparral
554408 acres (32%)

 Rural
378036 acres (21.8%)

 Transportation
14777 acres (0.9%)

Map 7.13

Regional Low-Density Future
500,000 New Residents
Land Use/Land Cover

0   2.5   5 Kilometers
0   2.5   5 Miles

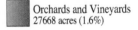 Fresh Water
8181 acres (0.5%)

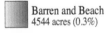 Barren and Beach
4544 acres (0.3%)

 Grassland
77540 acres (4.5%)

Orchards and Vineyards
27668 acres (1.6%)

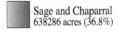 Sage and Chaparral
638286 acres (36.8%)

 Forest and Woodland
57831 acres (3.3%)

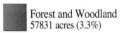 Riparian
41985 acres (2.4%)

 Agriculture and Golf Courses
10442 acres (0.6%)

 Ex-Urban
87518 acres (5.1%)

 Rural
326869 acres (18.9%)

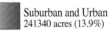 Suburban and Urban
241340 acres (13.9%)

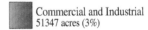 Commercial and Industrial
51347 acres (3%)

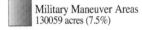 Military Maneuver Areas
130059 acres (7.5%)

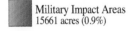 Military Impact Areas
15661 acres (0.9%)

 Transportation
14777 acres (0.9%)

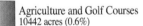

Map 7.15

Regional Low-Density Future
1,000,000 New Residents
Land Use/Land Cover

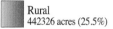
0 2.5 5 Kilometers
0 2.5 5 Miles

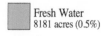

Fresh Water 8181 acres (0.5%)	**Forest and Woodland** 57006 acres (3.3%)	**Suburban and Urban** 290523 acres (16.8%)
Barren and Beach 2409 acres (0.1%)	**Riparian** 35461 acres (2%)	**Commercial and Industrial** 57391 acres (3.3%)
Grassland 50777 acres (2.9%)	**Agriculture and Golf Courses** 3929 acres (0.2%)	**Military Maneuver Areas** 130059 acres (7.5%)
Orchards and Vineyards 23269 acres (1.3%)	**Ex-Urban** 104555 acres (6%)	**Military Impact Areas** 15661 acres (0.9%)
Sage and Chaparral 497726 acres (28.7%)	**Rural** 442326 acres (25.5%)	**Transportation** 14777 acres (0.9%)

Map 7.18

Three-Centers Future
500,000 New Residents
Land Use/Land Cover

0 2.5 5 Kilometers
0   2.5   5 Miles

Fresh Water
8181 acres (0.5%)

Barren and Beach
3613 acres (0.2%)

Grassland
85284 acres (4.9%)

Orchards and Vineyards
28305 acres (1.6%)

Sage and Chaparral
690375 acres (39.8%)

Forest and Woodland
59887 acres (3.5%)

Riparian
43971 acres (2.5%)

Agriculture and Golf Courses
24614 acres (1.4%)

Ex-Urban
119680 acres (6.9%)

Rural
213455 acres (12.3%)

Suburban and Urban
244046 acres (14.1%)

Commercial and Industrial
52143 acres (3%)

Military Maneuver Areas
130059 acres (7.5%)

Military Impact Areas
15661 acres (0.9%)

Transportation
14777 acres (0.9%)

**Map
7.20**

Three-Centers Future
1,000,000 New Residents
Land Use/Land Cover

0 2,5 5 Kilometers
0 2,5 5 Miles

 Fresh Water
8181 acres (0.5%)

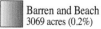 Barren and Beach
3069 acres (0.2%)

Grassland
72467 acres (4.2%)

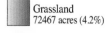 Orchards and Vineyards
27312 acres (1.6%)

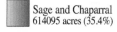 Sage and Chaparral
614095 acres (35.4%)

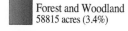 Forest and Woodland
58815 acres (3.4%)

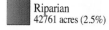 Riparian
42761 acres (2.5%)

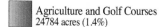 Agriculture and Golf Courses
24784 acres (1.4%)

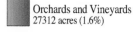 Ex-Urban
123592 acres (7.1%)

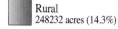 Rural
248232 acres (14.3%)

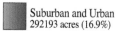 Suburban and Urban
292193 acres (16.9%)

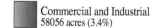 Commercial and Industrial
58056 acres (3.4%)

 Military Maneuver Areas
130059 acres (7.5%)

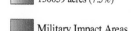 Military Impact Areas
15661 acres (0.9%)

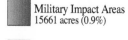 Transportation
14777 acres (0.9%)

Map 8.1

Region of MCB Camp Pendleton & MCAS Miramar
Existing Conditions 2000
Manning's n for the Santa Margarita River Basin

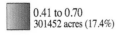 0 to 0.10
48387 acres (2.8%)

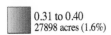 0.11 to 0.20
57851 acres (3.3%)

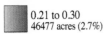 0.21 to 0.30
46477 acres (2.7%)

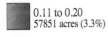 0.31 to 0.40
27898 acres (1.6%)

 0.41 to 0.70
301452 acres (17.4%)

Federal Lands

Map 10.11

Region of MCB Camp Pendleton & MCAS Miramar
Existing Conditions 2000
Landscape Ecological Pattern

0 2.5 5 Kilometers
0 2.5 5 Miles

 Contiguous Natural Vegetation
922634 acres (53.2%)

Built Land
533007 acres (30.7%)

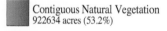 Isolated Natural Vegetation
35054 acres (2%)

 Federal Lands

Fragmented Natural Vegetation
103266 acres (6%)

 Stream Corridor
11122 acres (0.6%)

Disturbed Land
128149 acres (7.4%)

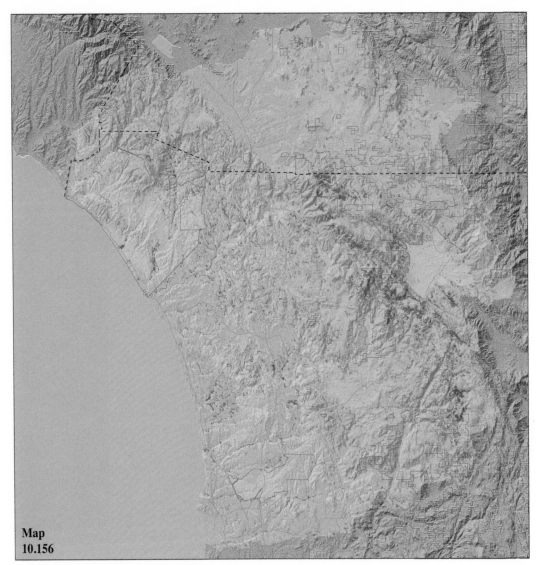

Map
10.156

Region of MCB Camp Pendleton & MCAS Miramar
Existing Conditions 2000
Native Species Richness

 36 to 60 Species
279473 acres (16.1%)

61 to 160 Species
144151 acres (8.3%)

161 to 190 Species
471837 acres (27.2%)

191 to 210 Species
746109 acres (43%)

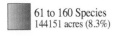 211 to 253 Species
92477 acres (5.3%)

Map 11.1

Coastal, Northern, and Regional Low-Density Futures
1,000,000 Allocation
Common Loss of Threatened and Endangered Species Potential Habitat

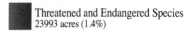 Threatened and Endangered Species
23993 acres (1.4%)

Map 11.2

All Four Futures
1,000,000 Allocation
Common Biological Impacts

0 2.5 5 Kilometers
0 2.5 5 Miles

Contiguous Natural Vegetation
64975 acres (3.8%)

Stream Corridor
1354 acres (0.1%)

Native Species Richness
42620 acres (2.5%)

Chapter 10—Overview

Biological Consequences Associated with the Alternative Futures

Scott D. Bassett

Policies reflecting attitudes on biological resources management can and do change.[1] In the United States a dramatic change took place in the early 1960s when the then dominant view of resource exploitation gave way to an emphasis on conservation.[2] One current school of thought on that subject is based on an ideal of sustainable management.[3] It calls for a land use model that allows for change, but retains sufficient protection for resources and the natural processes that support them; moreover, it seeks to provide a basis that will maintain these environmental features for generations to come. Implementing such a strategy requires the balancing of what can be competing needs. A piece of land must be assessed not only in terms of its biological potential, but also in light of other societal needs and goals. Such an assessment is not an either-or comparison between conservation and development. Even those values that appear to be biocentric are, indeed, part of society's larger value system. In attempting to achieve this balance over a region, some portion of a landscape may be conserved while locations exhibiting similar biological significance may be altered to accommodate other needs. Adjudicating which lands are to be changed and which are not can be a contentious process.[4] Thus, while sustainable management strategies offer a promising approach to resource conservation, achieving consensus on specific land use plans has proven to be difficult.

Often researchers in the natural resources focus their efforts on inventorying and monitoring.[5] These operations identify and chart the physical extent of a resource over time. However, meeting the goals of sustainability requires more than a knowledge of the past and the present; potential alterations that may take place in the future must be also be considered.[6]

Biological elements may be described in a variety of ways from simplistic overall vegetation or species inventories[7] to complex analyses, which include the population dynamics of individual species.[8] While more advanced approaches are preferable, the current level of scientific knowledge and/or limitations of available data may preclude some avenues of investigation. Restrictions of time, cost, and other resources may also limit the means of study. Regardless, it is not acceptable to simply ignore the biological consequences of policy decisions on the phenomena of interest for reasons of operational constraints. When such limitations occur, analogies, generalizations, and even educated guesses must be made. The models described in this chapter are not immune to these limitations and thus, may be missing specific information that some researchers and policy makers might consider valuable for an adequate assessment. With this caveat noted, the goal of the following assessments is to better understand the biological consequences of future changes to the landscape. To do this analysis, readily available data specific to the study were utilized. The models reflect biological concerns ranging from the connectivity of the natural landscape to an overall biodiversity assessment. All of the biological assessments stem from one of three major conservation assessment methodologies: landscape ecological pattern assessment, single species analysis, and species richness assessment. These are presented to determine relative impacts of each of the alternative futures.

The potential impacts of each future are summarized in tables that accompany each section and are shown in maps that depict change from the current conditions. Additionally, smaller maps are used to illustrate components of the analyses.

Chapter 10—Section I

Biological Consequences Associated with the Alternative Futures: Vegetation

David A. Mouat

An underlying issue of habitat conservation is vegetation. Not only is the total quantity of native vegetation important, but so is its relative quality and its spatial distribution. As discussed earlier, the physiographic characteristics of the region, the current pattern of vegetation is not simply a product of natural geographic conditions; it is also partly a result of human induced changes to the landscape. Future changes in land use will also influence the area's vegetation. While a vegetation transition model associated with the indirect impacts of development—primarily those associated with flood, fire, and the introduction of exotic species—was not a part of this study, the loss of specific vegetation through the development process is should be noted. These changes are summarized in Table 10.1 and shown in Maps 10.2-10.10. Map 10.1 allows a comparison of land use/land cover vs. vegetation mapping.

All four futures significantly impact the cover of Coastal Sage Scrub. With the addition of 500,000 new residents, they result in a decrease of Coastal Sage Scrub cover of from 5.3 to 9.5%. Both the Northern and Regional Low-Density Futures have the greatest loss. With the addition of 1,000,000 people, however, the decrease in cover due to the Regional Low-Density Future is nearly twice as high as that of the Northern Futures (30.2% vs. 16.3%, respectively). As stated in an earlier chapter, one possible factor for this change is ornamental landscaping around the large lots attendant in the Regional Low-Density Future. This landscaping removes the native vegetation.

All four futures also result in a significant decrease in Chaparral cover. The addition of 500,000 residents results in decreases ranging from 3.5% for the Coastal Future to 10.9% for the

Regional Low-Density Future. Decreases for the other two futures lie in between those two figures. The addition of 1,000,000 people results in losses of Chaparral cover, slightly more than double the figures for the lesser population increase. The increase in loss for the Regional Low-Density Future, however, is less than double, resulting in only a 17.9% loss in cover. On account of the large extent of Chaparral, the losses for the Northern and Regional Low-Density Futures are significant, approximately 143,000 acres and 125,000 acres, respectively. Those figures equal and exceed the extent of MCB Camp Pendleton. The decreases in cover of the Coastal Sage Scrub-Chaparral Mix lie roughly in between the figures for each of the futures affecting Coastal Sage Scrub and Chaparral.

Grassland is the vegetation type most affected by the futures in terms of percent change. The Regional Low-Density Future, with an increase of 1,000,000 additional residents, has a change of approximately 154,000 acres of Grassland cover. The Coastal and Three-Centers Futures have increases of approximately 20%. The Northern Future has an increase of nearly 37% and the Regional Low-Density Future has an increase of nearly 86%. Field inspection of areas undergoing development or which have recently been developed show that the composition of this type is largely comprised of annual grasses and forbs, most of which are both exotic (i.e., from outside the area) and invasive. Furthermore, it appears from field context that the increase in Grassland comes about as a result of a concomitant decrease in both the Coastal Sage Scrub and Chaparral (and Mixed) types. It is likely, although not analyzed, that the actual area of native perennial Grassland will decrease as a result of the futures.

Table 10.1

Vegetation (in acres)

	2000	Coastal		Northern		Reg. Low-Density		Three-Centers	
		500k	1,000k	500k	1,000k	500k	1,000k	500k	1,000k
Barren	15,132	12,045	10,102	11,214	9,606	12,118	10,639	10,294	8,634
Change from 2000		-3,087	-5,030	-3,918	-5,526	-3,014	-4,493	-4,838	-6,499
% Change from 2000		**-20.4**	**-33.3**	**-25.9**	**-36.5**	**-19.9**	**-29.7**	**-32.0**	**-43.0**
Beach	1,107	1,107	1,107	1,107	1,107	1,107	1,107	1,107	1,107
Change from 2000		0	0	0	0	0	0	0	0
% Change from 2000		**0.0**	**0.0**	**0.0**	**0.0**	**0.0**	**0.0**	**0.0**	**0.0**
Ag. and Golf Courses	58,790	54,449	50,743	59,527	59,658	50,167	43,629	58,790	58,790
Change from 2000		-4,341	-8,047	737	868	-8,623	-15,161	0	0
% Change from 2000		**-7.4**	**-13.7**	**1.3**	**1.5**	**-14.7**	**-25.8**	**0.0**	**0.0**
Orchard and Vineyard	57,871	51,785	45,502	80,322	106,798	57,871	57,871	57,871	57,871
Change from 2000		-6,086	-12,369	22,451	48,927	0	0	0	0
% Change from 2000		**-10.5**	**-21.4**	**38.8**	**85.0**	**0.0**	**0.0**	**0.0**	**0.0**
Grassland	179,155	193,653	214,246	208,934	245,017	253,863	332,951	199,201	217,097
Change from 2000		14,498	35,091	29,779	65,862	74,708	153,796	20,046	37,942
% Change from 2000		**8.1**	**19.6**	**16.6**	**36.8**	**41.7**	**85.9**	**11.2**	**21.2**
Coastal Sage Scrub	285,069	266,833	246,751	259,474	238,722	258,141	198,884	270,007	259,036
Change from 2000		-18,236	-38,318	-25,595	-46,347	-26,928	-86,185	-15,062	-26,033
% Change from 2000		**-6.4**	**-13.4**	**-9.0**	**-16.3**	**-9.5**	**-30.2**	**-5.3**	**-9.1**
Chaparral	696,029	671,400	641,280	632,734	553,012	620,055	571,312	654,525	608,539
Change from 2000		-24,629	-54,749	-63,295	-143,017	-75,974	-124,717	-41,504	-87,490
% Change from 2000		**-3.5**	**-7.9**	**-9.1**	**-20.6**	**-10.9**	**-17.9**	**-6.0**	**-12.6**
C.S.-Scrub-Chaparral Mix	26,516	25,606	24,539	23,970	22,034	24,079	18,962	24,995	23,333
Change from 2000		-910	-1,977	-2,546	-4,482	-2,437	-7,554	-1,521	-3,183
% Change from 2000		**-3.4**	**-7.5**	**-9.6**	**-16.9**	**-9.2**	**-28.5**	**-5.7**	**-12.0**
Meadow	2,002	2,000	1,997	1,996	1,959	1,997	1,997	1,998	1,998
Change from 2000		-2	-5	-6	-43	-5	-5	-4	-4
% Change from 2000		**-0.1**	**-0.2**	**-0.3**	**-2.1**	**-0.2**	**-0.2**	**-0.2**	**-0.2**
Wetland	1,216	1,216	1,216	1,216	1,216	1,216	1,216	1,216	1,216
Change from 2000		0	0	0	0	0	0	0	0
% Change from 2000		**0.0**	**0.0**	**0.0**	**0.0**	**0.0**	**0.0**	**0.0**	**0.0**
Riparian	61,271	61,270	61,271	61,271	61,271	61,271	61,271	61,271	61,271
Change from 2000		0	0	0	0	0	0	0	0
% Change from 2000		**0.0**	**0.0**	**0.0**	**0.0**	**0.0**	**0.0**	**0.0**	**0.0**
Oak Woodland	31,822	31,807	31,771	31,605	30,652	31,756	31,701	31,756	31,671
Change from 2000		-15	-51	-217	-1,170	-66	-121	-66	-151
% Change from 2000		**0.0**	**-0.2**	**-0.7**	**-3.7**	**-0.2**	**-0.4**	**-0.2**	**-0.5**
Mixed Forest	30,087	30,085	30,071	30,041	29,716	30,058	30,054	30,080	30,041
Change from 2000		-2	-16	-46	-371	-29	-33	-7	-46
% Change from 2000		**0.0**	**-0.1**	**-0.2**	**-1.2**	**-0.1**	**-0.1**	**0.0**	**-0.2**
Urban Lands	279,473	322,283	364,943	322,131	364,773	321,840	363,948	322,430	364,936
Change from 2000		42,810	85,470	42,658	85,300	42,367	84,475	42,957	85,463
% Change from 2000		**15.3**	**30.6**	**15.3**	**30.5**	**15.2**	**30.2**	**15.4**	**30.6**

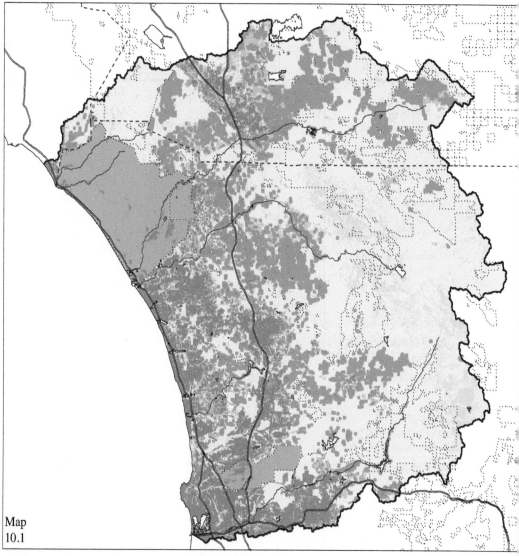

Map
10.1

Region of MCB Camp Pendleton & MCAS Miramar
Existing Conditions 2000
Land Use/Land Cover

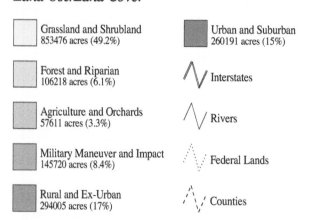

Grassland and Shrubland
853476 acres (49.2%)

Forest and Riparian
106218 acres (6.1%)

Agriculture and Orchards
57611 acres (3.3%)

Military Maneuver and Impact
145720 acres (8.4%)

Rural and Ex-Urban
294005 acres (17%)

Urban and Suburban
260191 acres (15%)

Interstates

Rivers

Federal Lands

Counties

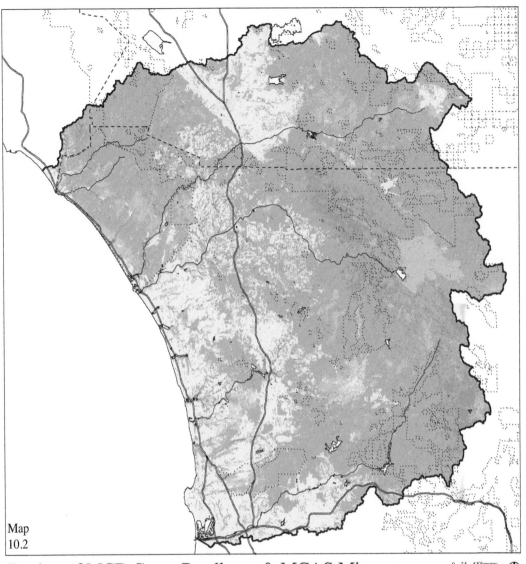

Map
10.2

Region of MCB Camp Pendleton & MCAS Miramar
Existing Conditions 2000
Vegetation

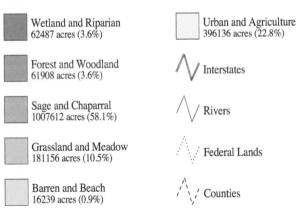

Wetland and Riparian
62487 acres (3.6%)

Forest and Woodland
61908 acres (3.6%)

Sage and Chaparral
1007612 acres (58.1%)

Grassland and Meadow
181156 acres (10.5%)

Barren and Beach
16239 acres (0.9%)

Urban and Agriculture
396136 acres (22.8%)

Interstates

Rivers

Federal Lands

Counties

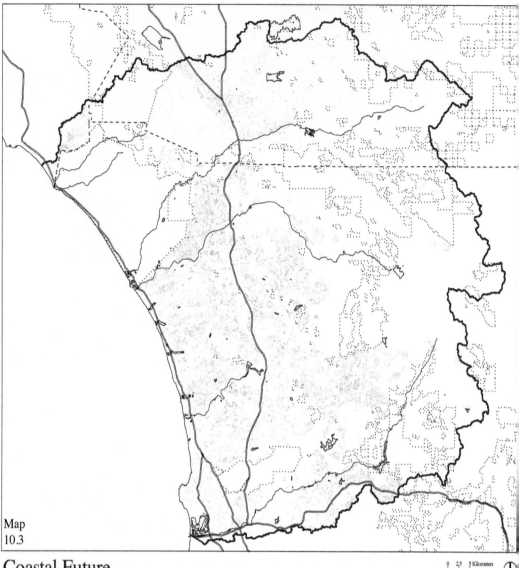

Map
10.3

Coastal Future
500,000 New Residents
Vegetation Change

0   2.5   5 Kilometers
0   2.5   5 Miles

 Vegetation Loss to Urban
42810 acres (2.5%)

 Counties

New Grassland
25894 acres (1.5%)

 Interstates

 Rivers

 Federal Lands

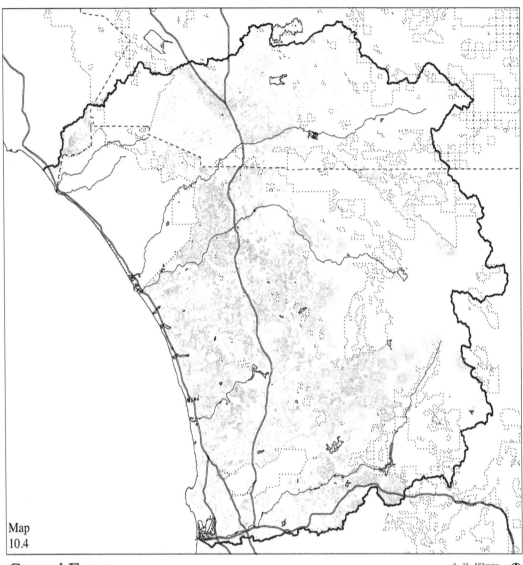

Map
10.4

Coastal Future
1,000,000 New Residents
Vegetation Change

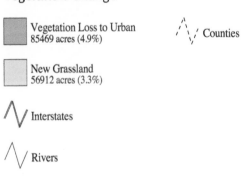

Vegetation Loss to Urban
85469 acres (4.9%)

New Grassland
56912 acres (3.3%)

Interstates

Rivers

Federal Lands

Counties

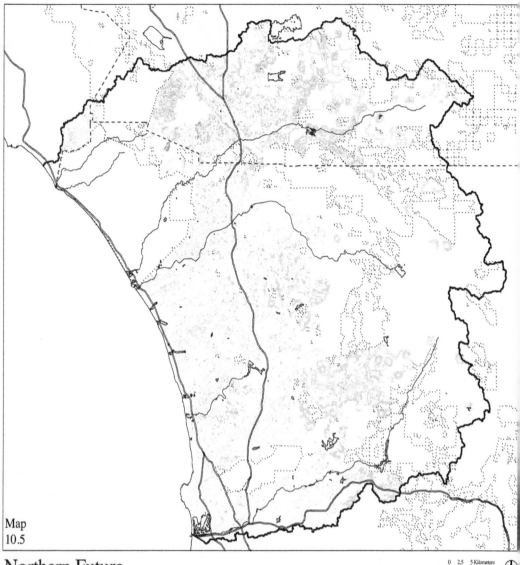

Map
10.5

Northern Future
500,000 New Residents
Vegetation Change

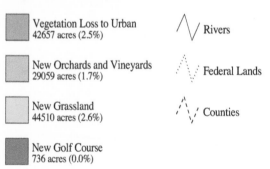

Vegetation Loss to Urban
42657 acres (2.5%)

New Orchards and Vineyards
29059 acres (1.7%)

New Grassland
44510 acres (2.6%)

New Golf Course
736 acres (0.0%)

Interstates

Rivers

Federal Lands

Counties

0   2.5   5 Kilometers
0   2.5   5 Miles

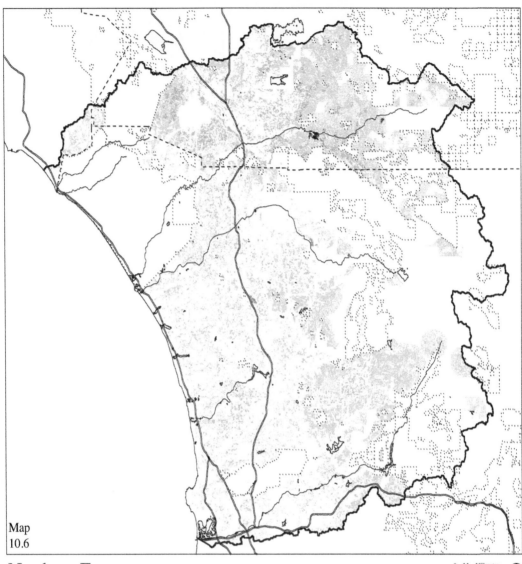

Map
10.6

Northern Future
1,000,000 New Residents
Vegetation Change

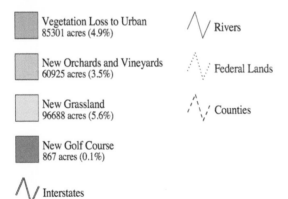

Vegetation Loss to Urban
85301 acres (4.9%)

New Orchards and Vineyards
60925 acres (3.5%)

New Grassland
96688 acres (5.6%)

New Golf Course
867 acres (0.1%)

Interstates

Rivers

Federal Lands

Counties

0   2.5   5 Kilometers
0   2.5   5 Miles

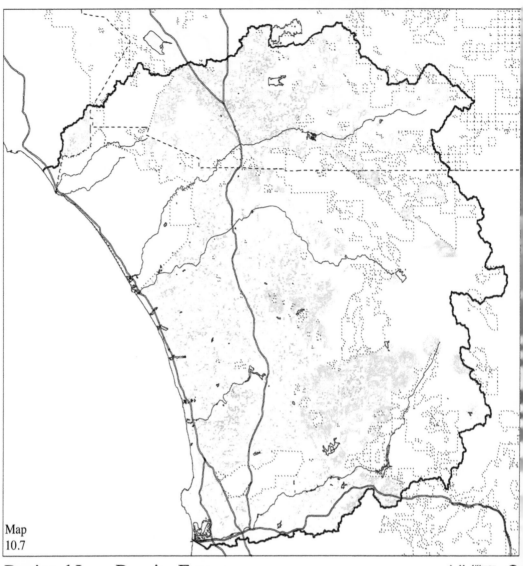

Map
10.7

Regional Low-Density Future
500,000 New Residents
Vegetation Change

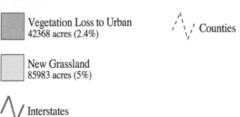

Vegetation Loss to Urban
42368 acres (2.4%)

New Grassland
85983 acres (5%)

Interstates

Rivers

Federal Lands

Counties

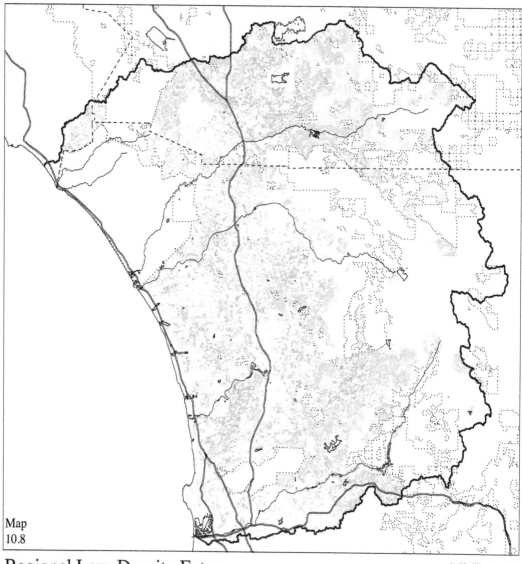

Map
10.8

Regional Low-Density Future
1,000,000 New Residents
Vegetation Change

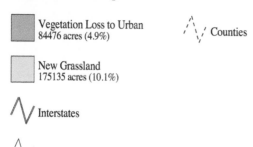

Vegetation Loss to Urban
84476 acres (4.9%)

Counties

New Grassland
175135 acres (10.1%)

Interstates

Rivers

Federal Lands

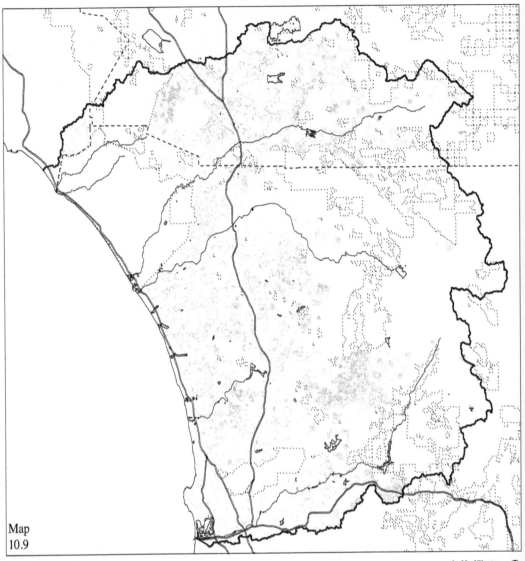

Map
10.9

Three-Centers Future
500,000 New Residents
Vegetation Change

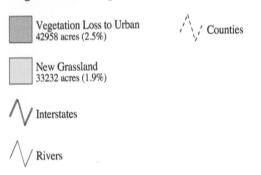

Vegetation Loss to Urban
42958 acres (2.5%)

Counties

New Grassland
33232 acres (1.9%)

Interstates

Rivers

Federal Lands

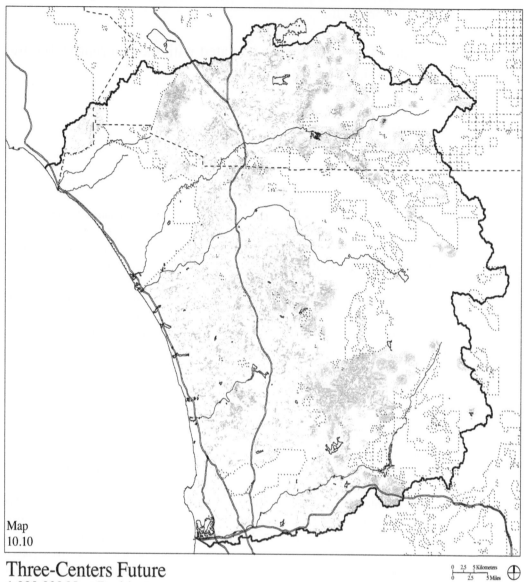

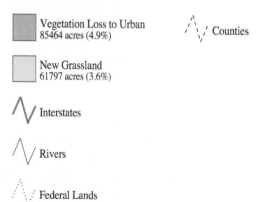

Map
10.10

Three-Centers Future
1,000,000 New Residents
Vegetation Change

Vegetation Loss to Urban
85464 acres (4.9%)

New Grassland
61797 acres (3.6%)

Interstates

Rivers

Federal Lands

Counties

0 2.5 5 Kilometers
0 2.5 5 Miles

Biological Consequences Associated with the Alternative Futures: Landscape Ecological Pattern

Scott D. Bassett

Analytical approaches based on landscape ecology emphasize the structural and functional patterns in the environment.[9] These patterns are comprised of elements including overall vegetation structure, connectivity among elements, and anthropogenically induced fragmentation.[10] Landscape ecological pattern includes the following four concepts: patch, corridor, edge, and mosaic.[11] Patches are contiguous areas that do not vary in composition. Corridors create connective links between patches allowing for movement. Edges represent areas located along the boundaries of patches. The entire pattern of patches, corridors, and edges is termed the mosaic.

Assessing the relative impacts each of the alternative futures has on the surrounding landscape ecological pattern requires an interpretation of how the mosaic might be altered. Alterations are generally described in terms of patch and corridor change. The ideal situation for maintaining the overall landscape ecological pattern is no loss in patch size or connectivity.

The landscape ecological pattern for this study has been defined in terms of six different elements.

1. **Contiguous Natural Vegetation**: A vegetated patch of land not used for agriculture or as a planted park which is at least 250 acres in area and attached by a corridor to a 5,000 acre patch or is itself 5,000 acres in area.

2. **Isolated Natural Vegetation**: A vegetated patch of land not used for agriculture or as a planted park which is at least 250 acres in area and is not attached by a corridor to a patch of contiguous natural vegetation.

3. **Fragmented Natural Vegetation**: A vegetated patch of land not used for agriculture or as a planted park which is less than 250 acres in area.

4. **Stream Corridor**: A vegetated piece of land along a stream 150 to 600 feet in width which connects vegetation patches at least 250 acres in area. Stream corridors that have over 50% of their width along their length comprised of development are not considered as connectivity links among vegetated patches.

5. **Disturbed Land**: Vegetated land where anthropogenic activities and plant production have a large influence on natural community structure. These lands include orchards, vineyards, row crops, military impact zones, golf courses, and other planted community parks.

6. **Built Land**: Land composed of urban development. Built land includes locations dominated by residential housing, commercial and industrial buildings, and other surfaces paved with asphalt or concrete.

Analysis of the landscape ecological pattern followed a specific sequence. It began by identifying the contiguous natural patches using the land use/land cover map. All natural vegetation patches of 250 or more acres in area were identified.

Next, stream corridors were identified as those places connecting natural patches at least 250 or more acres in area and having an undeveloped land of at least 150 feet on either side of the stream. Streams were identified using USGS digital elevation models with an approximately 100-foot pixel resolution (30 meters). A stream pixel is described if at least 200 acres of upstream area contributed to water runoff through the pixel. Streams flowing through a contiguous patch of 250 acres or more were identified as potential corridors. All other stream pixels that did not provide a connectivity link between patches were eliminated as potential corridors. Potential stream corridors are buffered by 300 feet on either side and subsequently traced up and down the stream to determine if land within 150 feet at any point along the watercourse was undeveloped. Those corridors not meeting the 150 feet width minimum were eliminated as potential stream corridors.

With the corridors and patches identified, patches of at least 250 acres that were connected by corridors and which, when summed, constituted an area greater than 5,000 acres in area, were identified as contiguous natural vegetation. Summed area patches 250 to 5,000 acres in area were identified as Isolated Natural Vegetation. All other natural vegetation was identified as Fragmented Natural Vegetation. The Disturbed and Built Lands were extracted from the Land Use/Land Cover map.

Most of the existing large natural patches of habitat are located in the higher mountains and in inland areas. The largest remaining coastal patch of natural habitat is located inside MCB Camp Pendleton. Thus, the best connections between coastal habitats and the uplands occur on MCB Camp Pendleton.

Other existing and notable artifacts of the landscape ecological pattern occur. Currently, a northwest–southeast connection exists within the Coastal Range, northeast of Fallbrook. Elimination of this connection would make it difficult for species attempting to move between large patches on either side of I-15. A second similar connection exists between MCAS Miramar and the southern part of the Cleveland National Forest. Any development to the east of MCAS Miramar could threaten the movement of species between the large amount of open space on that installation and other public lands. Currently, the Multiple Species Conservation Plan (MSCP) calls for the protection of areas

directly east of Miramar as open space, but any reluctance to implement the plan could isolate the base's landscape.

The alternative futures vary in their relative impact on landscape ecological pattern. Summary statistics of change associated with each future are provided in Table 10.2; however, while these figures do reflect change within the study area as a whole, they do not convey the spatial distributions of change within the region. Readers should, therefore, also refer to Maps 10.11-10.19. All the alternative futures see a reduction in Contiguous Natural Vegetation, Fragmented Natural Vegetation, and Stream Corridors. The Northern and Regional Low-Density Futures, which both assume a large amount of low-density residential development, cause the greatest impacts. The Coastal Future, which has the densest development, has the least amount of impact in terms of the landscape ecological pattern.

Two seemingly anomalous values appear in Table 10.2 and are the result of assumptions associated with housing allocation. First, the amount of Isolated Natural Vegetation decreases in the 1,000k Regional Low-Density Future, but increases in every other alternative. In all of the scenarios, some new development occurs on land currently classified as Contiguous Natural Vegetation. In part, this development transforms these lands into Built or Disturbed parcels. Also, in part, this development divides or splinters some natural vegetation from large patches and thus creates Isolated Natural Vegetation. As indicated by its name, the Regional Low-Density Scenario calls for a significant amount of low-density development. With the first increase of 500,000 new residents, housing needs are easily accommodated in the large open areas in the central and eastern parts of the study area. However, with the second 500,000 increase, buildable land becomes scarce and new residents who want large lots are forced to purchase land in the western part of the region. Parcels of Isolated Natural Vegetation that are currently within the urban-dominated landscape near the Pacific Ocean are subsequently developed. The result is the loss of the larger unconnected fragments of open space that are presently found between the coast and I-15.

The second apparent anomaly occurs in the Three-Centers Future. In the other three futures, Fragmented Natural Vegetation is lost as population

Table 10.2

Landscape Ecological Pattern (in acres)

	2000	Coastal		Northern		Reg. Low-Density		Three-Centers	
		500k	1,000k	500k	1,000k	500k	1,000k	500k	1,000k
Contiguous Natural Veg.	922,634	872,453	794,916	832,729	688,804	783,586	663,669	853,429	763,44:
Change from 2000		-50,181	-127,718	-89,905	-233,830	-139,048	-258,965	-69,205	-159,19.
% Change from 2000		**-5.4**	**-13.8**	**-9.7**	**-25.3**	**-15.1**	**-28.1**	**-7.5**	**-17.:**
Isolated Natural Veg.	35,054	36,722	41,243	38,530	49,010	46,467	20,848	44,849	44,34:
Change from 2000		1,668	6,189	3,476	13,056	11,413	-14,206	9.795	9,288
% Change from 2000		**4.8**	**17.7**	**9.9**	**39.8**	**32.6**	**-40.5**	**27.9**	**26.:**
Fragmented Natural Veg.	103,266	92,359	87,295	88,781	85,924	92,198	69,897	84,204	87,51●
Change from 2000		-10,907	-15,971	-14,485	-17,342	-11,068	-33,369	-19,062	-15,75●
% Change from 2000		**-10.6**	**-15.5**	**-14.0**	**-16.8**	**-10.7**	**-32.3**	**-18.5**	**-15.:**
Stream Corridor	11,122	9,187	8,564	7,015	6,121	7,695	5,626	8,115	7,38:
Change from 2000		-1,935	-2,558	-4,107	-5,001	-3,427	-5,496	-3,007	-3,73●
% Change from 2000		**-17.4**	**-23.0**	**-36.9**	**-45.0**	**-30.8**	**-49.4**	**-27.0**	**-33.●**
Disturbed Land	128,122	117,916	108,232	152,671	179,597	120,380	113,676	128,620	128,932
Change from 2000		-10,206	-19,890	24,495	51,475	-7,742	-14,446	498	81●
% Change from 2000		**-8.0**	**-15.5**	**19.2**	**40.2**	**-6.0**	**-11.3**	**0.4**	**0.●**
Built Land	533,007	582,328	648,012	594,305	685,124	676,557	846730	594,804	662,428
Change from 2000		49,321	115,005	61,298	152,117	143,550	313,723	61,797	129,42●
% Change from 2000		**9.3**	**21.6**	**11.5**	**28.5**	**26.9**	**58.9**	**11.6**	**24.:**

increases. But in the Three-Centers Future, the summary statistics report that less Fragmented Natural Vegetation is lost as the population increases. With the first 500,000 new residents, there is an 18.5% loss relative to the conditions in 2000; with the second 500,000 new residents, there is only a 15.3% loss relative to the conditions in 2000. This result occurs because of the spatial patchiness of the conservation efforts that are assumed under this scenario. The Three-Centers Future attempts to preserve species habitat and other issues of biological concern on a case-by-case basis. As a result, housing is less clumped at larger populations. After some smaller and less critical vegetation patches have been developed, Contiguous and Isolated Natural Vegetation is subject to some development that causes fragmentation.

With the addition of 1,000,000 people, all the futures effectively eliminate the northeast–southwest connection that crosses I-15 in the mountains south of Temecula. Only the Coastal Future with 500,000 new residents maintains this connection. The Regional Low-Density Future also eliminates the connection between the open land on MCAS Miramar and locations to its east.

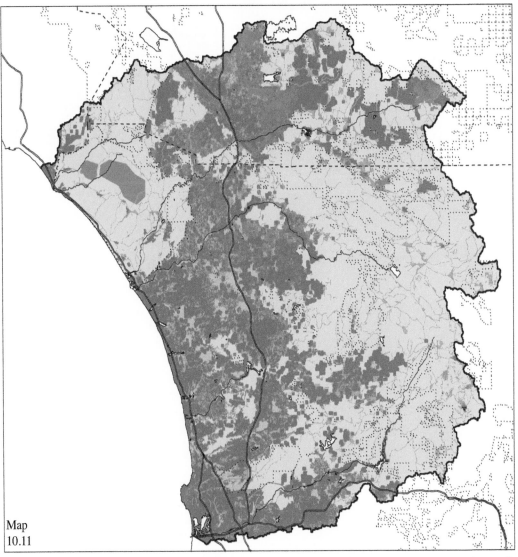

Region of MCB Camp Pendleton & MCAS Miramar
Existing Conditions 2000
Landscape Ecological Pattern

Map
10.11

Contiguous Natural Vegetation
922634 acres (53.2%)

Isolated Natural Vegetation
35054 acres (2%)

Fragmented Natural Vegetation
103266 acres (6%)

Stream Corridor
11122 acres (0.6%)

Disturbed Land
128149 acres (7.4%)

Built Land
533007 acres (30.7%)

Interstates

Rivers

Federal Lands

Counties

0   25   5 Kilometers
0   25   5 Miles

— See color insert following page 204 —

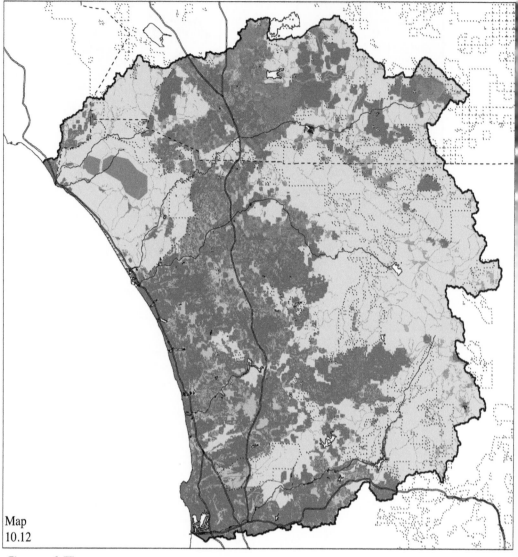

Map
10.12

Coastal Future
500,000 New Residents
Landscape Ecological Pattern

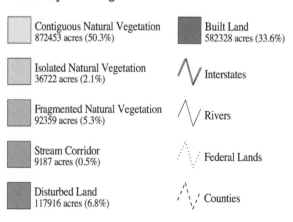

Contiguous Natural Vegetation
872453 acres (50.3%)

Isolated Natural Vegetation
36722 acres (2.1%)

Fragmented Natural Vegetation
92359 acres (5.3%)

Stream Corridor
9187 acres (0.5%)

Disturbed Land
117916 acres (6.8%)

Built Land
582328 acres (33.6%)

Interstates

Rivers

Federal Lands

Counties

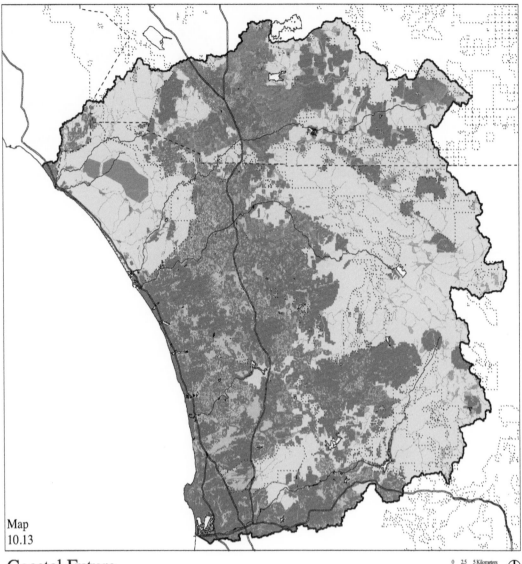

Map
10.13

Coastal Future
1,000,000 New Residents
Landscape Ecological Pattern

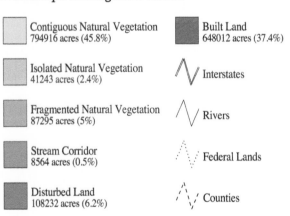

Contiguous Natural Vegetation 794916 acres (45.8%)		Built Land 648012 acres (37.4%)	
Isolated Natural Vegetation 41243 acres (2.4%)		Interstates	
Fragmented Natural Vegetation 87295 acres (5%)		Rivers	
Stream Corridor 8564 acres (0.5%)		Federal Lands	
Disturbed Land 108232 acres (6.2%)		Counties	

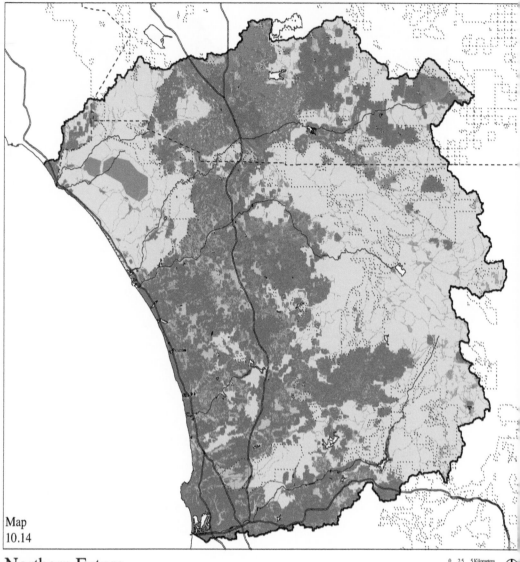

Map
10.14

Northern Future
500,000 New Residents
Landscape Ecological Pattern

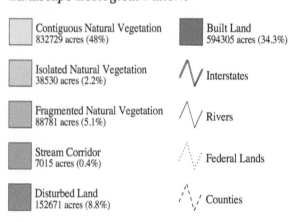

Contiguous Natural Vegetation
832729 acres (48%)

Built Land
594305 acres (34.3%)

Isolated Natural Vegetation
38530 acres (2.2%)

Interstates

Fragmented Natural Vegetation
88781 acres (5.1%)

Rivers

Stream Corridor
7015 acres (0.4%)

Federal Lands

Disturbed Land
152671 acres (8.8%)

Counties

0   2.5   5 Kilometers
0   2.5   5 Miles

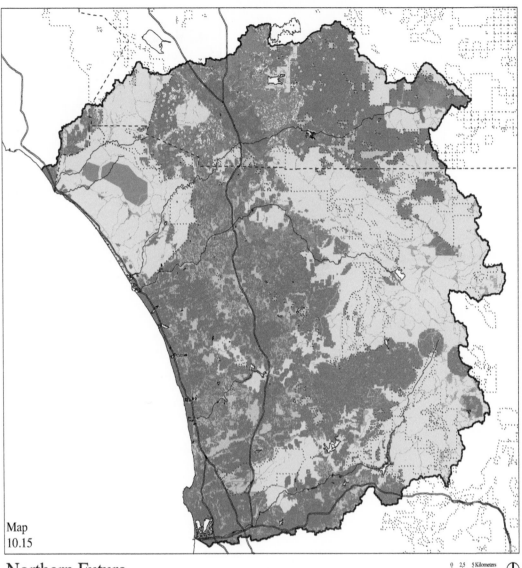

Map
10.15

Northern Future
1,000,000 New Residents
Landscape Ecological Pattern

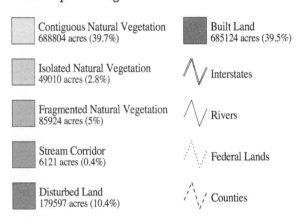

Contiguous Natural Vegetation
688804 acres (39.7%)

Isolated Natural Vegetation
49010 acres (2.8%)

Fragmented Natural Vegetation
85924 acres (5%)

Stream Corridor
6121 acres (0.4%)

Disturbed Land
179597 acres (10.4%)

Built Land
685124 acres (39.5%)

Interstates

Rivers

Federal Lands

Counties

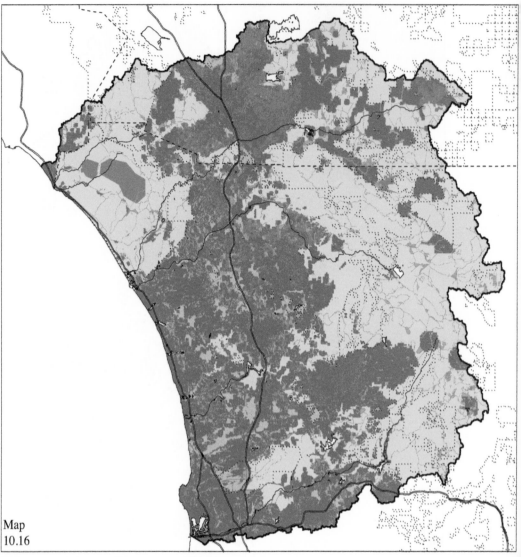

Map
10.16

Regional Low-Density Future
500,000 New Residents
Landscape Ecological Pattern

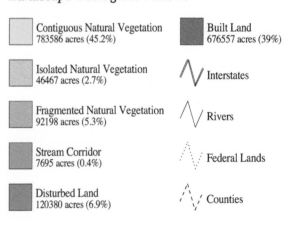

Contiguous Natural Vegetation 783586 acres (45.2%)	Built Land 676557 acres (39%)
Isolated Natural Vegetation 46467 acres (2.7%)	Interstates
Fragmented Natural Vegetation 92198 acres (5.3%)	Rivers
Stream Corridor 7695 acres (0.4%)	Federal Lands
Disturbed Land 120380 acres (6.9%)	Counties

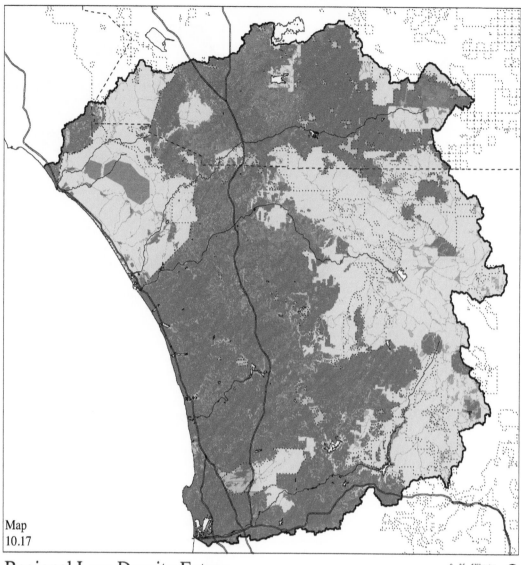

Map
10.17

Regional Low-Density Future
1,000,000 New Residents
Landscape Ecological Pattern

0   2.5   5 Kilometers
0   2.5   5 Miles

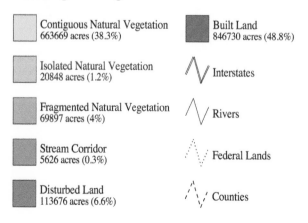

Contiguous Natural Vegetation
663669 acres (38.3%)

Isolated Natural Vegetation
20848 acres (1.2%)

Fragmented Natural Vegetation
69897 acres (4%)

Stream Corridor
5626 acres (0.3%)

Disturbed Land
113676 acres (6.6%)

Built Land
846730 acres (48.8%)

Interstates

Rivers

Federal Lands

Counties

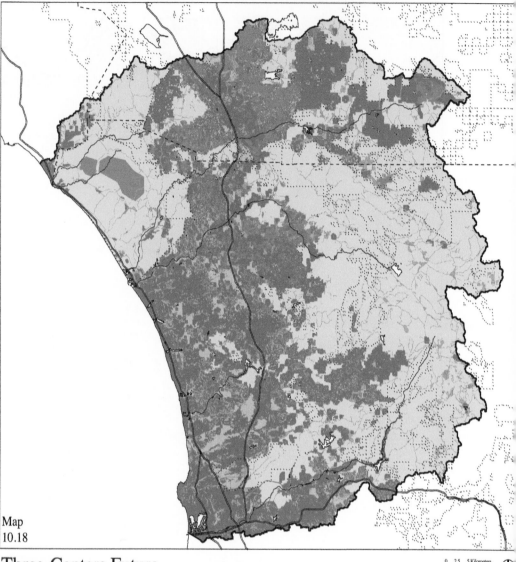

Map
10.18

Three-Centers Future
500,000 New Residents
Landscape Ecological Pattern

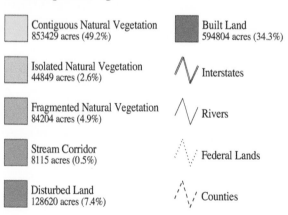

Contiguous Natural Vegetation
853429 acres (49.2%)

Isolated Natural Vegetation
44849 acres (2.6%)

Fragmented Natural Vegetation
84204 acres (4.9%)

Stream Corridor
8115 acres (0.5%)

Disturbed Land
128620 acres (7.4%)

Built Land
594804 acres (34.3%)

Interstates

Rivers

Federal Lands

Counties

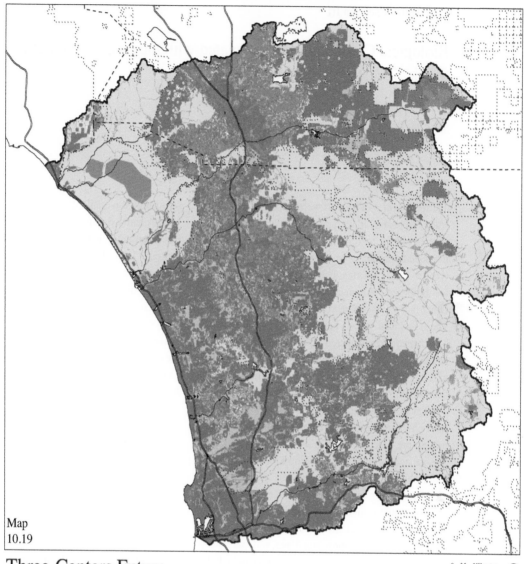

Map
10.19

Three-Centers Future
1,000,000 New Residents
Landscape Ecological Pattern

0 2.5 5 Kilometers
0 2.5 5 Miles

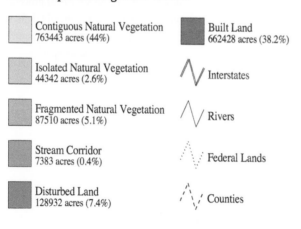

Contiguous Natural Vegetation
763443 acres (44%)

Isolated Natural Vegetation
44342 acres (2.6%)

Fragmented Natural Vegetation
87510 acres (5.1%)

Stream Corridor
7383 acres (0.4%)

Disturbed Land
128932 acres (7.4%)

Built Land
662428 acres (38.2%)

Interstates

Rivers

Federal Lands

Counties

Chapter 10—Section III

Biological Consequences Associated with the Alternative Futures: Single Species Potential Habitat

Scott D. Bassett and Craig W. Johnson

Thirteen species of interest were selected for detailed study. In general, they represent the biological diversity of the region. The selection process was intended to provide a representation of vegetation types that are important in their own right, or that are important components of habitat for local fauna. Other factors also influenced the selection process. These included species being 1) federally listed as Threatened or Endangered, 2) an indicator of the health of a specific ecological system, 3) a generalist in terms of required habitat, 4) rare in terms of occurrence, or 5) a combination of these reasons.

The potential habitat models follow the reasoning presented in the Habitat Suitability Index models presented by the U.S. Fish and Wildlife Service and its Habitat Evaluation Procedures.[12] The Habitat Suitability Index models focus on the identification of spatially explicit elements that contribute to an individual species' habitat preferences. These elements may be comprised of vegetation composition, stand age, soil, percent cover, patch size, patch distribution, and can include a host of other factors.

Each single species potential habitat model was derived from the available literature and personal communication with experts. The models vary in their relative complexity and reflect the amount of information available about each individual species and—significantly—the ability to model this information within the Geographic Information System. Thus, the inclusion or exclusion of any portion of information from one species potential habitat model does not indicate the relative importance of that information unless noted. Each species potential habitat map delineates the estimated distribution of potential habitat at the present time. Impacts relate the probable loss or gain in potential habitat in each of the alternative futures.

The single species potential habitat models presented and described here represent a simple version of how potential habitat can be predicted given new land uses. More complex techniques describing species distributions could be done using non-parametric statistical techniques.[13] This would require an adequate sampling design for abundance and more detailed knowledge about the physical attributes of known locations where specific species occur. Future research about individual species may be done in a manner to accommodate advances in modeling techniques, including new statistical procedures. These statistical procedures would allow for a more scientifically defensible rendition of potential habitat and the subsequent predicted distribution of individual species.

Willowy Monardella

Willowy monardella (*Monardella linoides ssp. viminea*) is a federally listed Endangered Species endemic to San Diego County and surrounding regions. This perennial herb belongs to the mint family and is characterized by a woody base and aromatic foliage. The leaves are linear to lance-shaped. When in bloom, the flowers are pale white to rose in coloration. The species can be distinguished from other species of the same genus by its waxy green coloration and hairy stems.[14]

Willowy monardella primarily inhabits washes located within coastal sage scrub or riparian vegetation communities. Although it is often associated with these two vegetation types, the willowy monardella may be found in association with a wide range of species that prefer sandy washes and floodplains. This species is often not the dominant plant at a given location. The actual geomorphological characteristics of the wash appear to be more important than the associated vegetation. Willowy monardella is located in higher, infrequently flooded areas along streams (Maps 10.20 and 10.21). These locations include benches, midchannel sandbars, and floodplain terraces. This association indicates that the willowy monardella requires a certain interval of disturbance induced by the rapid flow of water. Alterations to flow events and flow rate have the potential to impact this species. Land use alterations upstream from a population of willowy monardella could alter the flow of water in the ephemeral streams and thereby alter available habitat.[15]

Creating a potential habitat model for the willowy monardella required that emphasis be placed more on landform than community vegetation type. Identification of landform elements was done using location information obtained from MCAS Miramar. Although many of the locations occurred within the boundary of this installation, some points were located beyond the base. Known locations of willowy monardella within the San Diego, Miramar, and Poway watersheds were overlaid on landform information. Willowy monardella occurred in the elevation range of 160 to 850 feet and along ephemeral streams that had an upstream contributing area of 220 to 11,000 acres.

With this information, model algorithms were developed and executed to identify the potential habitat. Areas located within the three watersheds and in the 160 to 850 feet interval of elevation were identified. Locations within this area where the upstream contributing area was between 220 and 11,000 acres provided a mask for a more complex assessment of flood potential along ephemeral streams. An algorithm based on the available "wall-to-wall" landform information was developed and executed to determine which areas would be flooded within the three watersheds given a rise in the water level to 6 feet. Areas where the water level could rise to 6 feet were identified as potential habitat locations, thereby identifying locations with the potential to have sporadic and periodic flooding. The land use/land cover types identified as urban, water, agriculture, or wetland were eliminated as potential habitat. Thus, the potential habitat map for the willowy monardella was comprised of locations within the three watersheds, in the 160 to 850 foot elevation interval, with 220 to 11,000 acres of upstream contributing area, where the water level could rise up to 6 feet, and not located on uninhabitable land use/land cover types.

Potential habitat for this species follows the dendritic nature of the stream system. In most situations, the predictive model adequately identifies potential habitat locations. Possible errors in the potential habitat model occur in locations near reservoirs. The regulation of water levels within the reservoirs creates areas of recently disturbed land, and based on habitat descriptions, these locations are unlikely to provide adequate habitat for the plant. Many of these locations remained as potential habitat in the model due to difficulties of algorithmically removing them and because the literature did not fully support their removal.

Existing conditions and impacts associated with each future are summarized in Table 10.3 and shown in Maps 10.22-10.30. The Three-Centers Future is the only alternative not to negatively impact the willowy monardella's potential habitat. In the other futures, the difference between adding 500,000 and 1,000,000 people to the region results in a roughly

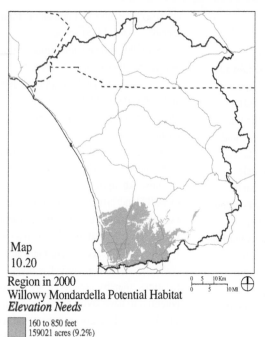

Map
10.20

Region in 2000
Willowy Mondardella Potential Habitat
Elevation Needs

160 to 850 feet
159021 acres (9.2%)

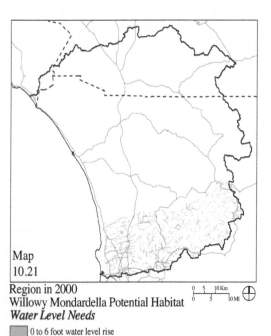

Map
10.21

Region in 2000
Willowy Mondardella Potential Habitat
Water Level Needs

0 to 6 foot water level rise
40557 acres (2.3%)

linear decrease in the percent loss with the Coastal and Regional Low-Density Futures resulting in slightly less than twice the impact and the Northern Future results are slightly more than twice the impact.

The Northern Future does not have as large a decrease in potential habitat when compared with the Regional Low-Density and Coastal Futures. This result is expected given that all willowy monardella potential habitat occurs in the southern portion of the study region. The higher impact of the Regional Low-Density Future is due to the large amount of land area required to accommodate the southern rural residential population influx. The impact of the Coastal Future on potential willowy monardella habitat is relatively high given the small land area required to accommodate the population gain which is placed mainly in urban development (0.2 acre housing lots).

In summary, low and high density housing occurring in southwestern San Diego County has the potential to reduce the amount of available land afforded willowy monardella. Conservation measures and land use planning efforts, as described in the Three Centers Future, can protect potential habitat for this species.

Table 10.3

Willowy Monardella Potential Habitat (in acres)

	2000	Coastal		Northern		Reg. Low-Density		Three-Centers	
		500k	1,000k	500k	1,000k	500k	1,000k	500k	1,000k
Potential Habitat	8,401	7,704	7,026	7,815	7,128	7,591	6,908	8,401	8,401
Change from 2000		-697	-1,376	-586	-1,273	-810	-1,493	0	0
% Change from 2000		**-8.3**	**-16.4**	**-7.0**	**-15.2**	**-9.7**	**-17.8**	**0**	**0**

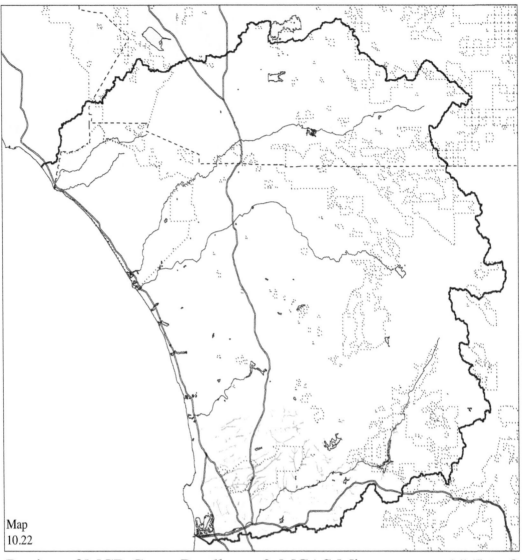

Map
10.22

Region of MCB Camp Pendleton & MCAS Miramar
Existing Conditions 2000
Willowy Monardella Potential Habitat

0   2.5   5 Kilometers
0   2.5   5 Miles

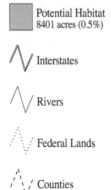

Potential Habitat
8401 acres (0.5%)

Interstates

Rivers

Federal Lands

Counties

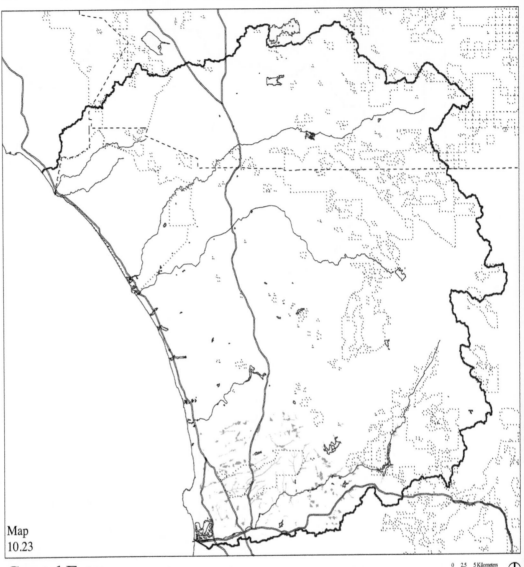

Map
10.23

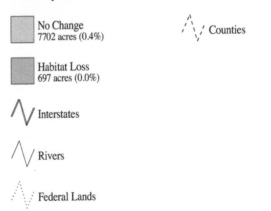

Coastal Future
500,000 New Residents
Willowy Monardella Potential Habitat Change

No Change
7702 acres (0.4%)

Habitat Loss
697 acres (0.0%)

Interstates

Rivers

Federal Lands

Counties

NOTE: Some impacted areas are too small to be depicted at the scale used in the map above. In order to better show these locations, they have been expanded by the drawing of a uniform line around their perimeters. Statistics reported in the legend and in Table 10.3 are based on the data used to create the map and are not influenced by this cartographic necessity.

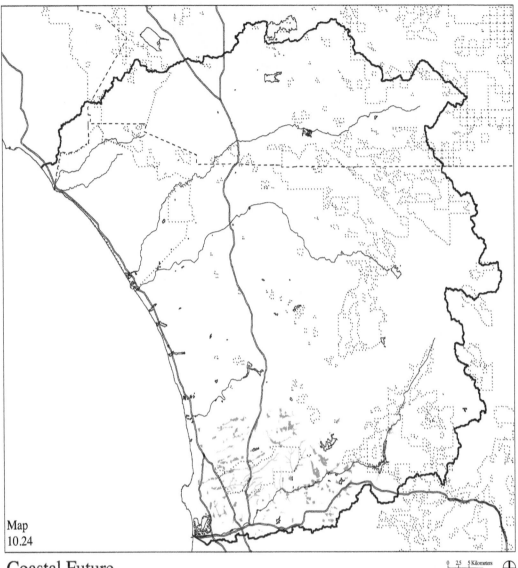

Map
10.24

Coastal Future
1,000,000 New Residents
Willowy Monardella Potential Habitat Change

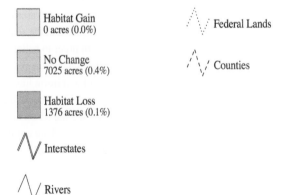

Habitat Gain
0 acres (0.0%)

No Change
7025 acres (0.4%)

Habitat Loss
1376 acres (0.1%)

Interstates

Rivers

Federal Lands

Counties

NOTE: Some impacted areas are too
small to be depicted at the scale used in the
map above. In order to better show these
locations, they have been expanded by
the drawing of a uniform line around their
perimeters. Statistics reported in the legend
and in Table 10.3 are based on the data used
to create the map and are not influenced by
this cartographic necessity.

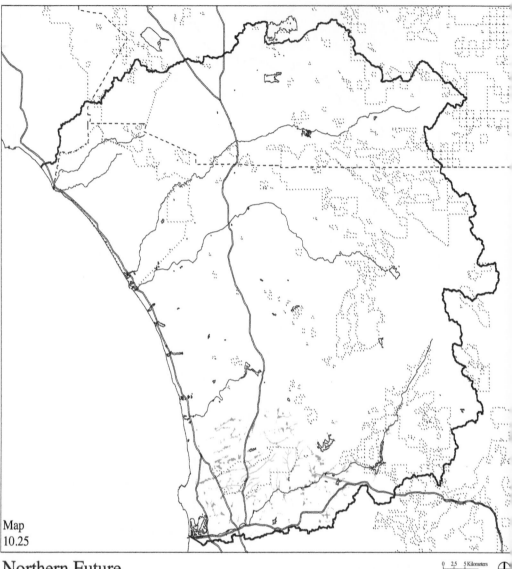

Map
10.25

Northern Future
500,000 New Residents
Willowy Monardella Potential Habitat Change

0   2.5   5 Kilometers
0   2.5   5 Miles

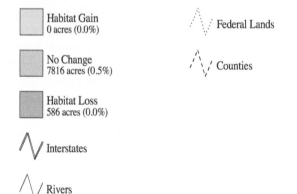

Habitat Gain
0 acres (0.0%)

No Change
7816 acres (0.5%)

Habitat Loss
586 acres (0.0%)

Interstates

Rivers

Federal Lands

Counties

NOTE: Some impacted areas are too
small to be depicted at the scale used in the
map above. In order to better show these
locations, they have been expanded by
the drawing of a uniform line around their
perimeters. Statistics reported in the legend
and in Table 10.3 are based on the data used
to create the map and are not influenced by
this cartographic necessity.

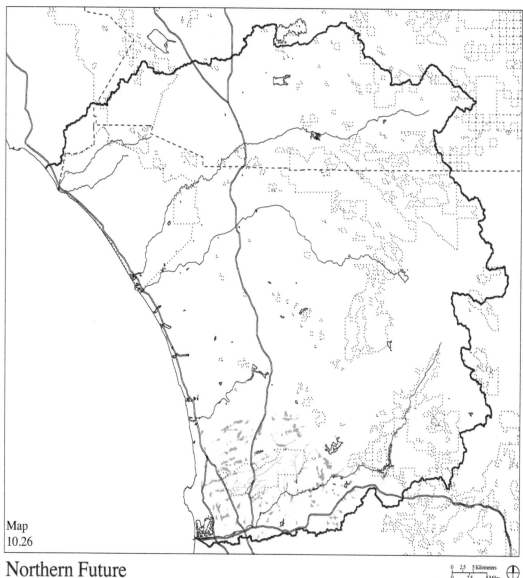

Map
10.26

Northern Future
1,000,000 New Residents
Willoy Monardella Potential Habitat Change

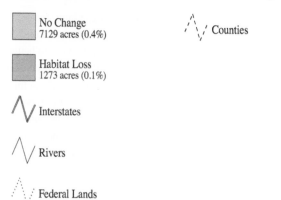

No Change
7129 acres (0.4%)

Habitat Loss
1273 acres (0.1%)

Interstates

Rivers

Federal Lands

Counties

NOTE: Some impacted areas are too
small to be depicted at the scale used in the
map above. In order to better show these
locations, they have been expanded by
the drawing of a uniform line around their
perimeters. Statistics reported in the legend
and in Table 10.3 are based on the data used
to create the map and are not influenced by
this cartographic necessity.

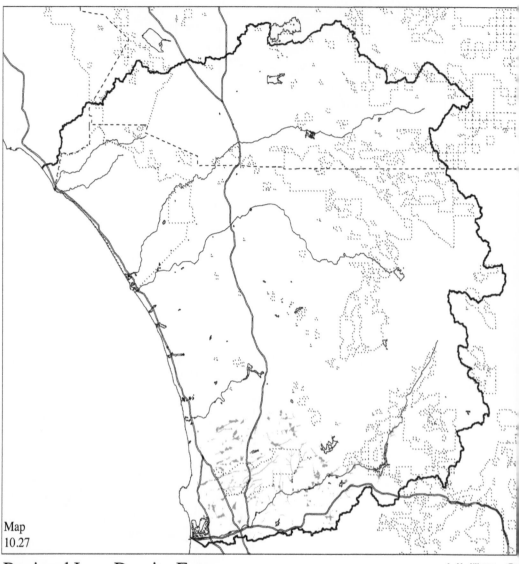

Map
10.27

Regional Low-Density Future
500,000 New Residents
Willowy Monardella Potential Habitat Change

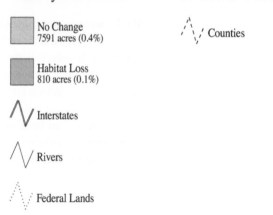

No Change
7591 acres (0.4%)

Habitat Loss
810 acres (0.1%)

Interstates

Rivers

Federal Lands

Counties

NOTE: Some impacted areas are too small to be depicted at the scale used in the map above. In order to better show these locations, they have been expanded by the drawing of a uniform line around their perimeters. Statistics reported in the legend and in Table 10.3 are based on the data used to create the map and are not influenced by this cartographic necessity..

0 2.5 5 Kilometers
0 2.5 5 Miles

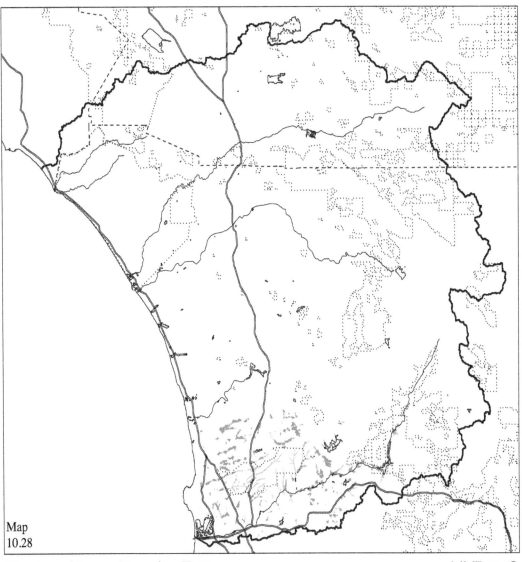

Map
10.28

Regional Low-Density Future
1,000,000 New Residents
Willowy Monardella Potential Habitat Change

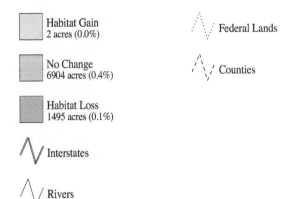

Habitat Gain
2 acres (0.0%)

No Change
6904 acres (0.4%)

Habitat Loss
1495 acres (0.1%)

Interstates

Rivers

Federal Lands

Counties

NOTE: Some impacted areas are too
small to be depicted at the scale used in the
map above. In order to better show these
locations, they have been expanded by
the drawing of a uniform line around their
perimeters. Statistics reported in the legend
and in Table 10.3 are based on the data used
to create the map and are not influenced by
this cartographic necessity.

0   2.5   5 Kilometers
0   2.5   5 Miles

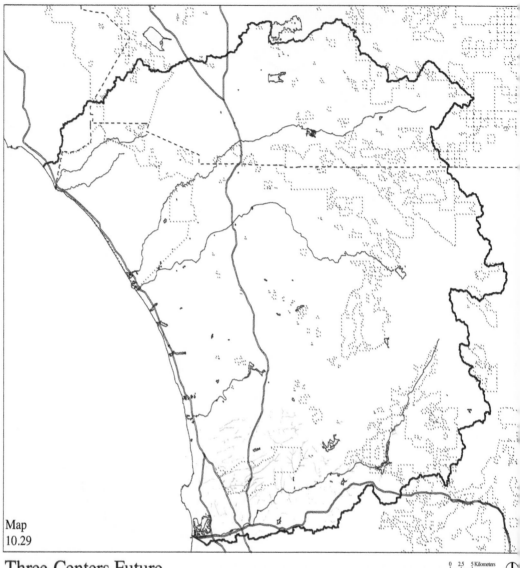

Map
10.29

Three-Centers Future
500,000 New Residents
Willowy Monardella Potential Habitat Change

0 2.5 5 Kilometers
0 2.5 5 Miles

No Change
8401 acres (0.5%)

Interstates

Rivers

Federal Lands

Counties

NOTE: Some impacted areas are too
small to be depicted at the scale used in the
map above. In order to better show these
locations, they have been expanded by
the drawing of a uniform line around their
perimeters. Statistics reported in the legend
and in Table 10.3 are based on the data used
to create the map and are not influenced by
this cartographic necessity.

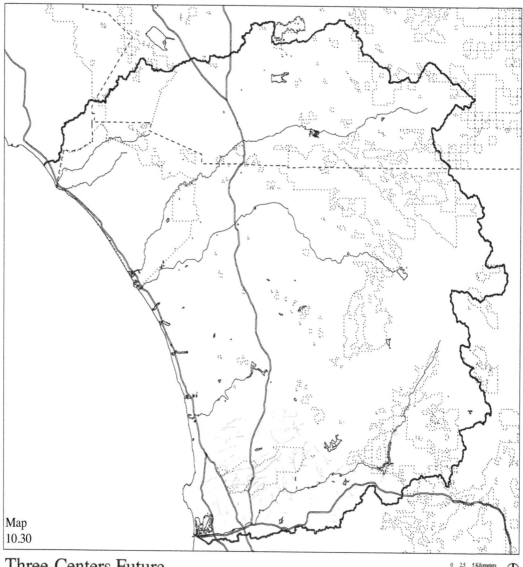

Map
10.30

Three-Centers Future
1,000,000 New Residents
Willowy Monardella Potential Habitat Change

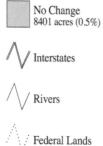

No Change
8401 acres (0.5%)

Interstates

Rivers

Federal Lands

Counties

0   2.5   5 Kilometers
0   2.5   5 Miles

NOTE: Some impacted areas are too
small to be depicted at the scale used in the
map above. In order to better show these
locations, they have been expanded by
the drawing of a uniform line around their
perimeters. Statistics reported in the legend
and in Table 10.3 are based on the data used
to create the map and are not influenced by
this cartographic necessity.

California Sycamore

California sycamore (*Platanus racemosa*) is a large cloning tree that reaches heights of 100 feet. The leaves are simple, five-lobed, and 6 to 10 inches long with the width being slightly greater than the length. The outer bark is brown exfoliating to expose the cream to light green bark beneath.[16]

The California sycamore is considered a mid-successional species that forms older groves on the higher terraces of riparian areas.[17] It has an obligate restriction to riparian zones indicating a need for access to ground water within its root zone. This species of sycamore prefers coarse, porous, gravelly unconsolidated, typically riparian deposits as a substrate. It is an important secondary component of mixed riparian forests where it is often associated with sites higher and drier than those occupied by the dominant Fremont cottonwood (*Populus fremontii*). The California sycamore is frequently the single dominant tree species along intermittent streams of the southern coast range of California indicating a better adaptation to drier sites than other riparian pioneer species.[18]

The relative importance of this tree is revealed through wildlife usage. Birds use the canopy area of the riparian system most heavily and especially depend on species such as the California sycamore. Many raptors, woodpeckers, and various species of warblers would be largely eliminated by the absence of the upper tree stratum provided by this species.[19] Some species of woodpecker show strong preference selection for the use of large dead sycamore limbs as nest locations.[20]

With the elimination of the natural water flow regimes in this region caused by the construction of reservoirs, changes in plant community structure are apparent. While the sycamore has adapted to intermittent streams with large pulses of water in winter and dry periods during summer, reservoirs create a perennial source of water allowing the establishment of other plant species. In addition to the presence of perennial water behind reservoirs, changes in the flood regime can also have an impact. Changes in the type of water flow caused by altering water use policies can result in an increase in finer sediments and a decrease in larger sediments

such as cobble or gravel that are a prerequisite for establishment of sycamores.[21]

Based on the literature, a sycamore model was created mainly using landform information. Intermittent and perennial streams for the region were identified using a digital elevation model to locate points through which water would likely flow. This analysis was overlaid onto the U.S. Geological Survey's 7.5-minute topographic maps to ensure positional accuracy. The identification of streams in this manner allowed potential flood algorithms to be developed using the digital elevation model. An algorithm based on the available region-wide landform information was developed and run to determine which areas would be flooded given a rise in the water level to 6 feet. Areas along the streams that could be flooded with a 3-feet rise in water level were also identified. Locations that flooded to a 3-feet level were eliminated from the 6-feet flood zone and not identified as potential sycamore habitat. Thus, areas along the intermittent and perennial streams where the water level could rise 3 to 6 feet were identified as potential habitat locations. The land use/land cover types of urban, water, beach, agriculture, or wetland were eliminated as potential habitat. Although the literature identified that the preferred habitat of sycamores should contain a cobble or gravel substrate, these data were not readily available at the time of this writing. A second item excluded from the model was the effect of dam and reservoir construction. As noted, water use policies which increase sedimentation over gravel and cobble substrates have been identified as a factor influencing potential habitat. Reservoirs and dams could be precisely identified, but their subsequent impact could not be quantified. The exclusion of this information could result in commission errors.

Existing California sycamore potential habitat occurs along intermittent and perennial streams in the region. Potential habitat appears to be concentrated in locations lacking a high degree of human infrastructure. This information indicates the need to preserve locations of low slope, characteristic of riparian areas in the region, located along streams. Again, errors of commission may appear due to the lack of adequate information about substrate type and the unknown quantifiable effects of sedimentation on potential sycamore habitat.

Existing conditions and impacts associated wtih each future are summarized in Table 10.4 and shown in Maps 10.31-10.39. The Northern Future has the greatest impact of the four alternative futures on sycamore potential habitat. The three other futures all reduce the amount of sycamore potential habitat at about the same level. Much of the available habitat occurs within flat locations in the northern part of the study region. Thus, the higher percentage impact of the Northern Future is not surprising.

The futures adding 500,000 people to the region all produce less significant impacts than the futures adding 1,000,000 people. Doubling the number of additional people in the region doubles the impact on California sycamore potential habitat.

The results show the importance of California sycamore within Western Riverside County. Housing increases in the northern portion of the study area, as indicated by the Northern Future's impacts, have a greater potential to reduce the habitat available for the sycamore. Efforts to conserve California sycamore should focus on intermittent and perennial streams and emphasize the localized need to preserve cobble and gravel substrates that were not modeled.

Table 10.4

California Sycamore Potential Habitat (in acres)

	2000	Coastal		Northern		Reg. Low-Density		Three-Centers	
		500k	1,000k	500k	1,000k	500k	1,000k	500k	1,000k
Potential Habitat	52,872	51,328	49,734	49,583	46,025	51,157	49,400	50,752	49,734
Change from 2000		-1,544	-3,138	-3,289	-6,847	-1,715	-3,472	-2,120	-3,138
% Change from 2000		**-2.9**	**-5.9**	**-6.2**	**-13.0**	**-3.2**	**-6.6**	**-4.0**	**-5.9**

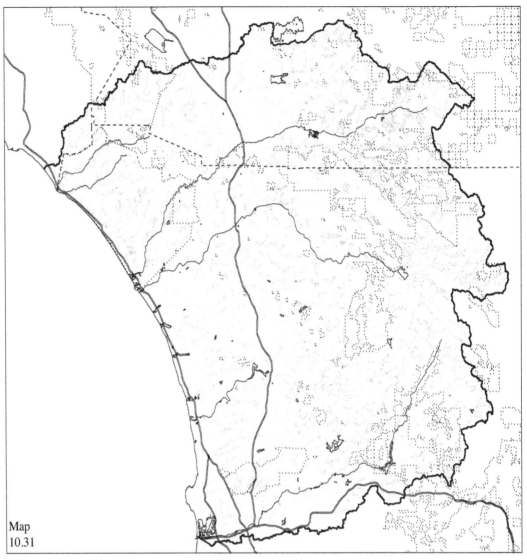

Map
10.31

Region of MCB Camp Pendleton & MCAS Miramar
Existing Conditions 2000
California Sycamore Potential Habitat

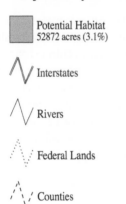

Potential Habitat
52872 acres (3.1%)

Interstates

Rivers

Federal Lands

Counties

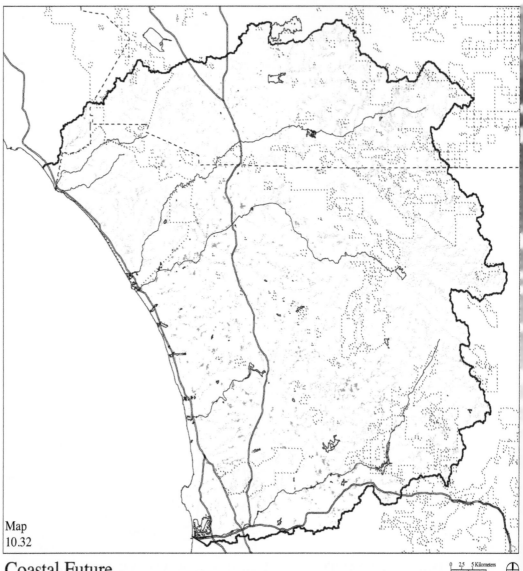

Map
10.32

Coastal Future
500,000 New Residents
California Sycamore Potential Habitat Change

0   2.5   5 Kilometers
0   2.5   5 Miles

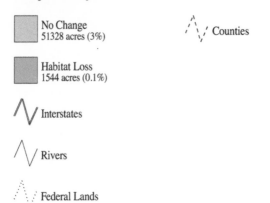

No Change
51328 acres (3%)

Habitat Loss
1544 acres (0.1%)

Interstates

Rivers

Federal Lands

Counties

NOTE: Some impacted areas are too
small to be depicted at the scale used in the
map above. In order to better show these
locations, they have been expanded by
the drawing of a uniform line around their
perimeters. Statistics reported in the legend
and in Table 10.4 are based on the data used
to create the map and are not influenced by
this cartographic necessity.

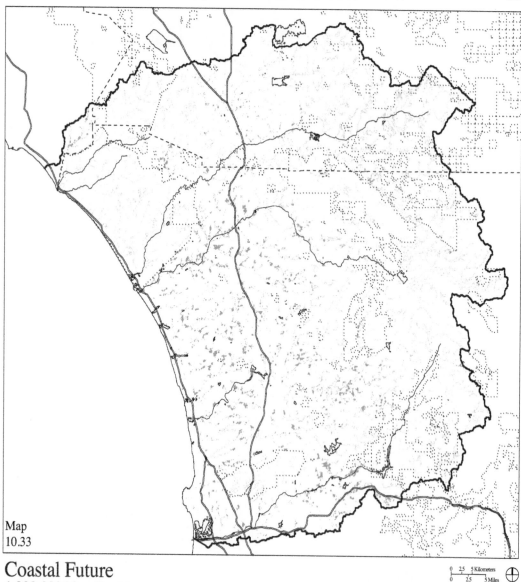

Map
10.33

Coastal Future
1,000,000 New Residents
California Sycamore Potential Habitat Change

0   2.5   5 Kilometers
0   2.5   5 Miles

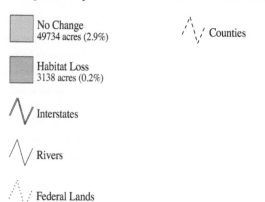

No Change
49734 acres (2.9%)

Habitat Loss
3138 acres (0.2%)

Interstates

Rivers

Federal Lands

Counties

NOTE: Some impacted areas are too
small to be depicted at the scale used in the
map above. In order to better show these
locations, they have been expanded by
the drawing of a uniform line around their
perimeters. Statistics reported in the legend
and in Table 10.4 are based on the data used
to create the map and are not influenced by
this cartographic necessity.

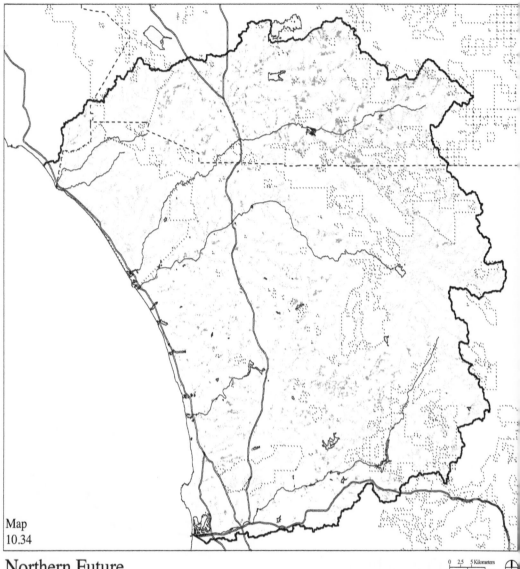

Map
10.34

Northern Future
500,000 New Residents
California Sycamore Potential Habitat Change

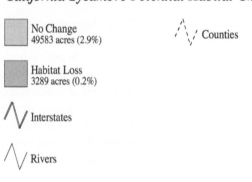

No Change
49583 acres (2.9%)

Habitat Loss
3289 acres (0.2%)

Interstates

Rivers

Federal Lands

Counties

NOTE: Some impacted areas are too
small to be depicted at the scale used in the
map above. In order to better show these
locations, they have been expanded by
the drawing of a uniform line around their
perimeters. Statistics reported in the legend
and in Table 10.4 are based on the data used
to create the map and are not influenced by
this cartographic necessity.

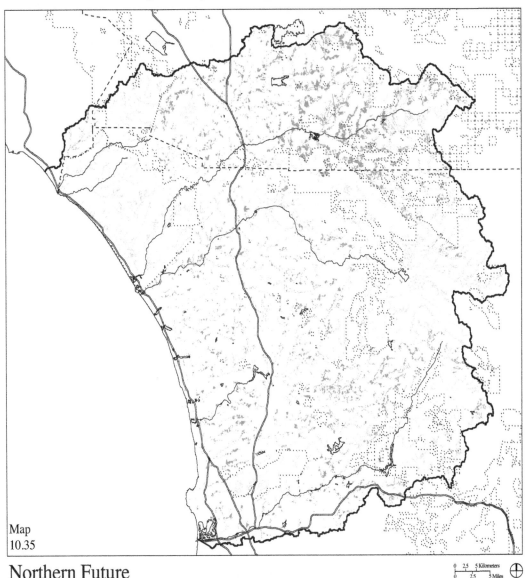

Map
10.35

Northern Future
1,000,000 New Residents
California Sycamore Potential Habitat Change

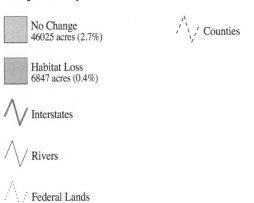

No Change
46025 acres (2.7%)

Habitat Loss
6847 acres (0.4%)

Interstates

Rivers

Federal Lands

Counties

NOTE: Some impacted areas are too
small to be depicted at the scale used in the
map above. In order to better show these
locations, they have been expanded by
the drawing of a uniform line around their
perimeters. Statistics reported in the legend
and in Table 10.4 are based on the data used
to create the map and are not influenced by
this cartographic necessity.

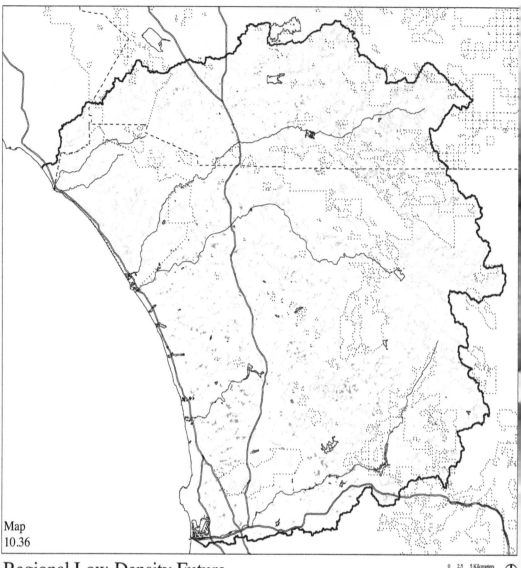

Map
10.36

Regional Low-Density Future
500,000 New Residents
California Sycamore Potential Habitat Change

0   2.5   5 Kilometers
0   2.5   5 Miles

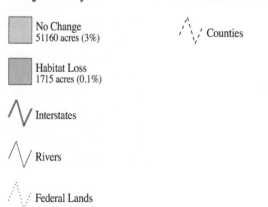

No Change
51160 acres (3%)

Counties

Habitat Loss
1715 acres (0.1%)

Interstates

Rivers

Federal Lands

NOTE: Some impacted areas are too
small to be depicted at the scale used in the
map above. In order to better show these
locations, they have been expanded by
the drawing of a uniform line around their
perimeters. Statistics reported in the legend
and in Table 10.4 are based on the data used
to create the map and are not influenced by
this cartographic necessity.

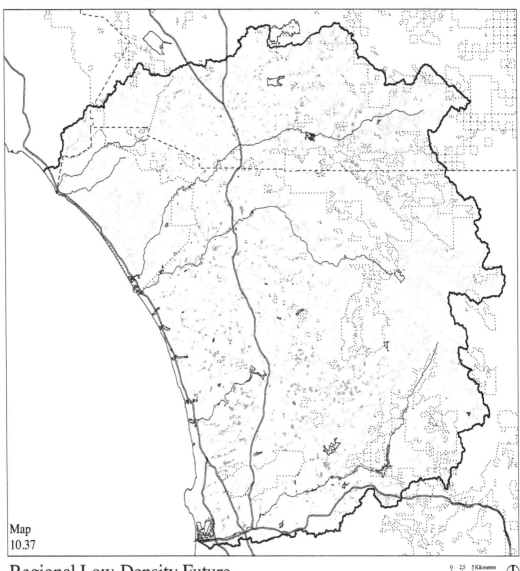

Map
10.37

Regional Low-Density Future
1,000,000 New Residents
California Sycamore Potential Habitat Change

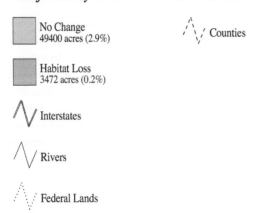

No Change
49400 acres (2.9%)

Habitat Loss
3472 acres (0.2%)

Interstates

Rivers

Federal Lands

Counties

0   2.5   5 Kilometers
0   2.5   5 Miles

NOTE: Some impacted areas are too
small to be depicted at the scale used in the
map above. In order to better show these
locations, they have been expanded by
the drawing of a uniform line around their
perimeters. Statistics reported in the legend
and in Table 10.4 are based on the data used
to create the map and are not influenced by
this cartographic necessity.

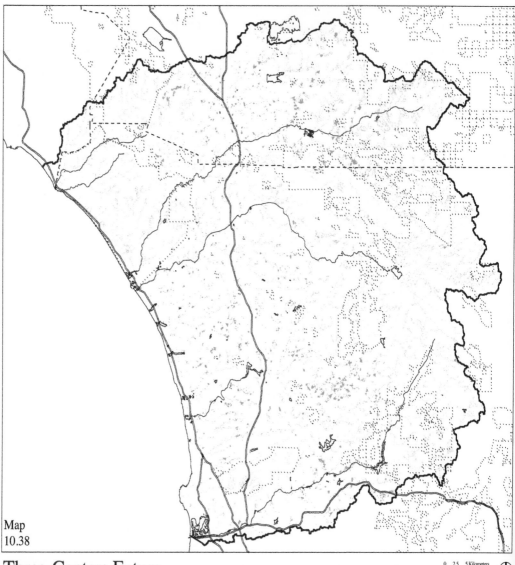

Map
10.38

Three-Centers Future
500,000 New Residents
California Sycamore Potential Habitat Change

0   2.5   5 Kilometers
0   2.5   5 Miles

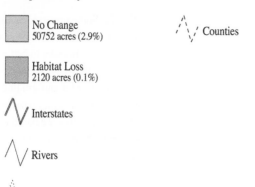

No Change
50752 acres (2.9%)

Habitat Loss
2120 acres (0.1%)

Interstates

Rivers

Federal Lands

Counties

NOTE: Some impacted areas are too
small to be depicted at the scale used in the
map above. In order to better show these
locations, they have been expanded by
the drawing of a uniform line around their
perimeters. Statistics reported in the legend
and in Table 10.4 are based on the data used
to create the map and are not influenced by
this cartographic necessity.

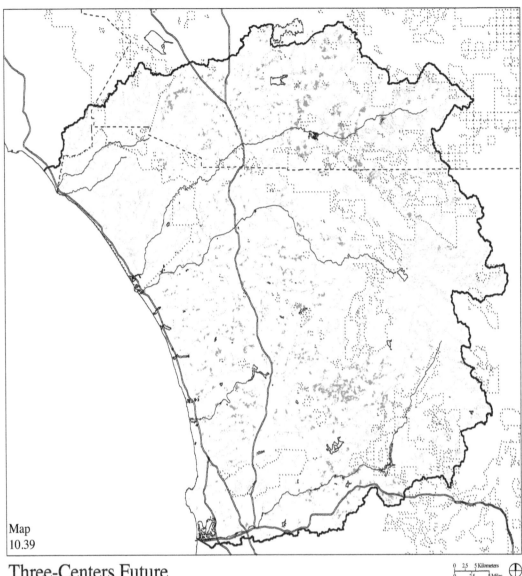

Map
10.39

Three-Centers Future
1,000,000 New Residents
California Sycamore Potential Habitat Change

0   2.5   5 Kilometers
0   2.5   5 Miles

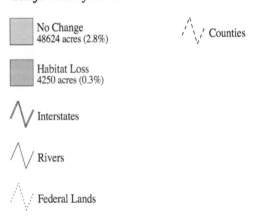

No Change
48624 acres (2.8%)

Habitat Loss
4250 acres (0.3%)

Interstates

Rivers

Federal Lands

Counties

NOTE: Some impacted areas are too
small to be depicted at the scale used in the
map above. In order to better show these
locations, they have been expanded by
the drawing of a uniform line around their
perimeters. Statistics reported in the legend
and in Table 10.4 are based on the data used
to create the map and are not influenced by
this cartographic necessity.

Fairy Shrimp

The San Diego fairy shrimp (*Branchinecta sandiegonensis*) and the Riverside fairy shrimp (*Streptocephalus wootoni*) are both federally listed endangered species. They are small freshwater crustaceans preferring to inhabit vernal pools in Southern California and Northern Baja California. Mature San Diego fairy shrimp range in size from a half an inch to sixth tenths of an inch in length while the Riverside fairy shrimp ranges from a half an inch to one inch in length. Both shrimp species are small, delicate, and may be confused with other fairy shrimp.[22] Although the San Diego and Riverside fairy shrimp inhabit vernal pools, other pools or areas modified to hold water may contain these species. The two species may co-occur in deeper pools; however, they have not been observed coexisting as adults.[23] The San Diego fairy shrimp prefers pools with a depth of 2 to 10 inches, while the Riverside fairy shrimp prefers pools greater than 10 inches in depth. These observations appear to be correlated to both shrimps' temperature preferences.[24] The distribution among these species of fairy shrimp appears to be linked in some way to the water chemistry of various pools with ion levels of sodium being linked to their ion regulatory abilities.[25] Vernal pools are perhaps the most notable component of both shrimps' habitat. Vernal pool declines, rareness, and Fish and Wildlife Service desires to restore historic locations add to the relative importance of this type of habitat.[26]

Vernal pools occur in heavy soil areas on mesas and terraces.[27] Areas on less than 1% slope where water may pool during the rainy season provide ideal habitat.[28] Vernal pools vary in size from a few square feet to that of small lakes and are dependent on water levels that may change from year to year as a function of rainfall.[29] In some years, no standing water may occur while in others the pools may overflow and resemble a lake. Seasonal variation in water, shallow depressions, soil type extent, and mesa location make the prediction of vernal pool locations difficult on a regional scale.

Lacking sufficiently detailed region-wide soils information and an elevation map at a one-foot resolution to discriminate shallow depressions, known vernal pool locations obtained from MCAS Miramar and MCB Camp Pendleton were used to determine the relative occurrence of vernal pools within existing digital information. These vernal pool sites were located in areas ranging in elevation from 50 to 550 feet (Map 10.40). Using a coarse slope map, over 90% of the vernal pool locations were identified on slopes of less than 5% (Map 10.41). Due to the coarse resolution of the slope map as identified from the 100-feet USGS digital elevation model, some points were placed on slopes of greater than 75% which had not been observed previously in the existing literature.

Utilizing information from the existing literature and the overlay of known vernal pool locations, potential fairy shrimp habitat was identified in a series of steps. First, areas located in the 50 to 500 foot range and on less than 5% slope were identified. After overlaying known locations of vernal pools, it became apparent that the resolution of the digital elevation model was too coarse to identify vernal pool complexes smaller than 3 acres in area and that most vernal pool complexes larger than 3 acres in the area existed adjacent to watersheds at least 200 acres in area. Thus, locations draining over 200 acres of area were identified. A small flood of up to 10 feet in depth was simulated at each of these locations. Although 10 feet appeared rather deep for a vernal pool, the characteristics of using the coarse digital elevation model dictated this figure. The subsequent outcome was compared with the extant locations of vernal pool complexes. Potential flood locations were masked to the elevation range and 5% slope map to create a layer to be filtered by area. A patch area filter across the masked potential flood map eliminated locations less than 5 acres in area. The remaining potential flood locations were identified as potential fairy shrimp habitat.

The potential fairy shrimp habitat map of existing conditions contains some errors of omission. A few locations appear to be missing in the potential habitat map and are most likely caused by the coarse grain thematic information utilized to identify potential habitat. Vernal pools are known to be located along the fringes of some smaller airports located in the region. More accurate and precise elevation information would provide for a more thorough identification of potential vernal pools.

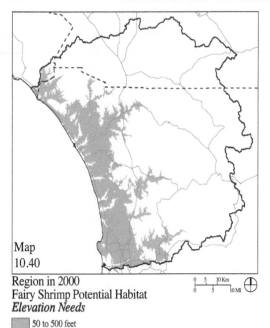

Map
10.40

Region in 2000
Fairy Shrimp Potential Habitat
Elevation Needs

50 to 500 feet
359133 acres (20.7%)

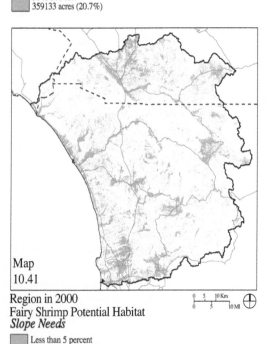

Map
10.41

Region in 2000
Fairy Shrimp Potential Habitat
Slope Needs

Less than 5 percent
287207 acres (16.6%)

Locations within the eastern part of the region are probably under-represented due to a lack of readily available information on the exact location of vernal pools.

Existing conditions and impacts associated with each future are summarized in Table 10.5 and shown in Maps 10.42-10.50. The overall impact on fairy shrimp potential habitat varies by alternative future. The Three-Centers Future has virtually no impact. The other three alternative futures vary in the amount that potential habitat is lost. The Coastal and Regional Low-Density Futures show the highest loss. This situation is expected given the lack of habitat identification in the eastern part of the study area. If potential errors in our ability to locate vernal pools could be rectified, the impacts of the Northern Future would be larger than the values indicated. For all the futures, any development on the mesas west of the Coastal Range reduces the amount of potential fairy shrimp habitat. When the addition of 1,000,000 versus the addition of 500,000 people is compared, an increase in impact occurs for the Coastal, Northern, and Regional Low-Density Futures. The increase is slightly lower than a doubling effect and reflects the initial preference for new housing lots to be placed in flat areas with a relatively large amount of buildable land. Thus, locations with vernal pool complexes represent prime areas for housing development.

Results indicate the need to conserve lands in Western San Diego County for fairy shrimp potential habitat. Any development occurring on the mesas located within the vicinity of potential vernal pools has the ability to impact fairy shrimp habitat. Conservation efforts, as indicated under the Three-Centers Future, would better ensure the maintenance of existing fairy shrimp potential habitat.

Table 10.5

Fairy Shrimp Potential Habitat (in acres)

	2000	Coastal		Northern		Reg. Low-Density		Three-Centers	
		500k	1,000k	500k	1,000k	500k	1,000k	500k	1,000k
Potential Habitat	11,633	10,635	9,991	10,842	10,264	10,588	9,809	11,633	11,633
Change from 2000		-988	-1,732	-791	-1,369	-1,045	-1,824	0	0
% Change from 2000		**-8.6**	**-14.9**	**-6.8**	**-11.8**	**-9.0**	**-15.7**	**0**	**0**

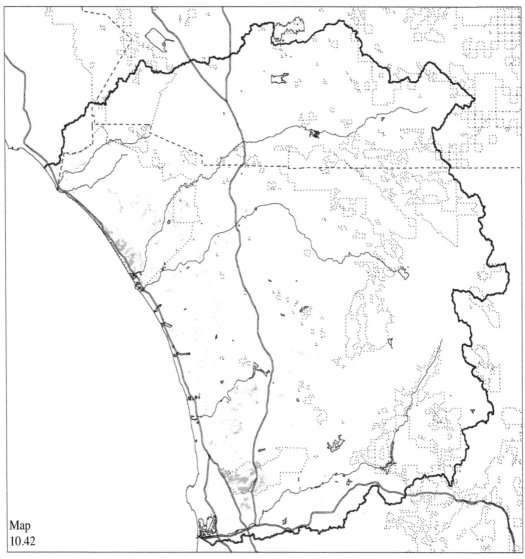

Map
10.42

Region of MCB Camp Pendleton & MCAS Miramar

Existing Conditions 2000

Fairy Shrimp Potential Habitat

Potential Habitat
11633 acres (0.7%)

Interstates

Rivers

Federal Lands

Counties

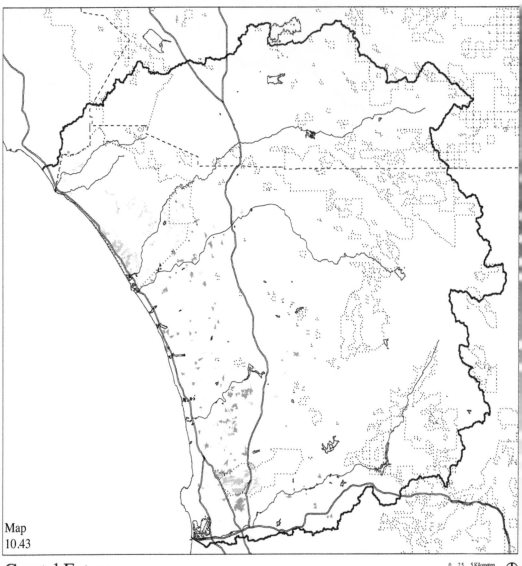

Map
10.43

Coastal Future
500,000 New Residents
Fairy Shrimp Potential Habitat Change

0   2.5   5 Kilometers
0   2.5   5 Miles

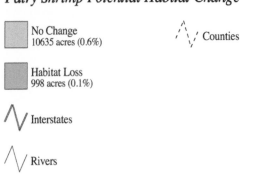

No Change
10635 acres (0.6%)

Habitat Loss
998 acres (0.1%)

Interstates

Rivers

Federal Lands

Counties

NOTE: Some impacted areas are too
small to be depicted at the scale used in the
map above. In order to better show these
locations, they have been expanded by
the drawing of a uniform line around their
perimeters. Statistics reported in the legend
and in Table 10.5 are based on the data used
to create the map and are not influenced by
this cartographic necessity.

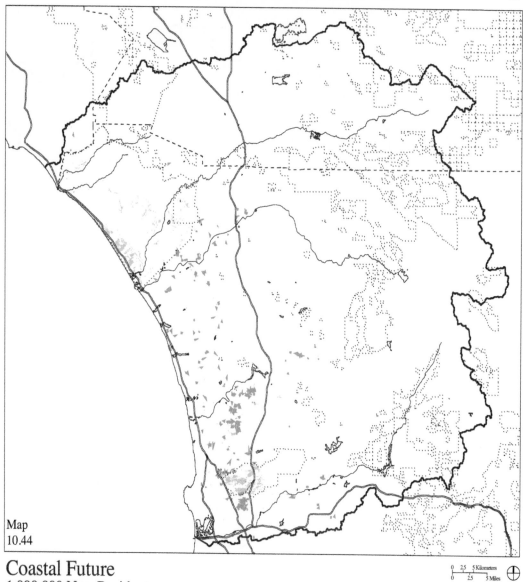

Map
10.44

Coastal Future
1,000,000 New Residents
Fairy Shrimp Potential Habitat Change

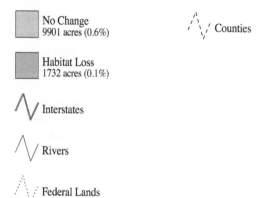

No Change
9901 acres (0.6%)

Habitat Loss
1732 acres (0.1%)

Interstates

Rivers

Federal Lands

Counties

NOTE: Some impacted areas are too
small to be depicted at the scale used in the
map above. In order to better show these
locations, they have been expanded by
the drawing of a uniform line around their
perimeters. Statistics reported in the legend
and in Table 10.5 are based on the data used
to create the map and are not influenced by
this cartographic necessity.

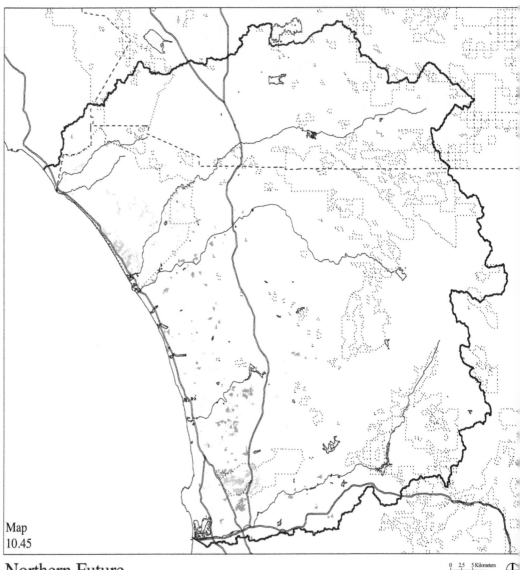

Map
10.45

Northern Future
500,000 New Residents
Fairy Shrimp Potential Habitat Change

0   2.5   5 Kilometers
0   2.5   5 Miles

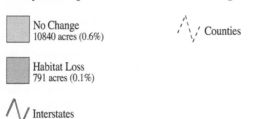

No Change
10840 acres (0.6%)

Habitat Loss
791 acres (0.1%)

Interstates

Rivers

Federal Lands

Counties

NOTE: Some impacted areas are too
small to be depicted at the scale used in the
map above. In order to better show these
locations, they have been expanded by
the drawing of a uniform line around their
perimeters. Statistics reported in the legend
and in Table 10.5 are based on the data used
to create the map and are not influenced by
this cartographic necessity.

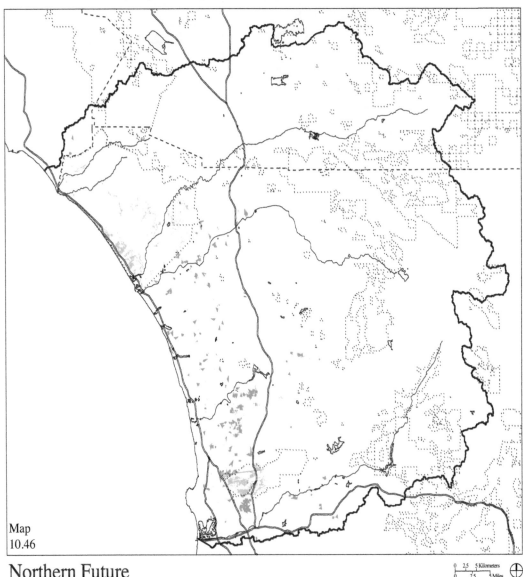

Map
10.46

Northern Future
1,000,000 New Residents
Fairy Shrimp Potential Habitat Change

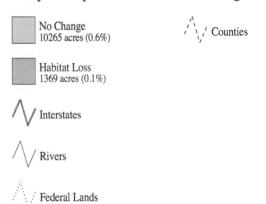

No Change
10265 acres (0.6%)

Habitat Loss
1369 acres (0.1%)

Interstates

Rivers

Federal Lands

Counties

0   2.5   5 Kilometers
0   2.5   5 Miles

NOTE: Some impacted areas are too
small to be depicted at the scale used in the
map above. In order to better show these
locations, they have been expanded by
the drawing of a uniform line around their
perimeters. Statistics reported in the legend
and in Table 10.5 are based on the data used
to create the map and are not influenced by
this cartographic necessity.

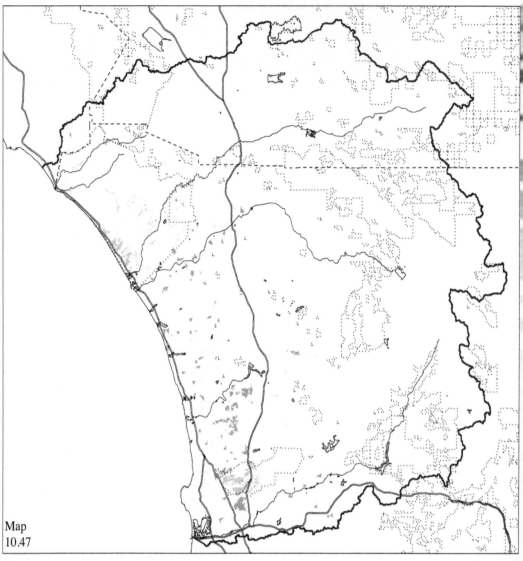

Map
10.47

Regional Low-Density Future
500,000 New Residents
Fairy Shrimp Potential Habitat Change

0 2.5 5 Kilometers
0 2.5 5 Miles

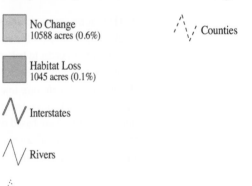

No Change
10588 acres (0.6%)

Habitat Loss
1045 acres (0.1%)

Interstates

Rivers

Federal Lands

Counties

NOTE: Some impacted areas are too small to be depicted at the scale used in the map above. In order to better show these locations, they have been expanded by the drawing of a uniform line around their perimeters. Statistics reported in the legend and in Table 10.5 are based on the data used to create the map and are not influenced by this cartographic necessity.

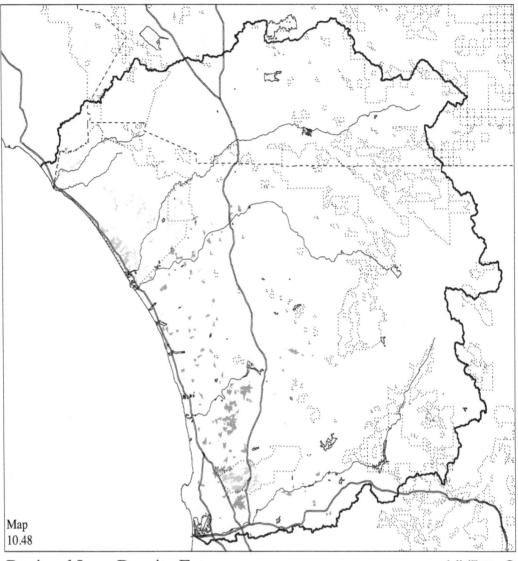

Map
10.48

Regional Low-Density Future
1,000,000 New Residents
Fairy Shrimp Potential Habitat Change

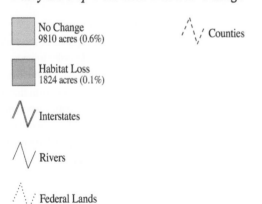

No Change
9810 acres (0.6%)

Habitat Loss
1824 acres (0.1%)

Interstates

Rivers

Federal Lands

Counties

NOTE: Some impacted areas are too small to be depicted at the scale used in the map above. In order to better show these locations, they have been expanded by the drawing of a uniform line around their perimeters. Statistics reported in the legend and in Table 10.5 are based on the data used to create the map and are not influenced by this cartographic necessity.

0   2.5   5 Kilometers
0   2.5   5 Miles

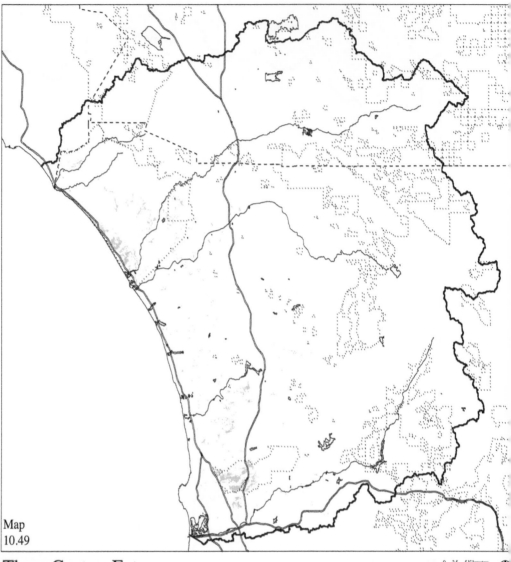

Map
10.49

Three-Centers Future
500,000 New Residents
Fairy Shrimp Potential Habitat Change

0 2.5 5 Kilometers
0 2.5 5 Miles

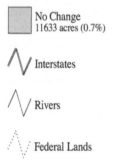 No Change
11633 acres (0.7%)

Interstates

Rivers

Federal Lands

Counties

NOTE: Some impacted areas are too small to be depicted at the scale used in the map above. In order to better show these locations, they have been expanded by the drawing of a uniform line around their perimeters. Statistics reported in the legend and in Table 10.5 are based on the data used to create the map and are not influenced by this cartographic necessity.

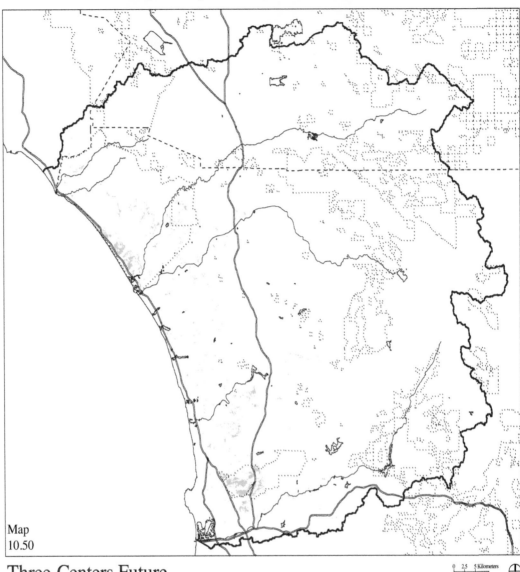

Map
10.50

Three-Centers Future
1,000,000 New Residents
Fairy Shrimp Potential Habitat Change

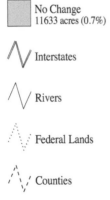

No Change
11633 acres (0.7%)

Interstates

Rivers

Federal Lands

Counties

0 2.5 5 Kilometers
0 2.5 5 Miles

NOTE: Some impacted areas are too
small to be depicted at the scale used in the
map above. In order to better show these
locations, they have been expanded by
the drawing of a uniform line around their
perimeters. Statistics reported in the legend
and in Table 10.5 are based on the data used
to create the map and are not influenced by
this cartographic necessity.

Argentine Ant

The Argentine ant (*Linepithema humile*)
measures, at most, an eighth of an inch in length.
The wingless worker ant, which is most commonly
encountered, has a honey brown coloration. Unlike
most other ants, an Argentine ant's movement
trail is up to five or more ants in width. They may
climb trees or buildings in search of food. Other
ant species tend not to climb obstacles in search of
food.[30]

Argentine ants are an invasive species native
to South America.[31] They are most successful
inhabiting Mediterranean and subtropical climates.[32]
In many climates they are directly associated with
human habitation and may use this association to
allow persistence in less favorable climates.[33] The
Argentine ant's association with human habitation
is even more interesting and more disconcerting
because the species is polydomous.[34] Polydomous
ants occupy multiple nests and this feature of the
colony allows it to readily adjust to alterations in
food supply.[35] Ants exhibiting this behavior can
redistribute workers, broods, and other resources
among nests giving them a competitive edge over
many other insects.

The invasive nature of the species causes a
variety of economic and ecological problems to
newly occupied lands.[36] Perhaps the most noticeable
impact is the displacement of native ant species.
Studies note that the displacement of native ant
populations by Argentine ant colonies may be due
to a number of factors with the most significant
being predation or competition.[37] Regardless of
the reason for the decline in native ant populations,
the observations in the literature note the empirical
relationship between presence of Argentine ants and
the reduction in the number of native ant species.
The reduction of native ant fauna could have drastic
effects on other species requiring these native
populations for subsistence. Information regarding
the extent to which Argentine ant invasions have
indirect effects on the native flora is limited, but
signs of decreasing populations of certain non-ant
species have appeared.[38]

Exact habitat descriptions for the Argentine ant
vary a great deal. Factors include soil, temperature,
and stream flow.[39] Specific vegetation types have

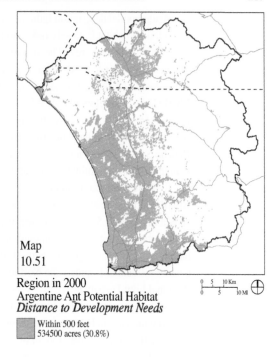

Map
10.51

Region in 2000
Argentine Ant Potential Habitat
Distance to Development Needs

Within 500 feet
534500 acres (30.8%)

0 5 10 Km
0 5 10 MI

not been identified as to their ability to support
ant populations; however, urban and other human
induced or disturbed landscapes are known to
support thriving populations.[40] The linkage to the
human landscape is simple with ant abundance
increasing as the distance to urban edge decreases.
At a distance of 325 to 650 feet from an urban edge,
the population of ants virtually disappears.[41] Natural
vegetation patches of less than 75 acres that are
surrounded by urban development are often entirely
inhabited by Argentine ants.[42] Such strong linkages
to the urban elements show the positive relationship
between human occupation and Argentine ant
invasion.

Modeling potential habitat for the Argentine ant
with the information above would be difficult due
to a lack of detailed spatial information, although it
appears that in California the driver for Argentine
ant invasion is linked to the expansion of the human
landscape. The algorithm for determining potential
Argentine ant habitat first identified areas within
500 feet of development (Map 10.51). Patches of
vegetation less than 75 acres in area were selected
and merged with the previous buffered development.
Water and wetlands were masked out of the merged
data. The remaining spatial data was identified as
potential Argentine ant habitat.

As the model description indicates, Argentine ant potential habitat is closely linked to the presence of anthropogenic features. The vast majority of potential habitat is located within close proximity to the coast. The largest (in area) exceptions are on MCB Camp Pendleton and MCAS Miramar. Other natural coastal inholdings lay slightly east of I-15 in close proximity to the cities of Escondido and Rancho Sante Fe, but their relative size is smaller than that of the two military installations.

Existing condidtions and impacts associated wtih the futures are summarized in Table 10.6 and shown in Maps 10.52-10.60. All the alternative futures show increases in the amount of potential habitat available for Argentine ants. The lower density residential futures—the Northern and Regional Low-Density alternatives—show the largest increases in potential habitat. The large potential increase in habitat under these two futures is the result of the relative amount of rural residential housing being allocated under these futures. The Coastal Future shows the lowest increase in potential habitat and is a direct result of the large amount of people being placed in high-density urban structures.

The addition of 500,000 and 1,000,000 people show varied results with the Coastal, Northern, and Three-Centers Futures. Argentine ant potential habitat increases in the Regional Low-Density Future over 40% with the addition of 500,000 people to 75% with the addition of 1,000,000 people. Doubling the population does not double the amount of potential ant habitat because of the need to place rural residential homes closer together. When the distance between house locations is low, ant dispersal area sources overlap, thus decreasing inhabited land.

These results indicate that with an increase in low-density residential development, an increase in Argentine ant potential habitat will occur. The juxtaposition of new rural residential housing with vegetation can result in increases of potential habitat. If an objective of the region is to control Argentine ants, then land managers should consider the relative density of new development and the juxtaposition of proposed development with current development. To control the indirect spread of Argentine ants, houses should be placed close together.

Table 10.6

Argentine Ant Potential Habitat (in acres)

	2000	Coastal		Northern		Reg. Low-Density		Three-Centers	
		500k	1,000k	500k	1,000k	500k	1,000k	500k	1,000k
Potential Habitat	530,766	620,016	719,358	712,696	884,448	746,882	930,485	649,594	782,136
Change from 2000		89,250	188,592	181,930	353,682	216,116	399,719	118,828	251,370
% Change from 2000		16.8	35.5	34.3	66.6	40.7	75.3	22.4	47.4

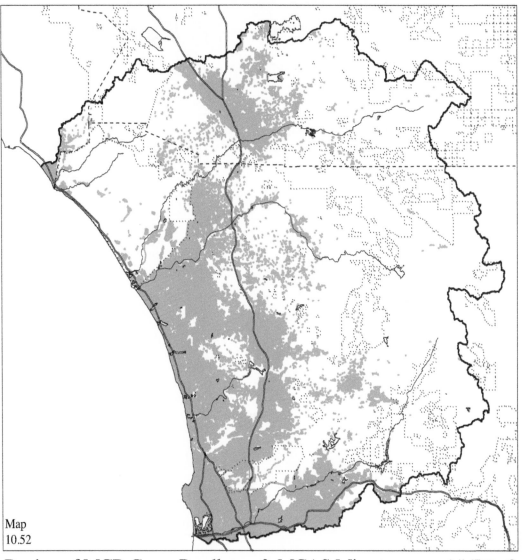

Map
10.52

Region of MCB Camp Pendleton & MCAS Miramar
Existing Conditions 2000
Argentine Ant Potential Habitat

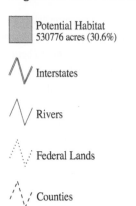

Potential Habitat
530776 acres (30.6%)

Interstates

Rivers

Federal Lands

Counties

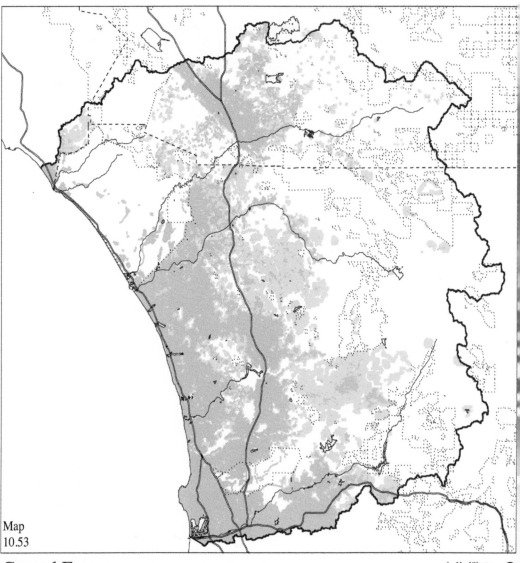

Map
10.53

Coastal Future
500,000 New Residents
Argentine Ant Potential Habitat Change

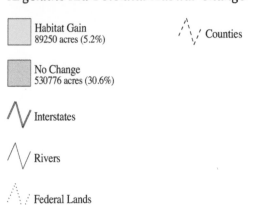

Habitat Gain
89250 acres (5.2%)

No Change
530776 acres (30.6%)

Interstates

Rivers

Federal Lands

Counties

NOTE: Some impacted areas are too small to be depicted at the scale used in the map above. In order to better show these locations, they have been expanded by the drawing of a uniform line around their perimeters. Statistics reported in the legend and in Table 10.6 are based on the data used to create the map and are not influenced by this cartographic necessity.

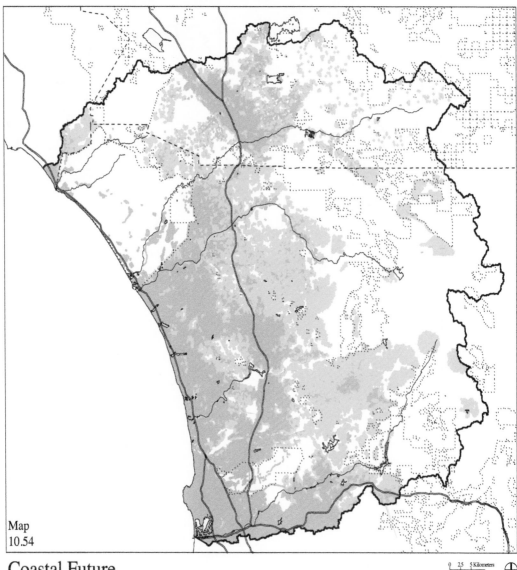

Map
10.54

Coastal Future
1,000,000 New Residents
Argentine Ant Potential Habitat Change

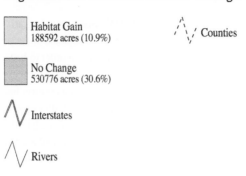

Habitat Gain
188592 acres (10.9%)

No Change
530776 acres (30.6%)

Interstates

Rivers

Federal Lands

Counties

NOTE: Some impacted areas are too
small to be depicted at the scale used in the
map above. In order to better show these
locations, they have been expanded by
the drawing of a uniform line around their
perimeters. Statistics reported in the legend
and in Table 10.6 are based on the data used
to create the map and are not influenced by
this cartographic necessity.

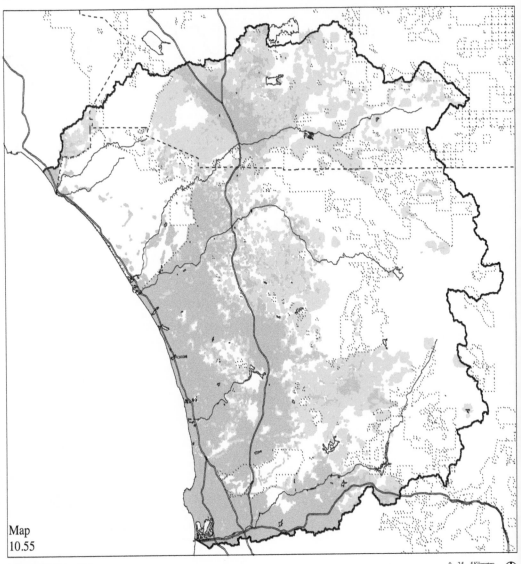

Map
10.55

Northern Future
500,000 New Residents
Argentine Ant Potential Habitat Change

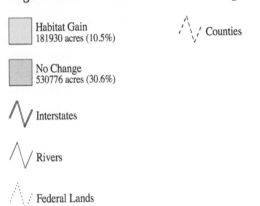

Habitat Gain
181930 acres (10.5%)

No Change
530776 acres (30.6%)

Interstates

Rivers

Federal Lands

Counties

NOTE: Some impacted areas are too
small to be depicted at the scale used in the
map above. In order to better show these
locations, they have been expanded by
the drawing of a uniform line around their
perimeters. Statistics reported in the legend
and in Table 10.6 are based on the data used
to create the map and are not influenced by
this cartographic necessity.

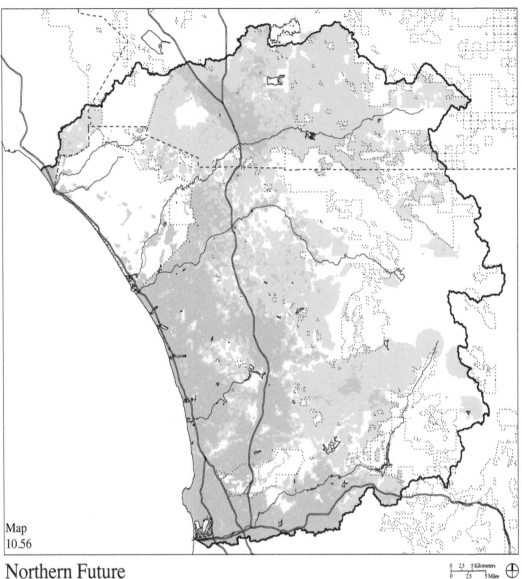

Map
10.56

Northern Future
1,000,000 New Residents
Argentine Ant Potential Habitat Change

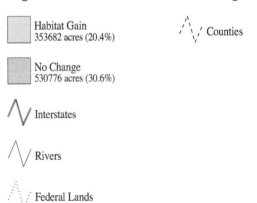

Habitat Gain
353682 acres (20.4%)

No Change
530776 acres (30.6%)

Interstates

Rivers

Federal Lands

Counties

NOTE: Some impacted areas are too
small to be depicted at the scale used in the
map above. In order to better show these
locations, they have been expanded by
the drawing of a uniform line around their
perimeters. Statistics reported in the legend
and in Table 10.6 are based on the data used
to create the map and are not influenced by
this cartographic necessity.

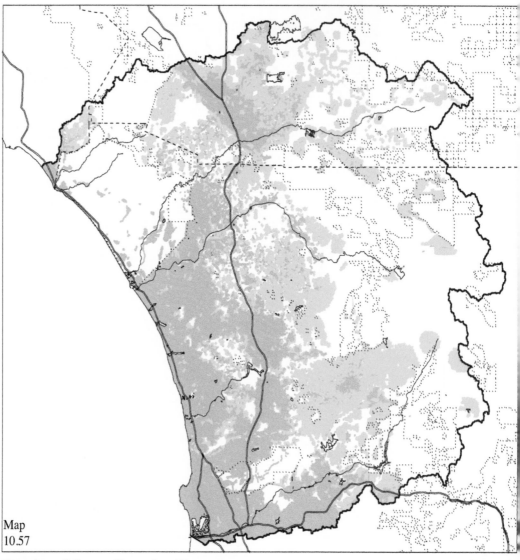

Map
10.57

Regional Low-Density Future
500,000 New Residents
Argentine Ant Potential Habitat Change

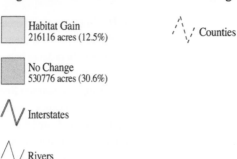

Habitat Gain
216116 acres (12.5%)

No Change
530776 acres (30.6%)

Interstates

Rivers

Federal Lands

Counties

NOTE: Some impacted areas are too
small to be depicted at the scale used in the
map above. In order to better show these
locations, they have been expanded by
the drawing of a uniform line around their
perimeters. Statistics reported in the legend
and in Table 10.6 are based on the data used
to create the map and are not influenced by
this cartographic necessity.

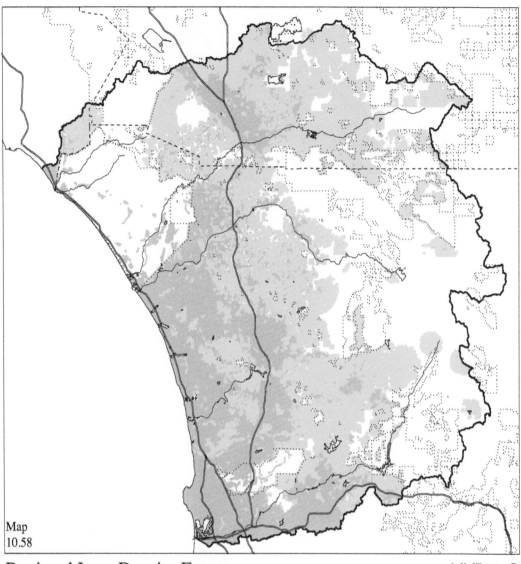

Map
10.58

Regional Low-Density Future
1,000,000 New Residents
Argentine Ant Potential Habitat Change

0   2.5   5 Kilometers
0   2.5   5 Miles

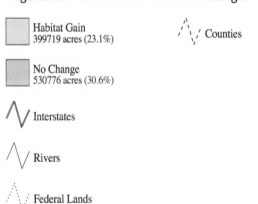

Habitat Gain
399719 acres (23.1%)

No Change
530776 acres (30.6%)

Interstates

Rivers

Federal Lands

Counties

NOTE: Some impacted areas are too small to be depicted at the scale used in the map above. In order to better show these locations, they have been expanded by the drawing of a uniform line around their perimeters. Statistics reported in the legend and in Table 10.6 are based on the data used to create the map and are not influenced by this cartographic necessity.

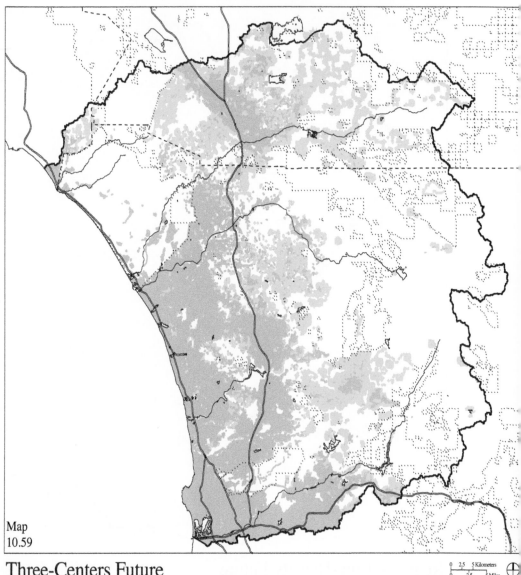

Map
10.59

Three-Centers Future
500,000 New Residents
Argentine Ant Potential Habitat Change

0   2.5   5 Kilometers
0   2.5   5 Miles

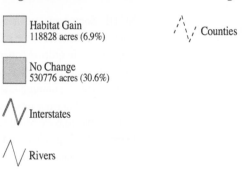

Habitat Gain
118828 acres (6.9%)

No Change
530776 acres (30.6%)

Interstates

Rivers

Federal Lands

Counties

NOTE: Some impacted areas are too
small to be depicted at the scale used in the
map above. In order to better show these
locations, they have been expanded by
the drawing of a uniform line around their
perimeters. Statistics reported in the legend
and in Table 10.6 are based on the data used
to create the map and are not influenced by
this cartographic necessity.

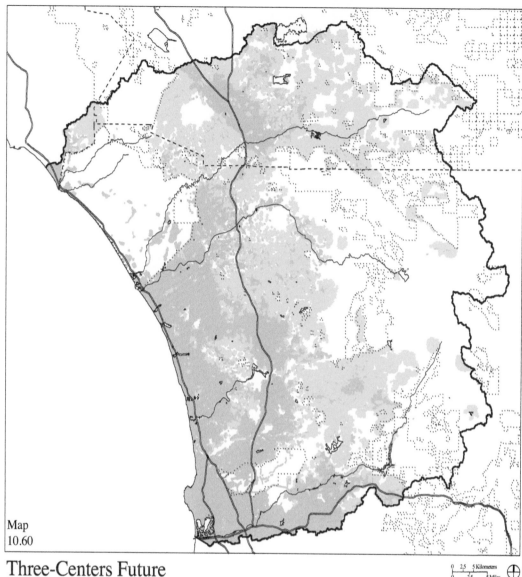

Map
10.60

Three-Centers Future
1,000,000 New Residents
Argentine Ant Potential Habitat Change

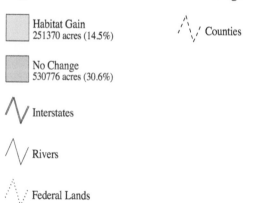

Habitat Gain
251370 acres (14.5%)

No Change
530776 acres (30.6%)

Interstates

Rivers

Federal Lands

Counties

NOTE: Some impacted areas are too
small to be depicted at the scale used in the
map above. In order to better show these
locations, they have been expanded by
the drawing of a uniform line around their
perimeters. Statistics reported in the legend
and in Table 10.6 are based on the data used
to create the map and are not influenced by
this cartographic necessity.

San Diego Coast Horned Lizard

The San Diego coast horned lizard (*Phrynosoma coronatum blainvillii*) is 2.5 to 4 inches long with a distinctive flat body. As the name would indicate, the coast horned lizard has two horns on the rear of its head; these horns are distinctively longer than the surrounding body spines. The body coloration may be yellowish, brown, or gray, and contains wavy blotches of darker color.[43]

The coast horned lizard has declined due to extensive habitat destruction caused by recent urbanization in Southern California.[44] Coast horned lizard populations are also impacted by the spread of Argentine ants—a topic described in the previous section. Over 90% of the insects consumed by coast horned lizards are native ants, and ants comprise around 45% of the total volume of food eaten.[45] With ants being a primary food source, an obvious concern is the persistence of native ant populations to sustain lizard numbers. Researchers have found that introduced Argentine ants have the ability to displace native ant populations through competition.[46] At first glance the exchange of ant species would appear to have little effect on the survivability of the San Diego coast horned lizard; however, information on prey selection of the San Diego coast horned lizard has revealed that Argentine ants are not eaten and where they occur coast horned lizard abundance has declined.[47]

The San Diego coast horned lizard prefers vegetation types with chaparral floristic components and sandy substrates although other vegetation communities may be inhabited especially where the sand substrate is present.[48] Also, this species is associated with locations near the coast showing a strong preference for open areas at elevations below 3,000 feet.[49]

The creation of the potential habitat model for the San Diego coast horned lizard required identification of the inhabited land cover types. All chaparral, grassland, and barren areas were extracted from the land cover map. The inhabited cover types were masked to select locations below 3,000 feet. All the masked locations that were not Argentine ant potential habitat, as identified from the Argentine ant potential habitat model, were eliminated. Thus, the potential habitat for the San Diego coast

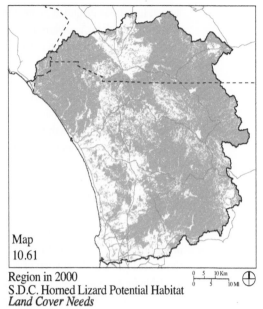

Map 10.61

Region in 2000
S.D.C. Horned Lizard Potential Habitat
Land Cover Needs

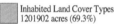

Inhabited Land Cover Types
1201902 acres (69.3%)

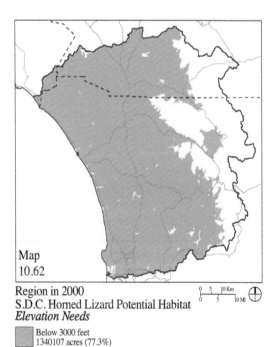

Map 10.62

Region in 2000
S.D.C. Horned Lizard Potential Habitat
Elevation Needs

Below 3000 feet
1340107 acres (77.3%)

horned lizard was identified as areas dominated by chaparral, grassland, and barren land covers below 3,000 feet and not within the potential habitat of the Argentine ant.

Most of the existing potential habitat for the San Diego coast horned lizard occurs west of the Coastal Range and in western Riverside County. One of the largest areas of potential habitat is located on MCB Camp Pendleton and within the Cleveland National Forest to the east of the base. A second location includes MCAS Miramar and public lands located to the east of the installation.

Existing conditions and impacts associated with the futures are summarized in Table 10.7 and shown in Mpas 10.61-10. 71. All of the alternative futures show a reduction in the amount of potential habitat available for the San Diego coast horned lizard. The increase of 500,000 to 1,000,000 people cause similar proportional decreases in lizard potential

habitat. The relative amount of loss is reflected in the Argentine ants gain in potential habitat. The Coastal Future is associated with the least amount of loss. The Northern and Regional Low-Density Futures have the highest loss. This situation is attributed to the large increase in Argentine ant potential habitat that is associated with the amount of low-density housing development. The potential habitat loss of the Three-Centers Future is in between that of the other futures.

The impacts associated with the alternative futures indicate that with lower density development comes a reduction in the amount of potential habitat available for the San Diego coast horned lizard. Thus, any future decisions surrounding the San Diego coast horned lizard should take into consideration the amount of low-density housing and the associated increase in Argentine ant potential habitat.

Table 10.7

San Diego Coast Horned Lizard Potential Habitat (in acres)

	2000	Coastal		Northern		Reg. Low-Density		Three-Centers	
		500k	1,000k	500k	1,000k	500k	1,000k	500k	1,000k
Potential Habitat	721,305	651,479	577,957	580,554	462,700	554,663	410,209	626,933	522,011
Change from 2000		-69,826	-143,348	-140,751	-258,605	-166,642	-311,096	-94,372	-199,294
% Change from 2000		**-9.7**	**-19.9**	**-19.5**	**-35.9**	**-23.1**	**-43.1**	**-13.1**	**-27.6**

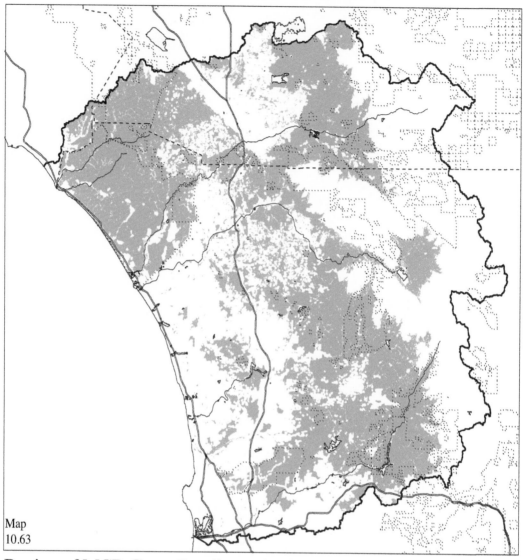

Map
10.63

Region of MCB Camp Pendleton & MCAS Miramar
Existing Conditions 2000
San Diego Coast Horned Lizard Potential Habitat

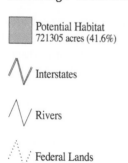

Potential Habitat
721305 acres (41.6%)

Interstates

Rivers

Federal Lands

Counties

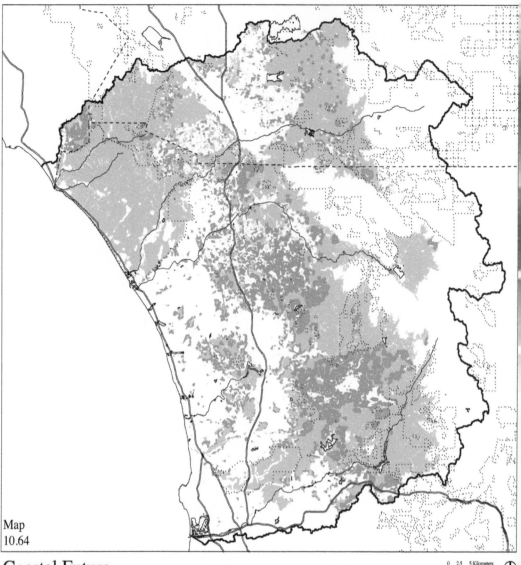

Map
10.64

Coastal Future
500,000 New Residents
San Diego Coast Horned Lizard Potential Habitat Change

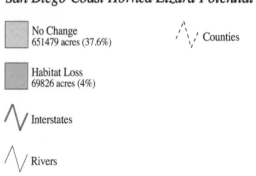

No Change
651479 acres (37.6%)

Habitat Loss
69826 acres (4%)

Interstates

Rivers

Federal Lands

Counties

NOTE: Some impacted areas are too small to be depicted at the scale used in the map above. In order to better show these locations, they have been expanded by the drawing of a uniform line around their perimeters. Statistics reported in the legend and in Table 10.7 are based on the data used to create the map and are not influenced by this cartographic necessity.

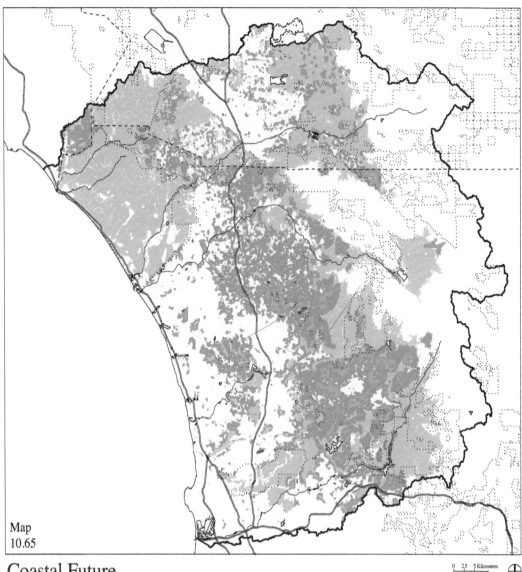

Map
10.65

Coastal Future
1,000,000 New Residents
San Diego Coast Horned Lizard Potential Habitat Change

0 2.5 5 Kilometers
0 2.5 5 Miles

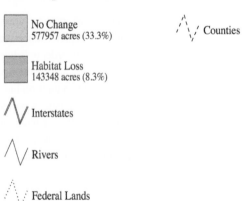

No Change
577957 acres (33.3%)

Habitat Loss
143348 acres (8.3%)

Interstates

Rivers

Federal Lands

Counties

NOTE: Some impacted areas are too
small to be depicted at the scale used in the
map above. In order to better show these
locations, they have been expanded by
the drawing of a uniform line around their
perimeters. Statistics reported in the legend
and in Table 10.7 are based on the data used
to create the map and are not influenced by
this cartographic necessity.

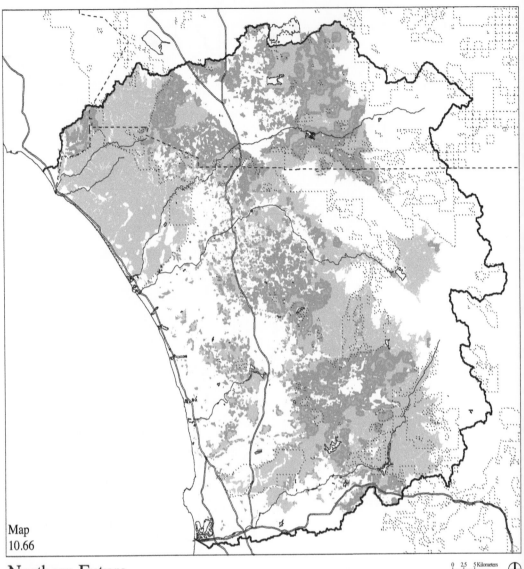

Map
10.66

Northern Future
500,000 New Residents
San Diego Coast Horned Lizard Potential Habitat Change

0   2.5   5 Kilometers
0   2.5   5 Miles

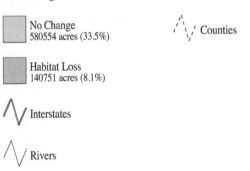

No Change
580554 acres (33.5%)

Habitat Loss
140751 acres (8.1%)

Interstates

Rivers

Federal Lands

Counties

NOTE: Some impacted areas are too small to be depicted at the scale used in the map above. In order to better show these locations, they have been expanded by the drawing of a uniform line around their perimeters. Statistics reported in the legend and in Table 10.7 are based on the data used to create the map and are not influenced by this cartographic necessity.

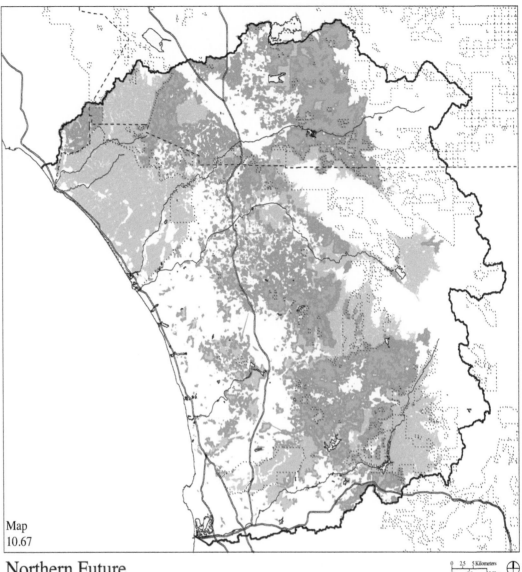

Map
10.67

Northern Future
1,000,000 New Residents
San Diego Coast Horned Lizard Potential Habitat Change

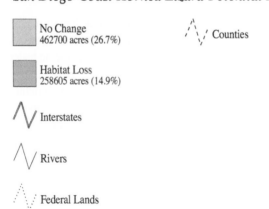

No Change
462700 acres (26.7%)

Habitat Loss
258605 acres (14.9%)

Interstates

Rivers

Federal Lands

Counties

NOTE: Some impacted areas are too small to be depicted at the scale used in the map above. In order to better show these locations, they have been expanded by the drawing of a uniform line around their perimeters. Statistics reported in the legend and in Table 10.7 are based on the data used to create the map and are not influenced by this cartographic necessity.

0 2.5 5 Kilometers
0 2.5 5 Miles

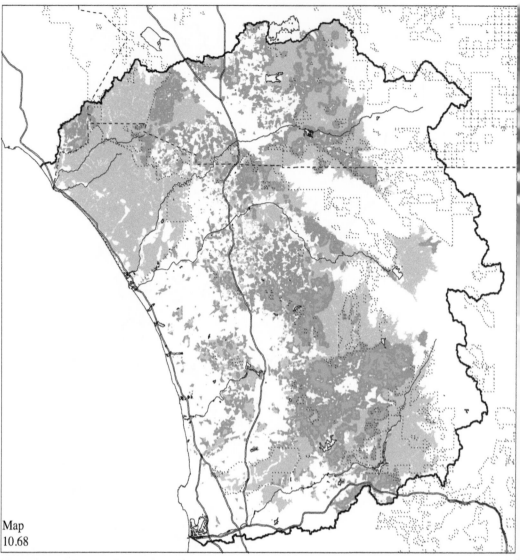

Map
10.68

Regional Low-Density Future
500,000 New Residents
San Diego Coast Horned Lizard Potential Habitat Change

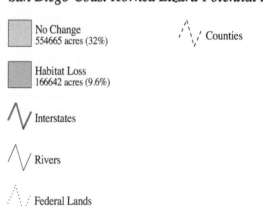

No Change
554665 acres (32%)

Habitat Loss
166642 acres (9.6%)

Interstates

Rivers

Federal Lands

Counties

NOTE: Some impacted areas are too small to be depicted at the scale used in the map above. In order to better show these locations, they have been expanded by the drawing of a uniform line around their perimeters. Statistics reported in the legend and in Table 10.7 are based on the data used to create the map and are not influenced by this cartographic necessity.

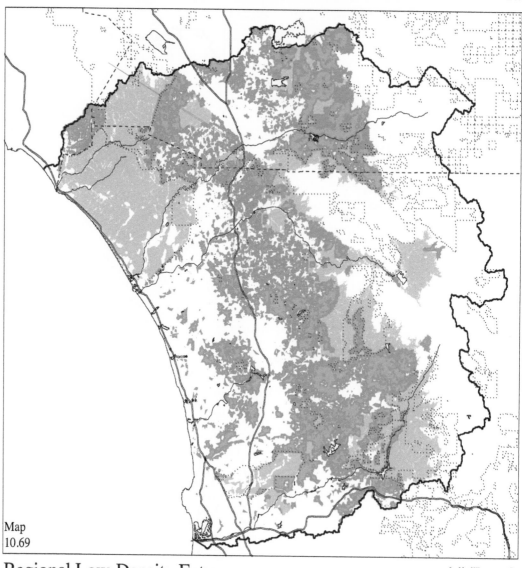

Map
10.69

Regional Low-Density Future
1,000,000 New Residents
San Diego Coast Horned Lizard Potential Habitat Change

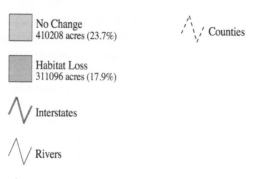

No Change
410208 acres (23.7%)

Habitat Loss
311096 acres (17.9%)

Interstates

Rivers

Federal Lands

Counties

NOTE: Some impacted areas are too
small to be depicted at the scale used in the
map above. In order to better show these
locations, they have been expanded by
the drawing of a uniform line around their
perimeters. Statistics reported in the legend
and in Table 10.7 are based on the data used
to create the map and are not influenced by
this cartographic necessity.

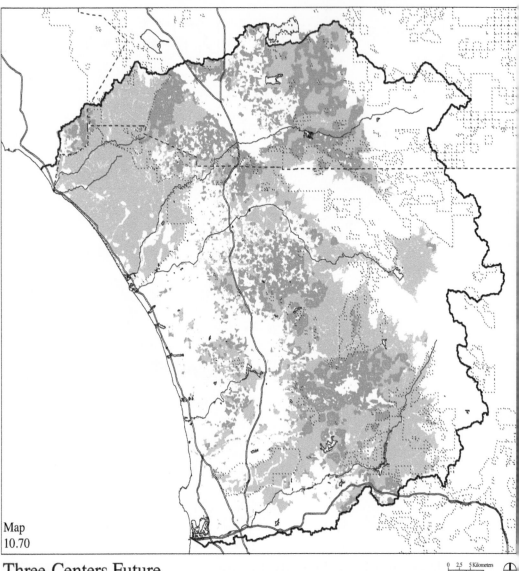

Map
10.70

Three-Centers Future
500,000 New Residents
San Diego Coast Horned Lizard Potential Habitat Change

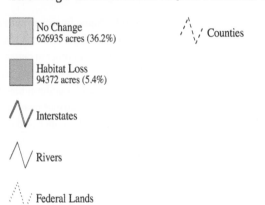

No Change
626935 acres (36.2%)

Habitat Loss
94372 acres (5.4%)

Interstates

Rivers

Federal Lands

Counties

NOTE: Some impacted areas are too small to be depicted at the scale used in the map above. In order to better show these locations, they have been expanded by the drawing of a uniform line around their perimeters. Statistics reported in the legend and in Table 10.7 are based on the data used to create the map and are not influenced by this cartographic necessity.

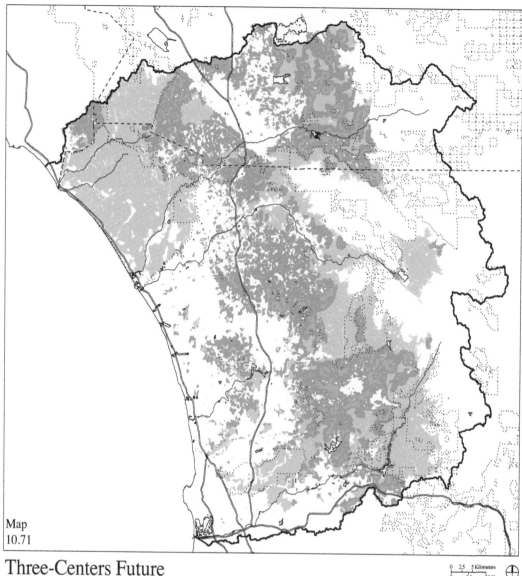

Map
10.71

Three-Centers Future
1,000,000 New Residents
San Diego Coast Horned Lizard Potential Habitat Change

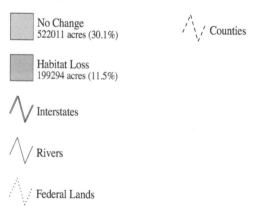

No Change
522011 acres (30.1%)

Habitat Loss
199294 acres (11.5%)

Interstates

Rivers

Federal Lands

Counties

NOTE: Some impacted areas are too
small to be depicted at the scale used in the
map above. In order to better show these
locations, they have been expanded by
the drawing of a uniform line around their
perimeters. Statistics reported in the legend
and in Table 10.7 are based on the data used
to create the map and are not influenced by
this cartographic necessity.

Rufous-Crowned Sparrow

The Rufous-crowned sparrow (*Aimophila ruficeps*) is widely distributed throughout Southern California. The species is a medium-sized sparrow with an overall length of from 5 to 6 inches. Dark in overall coloration, it has a plain dusky breast, rufous cap, and rounded tail. Its nasal call, punctuated with what sounds like the words "dear, dear, dear" may be the most distinguishing characteristics for locating individuals.[50]

This sparrow is most commonly found on dry and rocky slopes with a scattering of shrubs. In Southern California, this species prefers mixed chaparral and coastal sage scrub habitats. Although habitat preference is given to these vegetation communities, the Rufous-crowned sparrow has been seen in locations dominated by forbs and grass in association with rock outcrops. The species avoids continuous stands of single shrubs or trees. It inhabits elevations below 4,500 feet and the majority of rufous-crowned sparrow habitat (89%) indicates a preference for slopes in the range of 15–60 degrees.[51]

Home ranges average 4 acres and territories average 2 acres.[52] Typically, the home range or territory size for a species provides a reference for the amount of area required for an individual of the species to survive. The Rufous-crowned sparrow does require a certain amount of area for foraging, but more importantly researchers have found the abundance of Rufous-crowned sparrows to be low to zero in habitat patches less than 250 acres in area. Initially, speculation for the lack of sparrows in small habitat patches centered on edge effects from urbanization.[53] The reason for the loss of individuals in habitat patches less than 250 acres is unknown, but the knowledge about abundance levels of sparrows in patches at various sizes is, nevertheless, valuable for species management purposes.[54]

Identification of potential habitat for the Rufous-crowned sparrow began by identifying all coastal sage scrub and chaparral land cover (Map 10.72). Locations containing these vegetation communities were selected for areas below 4,500 feet (Map 10.73). Patches less than 250 acres in area were eliminated. Slopes less than 15 degrees were also eliminated. The resultant potential habitat

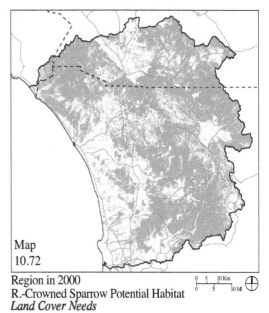

Map
10.72

Region in 2000
R.-Crowned Sparrow Potential Habitat
Land Cover Needs

 Inhabited Land Cover Types
1007612 acres (58.1%)

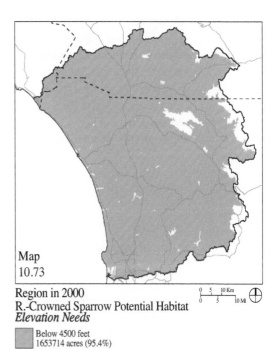

Map
10.73

Region in 2000
R.-Crowned Sparrow Potential Habitat
Elevation Needs

Below 4500 feet
1653714 acres (95.4%)

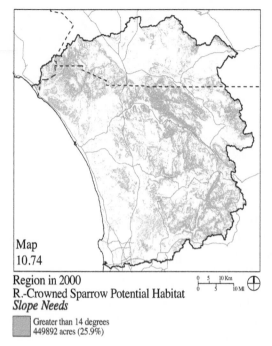

Map
10.74

Region in 2000
R.-Crowned Sparrow Potential Habitat
Slope Needs

Greater than 14 degrees
449892 acres (25.9%)

of the study area, other contiguous patches do occur on the steep slopes of the south. These other patches mainly extend off valley floors and canyon walls.

Existing conditions and impacts associated with the futures summarized in Table 10.8 and shown in Maps 10.75-10.83. All the alternative futures result in a reduction in the amount of rufous-crowned sparrow potential habitat. The increase in people from 500,000 to 1,000,000 results in increased habitat loss for all the futures. The relative amount of decrease in habitat is similar for the Three-Center and Coastal Futures. The Northern and Regional Low-Density Futures cause a larger decrease in potential habitat. This situation results from the amount of low density residential development and the associated vegetation alterations. The potential habitat loss is located throughout the study region for all the futures with the majority of the impact resulting in the loss of small isolated chaparral patches or loss of chaparral along the edge of valley locations. Most of the large contiguous parcels of potential habitat are maintained regardless of the alternative future assessed.

The clearing of chaparral and other vegetation types for firebreaks around residential structures causes the greatest impact on rufous-crowned sparrow potential habitat. Thus, with a decrease in housing density per unit land area, a like decrease in potential habitat would occur. Increased housing densities or the dismissal of shrub removal practices around houses would reduce the impacts on the potential habitat for the Rufous-crowned sparrow.

map consisted of patches of coastal sage scrub and chaparral land cover, below 4,500 feet, at least 250 acres in area, and on slopes greater than 14 degrees (Map 10.74).

The pattern of rufous-crowned sparrow potential habitat appears somewhat fragmented due to the preference for terrain featuring high slopes. The greatest area of potential habitat occurs within MCB Camp Pendleton and in the northern portion of the Cleveland National Forest. Although large patches of potential habitat dominate the northwest portion

Table 10.8

Rufous-Crowned Sparrow Potential Habitat (in acres)

	2000	Coastal		Northern		Reg. Low-Density		Three-Centers	
		500k	1,000k	500k	1,000k	500k	1,000k	500k	1,000k
Potential Habitat	343,882	340,403	334,191	334,964	317,807	331,511	310,183	339,924	331,315
Change from 2000		-3,479	-9,691	-8,918	-26,076	-12,372	-33,699	-3,959	-12,568
% Change from 2000		**-1.0**	**-2.8**	**-2.6**	**-7.6**	**-3.6**	**-9.8**	**-1.2**	**-3.7**

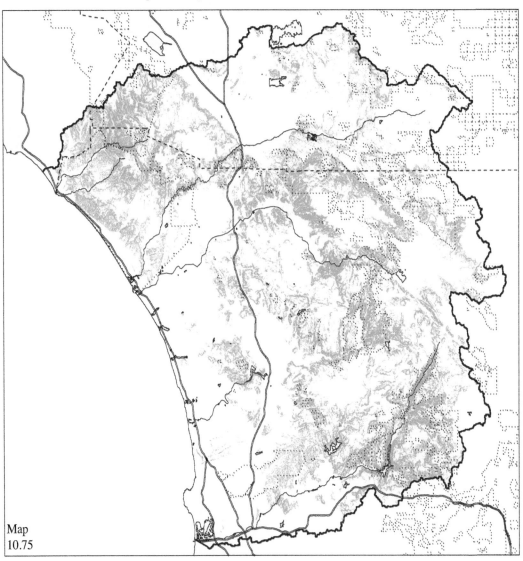

Map
10.75

Region of MCB Camp Pendleton & MCAS Miramar
Existing Conditions 2000
Rufous-Crowned Sparrow Potential Habitat

0   2.5   5 Kilometers
0   2.5   5 Miles

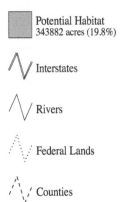

Potential Habitat
343882 acres (19.8%)

Interstates

Rivers

Federal Lands

Counties

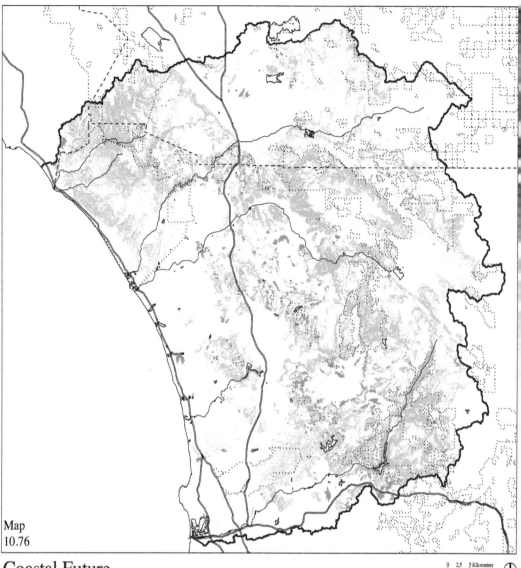

Map
10.76

Coastal Future
500,000 New Residents
Rufous-Crowned Sparrow Potential Habitat Change

0   2.5   5 Kilometers
0   2.5   5 Miles

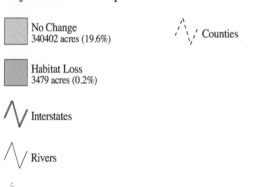

No Change
340402 acres (19.6%)

Counties

Habitat Loss
3479 acres (0.2%)

Interstates

Rivers

Federal Lands

NOTE: Some impacted areas are too small to be depicted at the scale used in the map above. In order to better show these locations, they have been expanded by the drawing of a uniform line around their perimeters. Statistics reported in the legend and in Table 10.8 are based on the data used to create the map and are not influenced by this cartographic necessity.

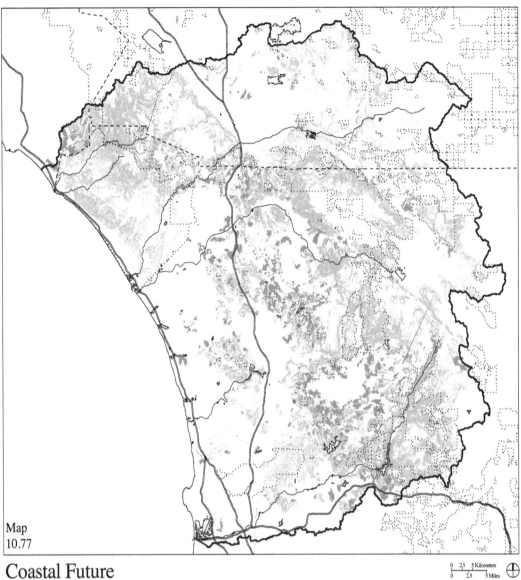

Map
10.77

Coastal Future
1,000,000 New Residents
Rufous-Crowned Sparrow Potential Habitat Change

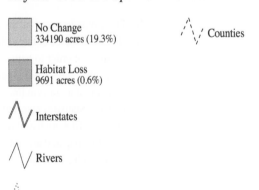

No Change
334190 acres (19.3%)

Habitat Loss
9691 acres (0.6%)

Interstates

Rivers

Federal Lands

Counties

NOTE: Some impacted areas are too
small to be depicted at the scale used in the
map above. In order to better show these
locations, they have been expanded by
the drawing of a uniform line around their
perimeters. Statistics reported in the legend
and in Table 10.8 are based on the data used
to create the map and are not influenced by
this cartographic necessity.

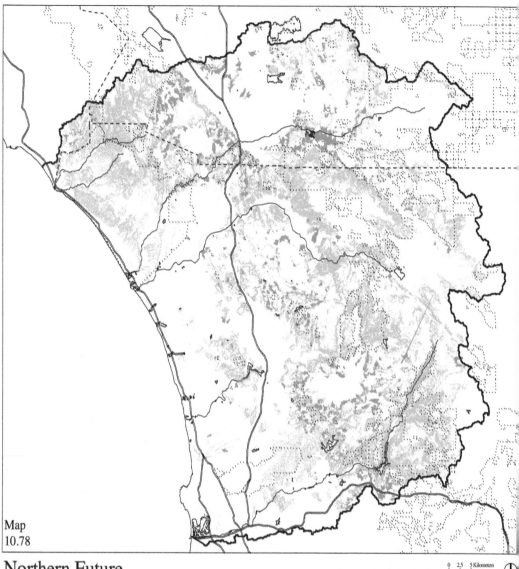

Map
10.78

Northern Future
500,000 New Residents
Rufous-Crowned Sparrow Potential Habitat Change

0   2.5   5 Kilometers
0   2.5   5 Miles

No Change
334966 acres (19.3%)

Habitat Loss
8918 acres (0.5%)

Counties

Interstates

Rivers

Federal Lands

NOTE: Some impacted areas are too
small to be depicted at the scale used in the
map above. In order to better show these
locations, they have been expanded by
the drawing of a uniform line around their
perimeters. Statistics reported in the legend
and in Table 10.8 are based on the data used
to create the map and are not influenced by
this cartographic necessity.

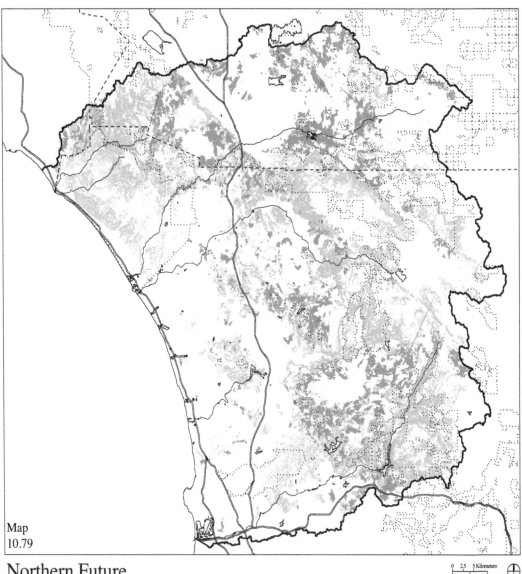

Map
10.79

Northern Future
1,000,000 New Residents
Rufous-Crowned Sparrow Potential Habitat Change

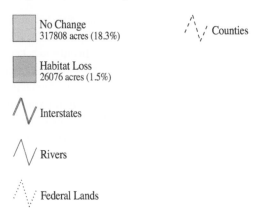

No Change
317808 acres (18.3%)

Habitat Loss
26076 acres (1.5%)

Interstates

Rivers

Federal Lands

Counties

NOTE: Some impacted areas are too small to be depicted at the scale used in the map above. In order to better show these locations, they have been expanded by the drawing of a uniform line around their perimeters. Statistics reported in the legend and in Table 10.8 are based on the data used to create the map and are not influenced by this cartographic necessity.

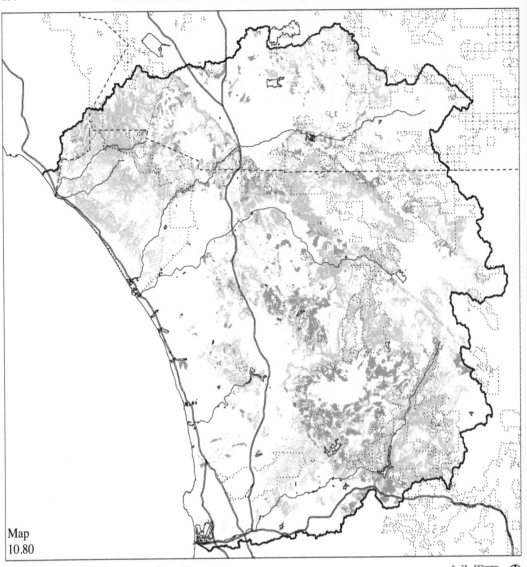

Map
10.80

Regional Low-Density Future
500,000 New Residents
Rufous-Crowned Sparrow Potential Habitat Change

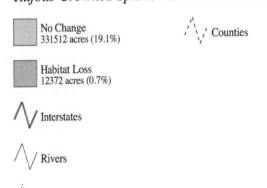

No Change
331512 acres (19.1%)

Habitat Loss
12372 acres (0.7%)

Interstates

Rivers

Federal Lands

Counties

NOTE: Some impacted areas are too
small to be depicted at the scale used in the
map above. In order to better show these
locations, they have been expanded by
the drawing of a uniform line around their
perimeters. Statistics reported in the legend
and in Table 10.8 are based on the data used
to create the map and are not influenced by
this cartographic necessity.

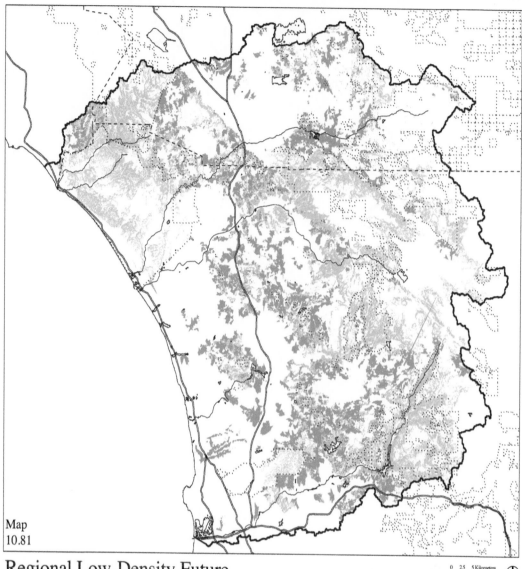

Map
10.81

Regional Low-Density Future
1,000,000 New Residents
Rufous-Crowned Sparrow Potential Habitat Change

0 2.5 5 Kilometers
0 2.5 5 Miles

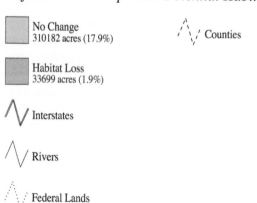

No Change
310182 acres (17.9%)

Habitat Loss
33699 acres (1.9%)

Interstates

Rivers

Federal Lands

Counties

NOTE: Some impacted areas are too small to be depicted at the scale used in the map above. In order to better show these locations, they have been expanded by the drawing of a uniform line around their perimeters. Statistics reported in the legend and in Table 10.8 are based on the data used to create the map and are not influenced by this cartographic necessity.

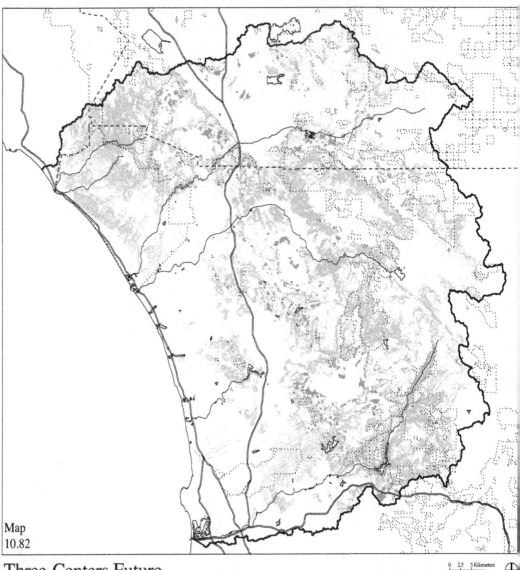

Map
10.82

Three-Centers Future
500,000 New Residents
Rufous-Crowned Sparrow Potential Habitat Change

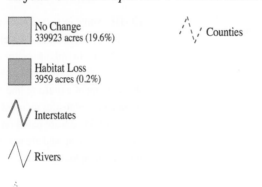

No Change
339923 acres (19.6%)

Habitat Loss
3959 acres (0.2%)

Interstates

Rivers

Federal Lands

Counties

NOTE: Some impacted areas are too
small to be depicted at the scale used in the
map above. In order to better show these
locations, they have been expanded by
the drawing of a uniform line around their
perimeters. Statistics reported in the legend
and in Table 10.8 are based on the data used
to create the map and are not influenced by
this cartographic necessity.

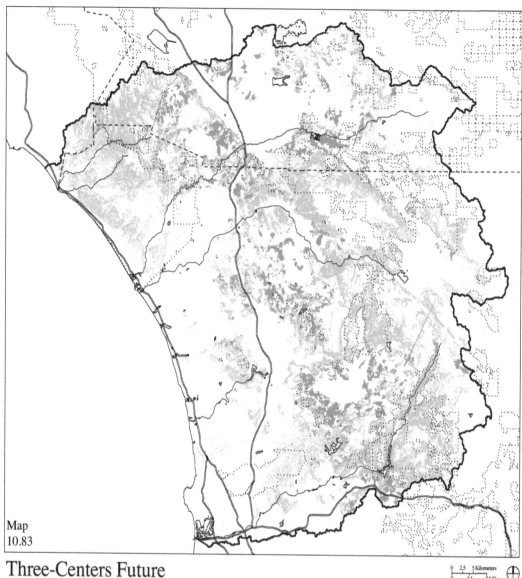

Map
10.83

Three-Centers Future
1,000,000 New Residents
Rufous-Crowned Sparrow Potential Habitat Change

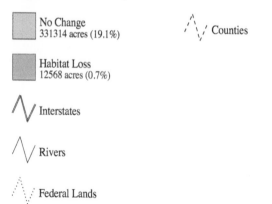

No Change
331314 acres (19.1%)

Habitat Loss
12568 acres (0.7%)

Interstates

Rivers

Federal Lands

Counties

NOTE: Some impacted areas are too
small to be depicted at the scale used in the
map above. In order to better show these
locations, they have been expanded by
the drawing of a uniform line around their
perimeters. Statistics reported in the legend
and in Table 10.8 are based on the data used
to create the map and are not influenced by
this cartographic necessity.

California Gnatcatcher

The California gnatcatcher (*Polioptila californica californica*) is a resident species listed as endangered by the federal government. This small bird is 4–5 inches long and has a dark gray back with white to gray underbelly coloration. A black cap is present on the head of males during the breeding season. The California gnatcatcher is similar in appearance to the black-tailed gnatcatcher, which has a whiter underbelly and tail.[55]

The species is found in association with semi-arid coastal scrub vegetation. Concern over the diminishing coastal sage scrub vegetation community has created a situation where this species is often viewed as a flagship species in debates on development and conservation goals.[56] Although many believe that preservation of the California gnatcatcher ensures the survivability of other species, some wildlife researchers have shown this not to be the case.[57] Regardless of the interpretation about the gnatcatcher's ability to be utilized as a species for protection, there is little doubt over the threat to the gnatcatcher's coastal sage scrub habitat.

Beyond loss of habitat, predators and other species are also threats. One of the most significant species to threaten the gnatcatchers survival is the brown-headed cowbird (*Molothrus ater*). The brown-headed cowbird practices nest parasitism by placing its own eggs in the nests of other bird species. The brown-headed cowbird chicks grow faster and outcompete the host's nestlings. Controlling cowbird numbers in gnatcatcher nesting locations has resulted in reduced impacts by nest parasitism and subsequently increased the nesting success of California gnatcatchers.[58]

This gnatcatcher is most numerous in dense coastal scrub in washes, on mesas, and on sloping coastal hills.[59] The shrub vegetation provides adequate habitat for roosting, nesting, and cover. Inhabited areas are generally located below 1,500 feet in elevation with the vast majority of individuals being found below 800 feet.[60] Patches of coastal sage scrub within Southern California may vary in habitat quality for gnatcatchers. This situation is reflected in the range of territory sizes from 4 to 39 acres represented by breeding pairs.[61] Territories comprised of smaller areas are located in closer

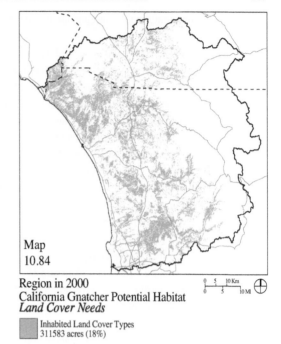

Map 10.84

Region in 2000
California Gnatcher Potential Habitat
Land Cover Needs

0 5 10 Km
0 5 10 Mi

Inhabited Land Cover Types
311583 acres (18%)

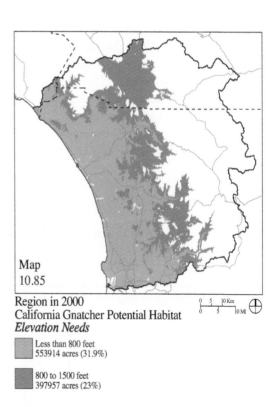

Map 10.85

Region in 2000
California Gnatcher Potential Habitat
Elevation Needs

0 5 10 Km
0 5 10 Mi

Less than 800 feet
553914 acres (31.9%)

800 to 1500 feet
397957 acres (23%)

proximity to the coast and larger territories are found farther inland.

The potential habitat model for the California gnatcatcher was based on vegetation, elevation, patch size, and distance from the coast. First, all the locations containing coastal sage scrub or a mix of coastal sage scrub and chaparral were selected (Map 10.84). The subsequent map was masked to eliminate areas above 1,500 feet (Map 10.85). The remaining land was masked according to patch size as a function of distance from the coast. Using the territory range of 4 to 39 acres and knowing that the distance from the coast to the furthest portion of gnatcatcher habitat was 44 miles, a scheme of 4-mile intervals from the coast was used. Thus, the range of territory sizes was broken into 11 values with a 4-acre territory area corresponding to locations from 0 to 4 miles of the coast and the maximum value of 39 acres corresponding to locations from 40 to 44 miles from the coastline. Vegetation patches under 1,500 feet in elevation that did not have adequate area as a function of territory size and distance from the coast were eliminated as potential habitat. The resultant map was classified into two categories identified as high quality habitat and other habitat. High quality habitat represented locations of exclusively coastal sage scrub vegetation below 800 feet in elevation. The remaining potential habitat was categorized as "other habitat."

Two of the largest patches of high quality habitat exist on MCB Camp Pendleton and MCAS Miramar. Another large patch of potential gnatcatcher habitat exists southeast of Escondido. Most other habitat occurs within inholdings of development and within the lower elevations of the mountains along the fringes of residential development.

Existing condidtions and impacts for the futures are summarized in Table 10.9 and shown in Maps 10.86-10.94. All the alternative futures reduce the amount of potential California gnatcatcher habitat. The relative impact with the addition of 500,000 or 1,000,000 people varies by future. In the Coastal scenario, doubling the population increase results in roughly a doubling in impact. The Regional Low-Density Future impact increases 3 to 4 times with twice the increase in population. This increase is caused by the need to find adequate land area for low density residential development in San Diego County which results in more development and land cover alterations closer to the coast within gnatcatcher habitat. Although the Northern Future contains a considerable amount of low-density residential development, the majority of it occurs in Western Riverside County on agricultural lands thereby resulting in proportionally less habitat loss. The Three-Centers 1,000,000 Future has a 1.3% loss of potential high quality gnatcatcher habitat. The low loss results from land conservation measures applied under this scenario for coastal sage scrub.

The model results show that development to the north has a relatively low impact on gnatcatcher habitat, but coastal development is significant. Prime potential gnatcatcher habitat can be maintained with appropriate conservation efforts.

Table 10.9

California Gnatcatcher Potential Habitat (in acres)

	2000	Coastal		Northern		Reg. Low-Density		Three-Centers	
		500k	1,000k	500k	1,000k	500k	1,000k	500k	1,000k
Hi-Qual. Potential Habitat	147,015	136,595	125,295	137,771	130,580	135,661	99,256	146,088	145,112
Change from 2000		-10,420	-21,720	-9,244	-16,435	-11,354	-47,759	-927	-1,903
% Change from 2000		**-7.1**	**-14.8**	**-6.3**	**11.2**	**-7.7**	**-32.5**	**-0.6**	**-1.3**
Other Habitat	88,237	82,959	76,806	78,368	73,987	78,222	51,046	86,875	82,371
Change from 2000		-5,278	-11,431	-9,869	-14,250	-10,015	-37,191	-1,362	-5,866
% Change from 2000		**-6.0**	**-11.2**	**-11.2**	**-16.1**	**-11.4**	**-42.1**	**-1.5**	**-6.6**
Total Habitat	235,252	219,554	202,101	216,139	204,567	213,883	150,302	232,963	227,483
Change from 2000		-15,698	-33,151	-19,113	-30,685	-21,369	-84,950	-2,289	-7,769
% Change from 2000		**-6.7**	**-14.1**	**-8.1**	**-13.0**	**-9.1**	**-36.1**	**-1.0**	**-3.3**

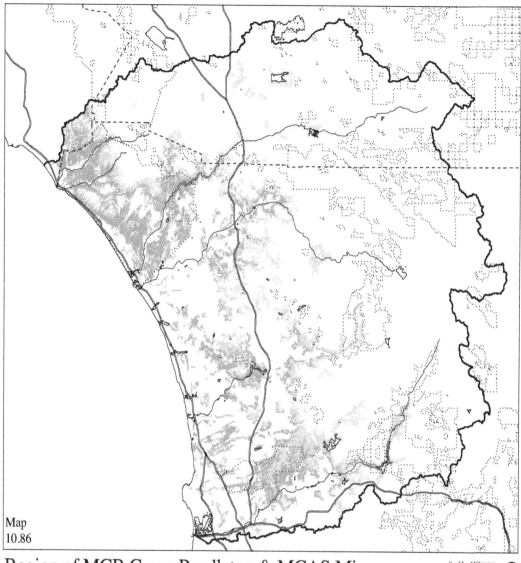

Map
10.86

Region of MCB Camp Pendleton & MCAS Miramar
Existing Conditions 2000
California Gnatcatcher Potential Habitat

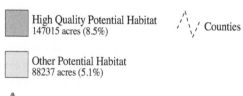

High Quality Potential Habitat
147015 acres (8.5%)

Other Potential Habitat
88237 acres (5.1%)

Counties

Interstates

Rivers

Federal Lands

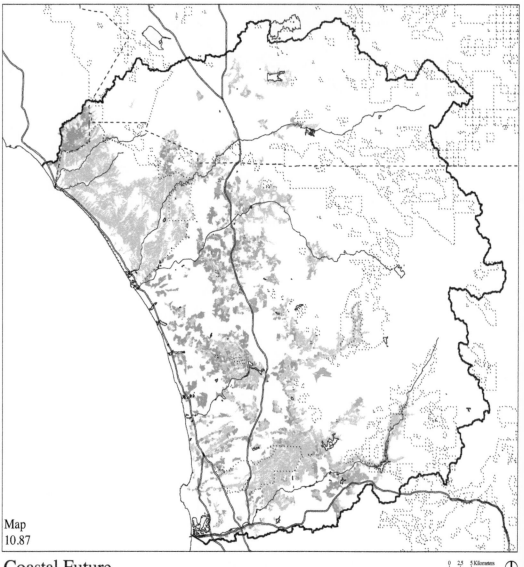

Map
10.87

Coastal Future
500,000 New Residents
California Gnatcatcher Potential Habitat Change

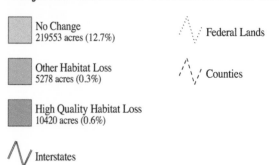

No Change
219553 acres (12.7%)

Other Habitat Loss
5278 acres (0.3%)

High Quality Habitat Loss
10420 acres (0.6%)

Interstates

Rivers

Federal Lands

Counties

NOTE: Some impacted areas are too small to be depicted at the scale used in the map above. In order to better show these locations, they have been expanded by the drawing of a uniform line around their perimeters. Statistics reported in the legend and in Table 10.9 are based on the data used to create the map and are not influenced by this cartographic necessity.

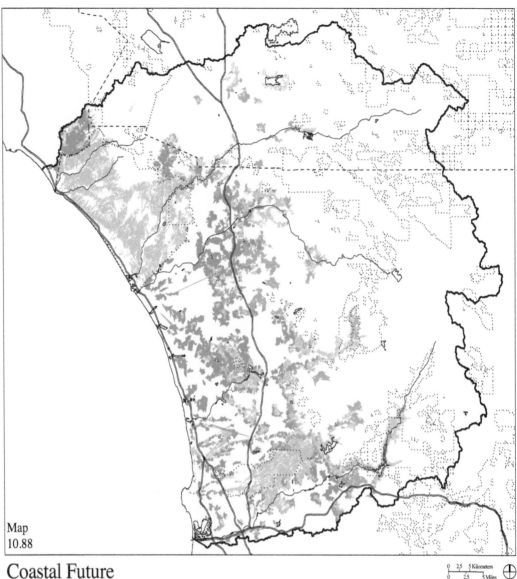

Map
10.88

Coastal Future
1,000,000 New Residents
California Gnatcatcher Potential Habitat Change

0   25   5 Kilometers
0   25   5 Miles

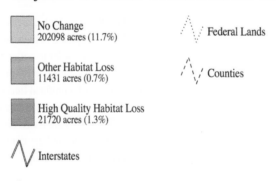

No Change
202098 acres (11.7%)

Other Habitat Loss
11431 acres (0.7%)

High Quality Habitat Loss
21720 acres (1.3%)

Interstates

Rivers

Federal Lands

Counties

NOTE: Some impacted areas are too
small to be depicted at the scale used in the
map above. In order to better show these
locations, they have been expanded by
the drawing of a uniform line around their
perimeters. Statistics reported in the legend
and in Table 10.9 are based on the data used
to create the map and are not influenced by
this cartographic necessity.

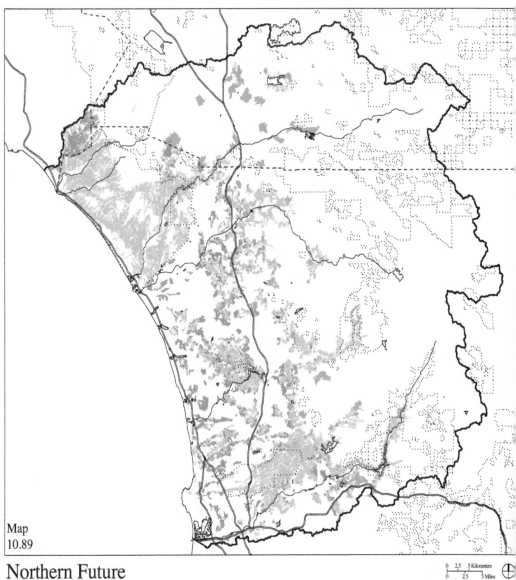

Map
10.89

Northern Future
500,000 New Residents
California Gnatcatcher Potential Habitat Change

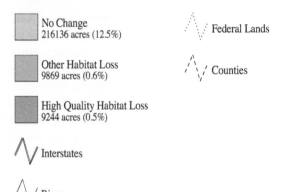

No Change
216136 acres (12.5%)

Other Habitat Loss
9869 acres (0.6%)

High Quality Habitat Loss
9244 acres (0.5%)

Interstates

Rivers

Federal Lands

Counties

NOTE: Some impacted areas are too small to be depicted at the scale used in the map above. In order to better show these locations, they have been expanded by the drawing of a uniform line around their perimeters. Statistics reported in the legend and in Table 10.9 are based on the data used to create the map and are not influenced by this cartographic necessity.

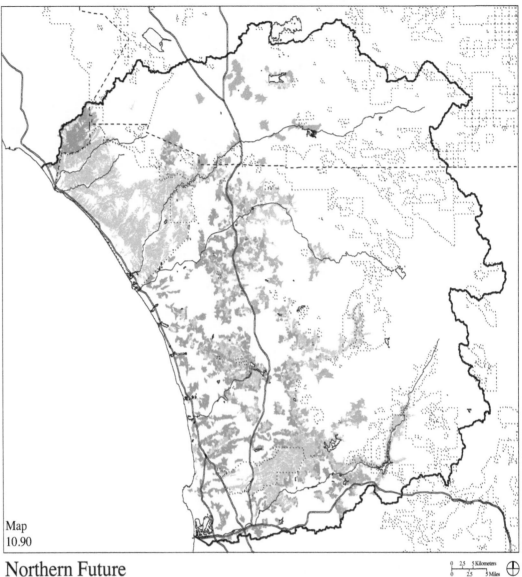

Map
10.90

Northern Future
1,000,000 New Residents
California Gnatcatcher Potential Habitat Change

0 2.5 5 Kilometers
0 2.5 5 Miles

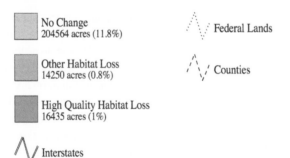

No Change
204564 acres (11.8%)

Other Habitat Loss
14250 acres (0.8%)

High Quality Habitat Loss
16435 acres (1%)

Interstates

Rivers

Federal Lands

Counties

NOTE: Some impacted areas are too small to be depicted at the scale used in the map above. In order to better show these locations, they have been expanded by the drawing of a uniform line around their perimeters. Statistics reported in the legend and in Table 10.9 are based on the data used to create the map and are not influenced by this cartographic necessity.

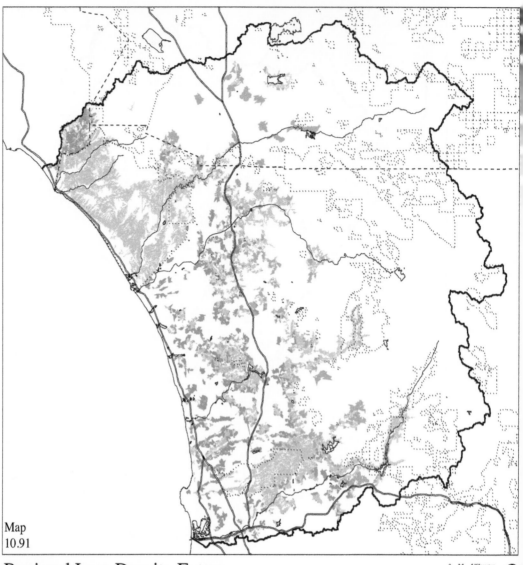

Map
10.91

Regional Low-Density Future
500,000 New Residents
California Gnatcatcher Potential Habitat Change

0   2.5   5 Kilometers
0   2.5   5 Miles

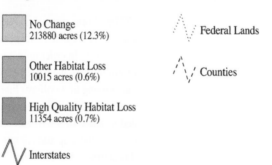

No Change
213880 acres (12.3%)

Other Habitat Loss
10015 acres (0.6%)

High Quality Habitat Loss
11354 acres (0.7%)

Interstates

Rivers

Federal Lands

Counties

NOTE: Some impacted areas are too
small to be depicted at the scale used in the
map above. In order to better show these
locations, they have been expanded by
the drawing of a uniform line around their
perimeters. Statistics reported in the legend
and in Table 10.9 are based on the data used
to create the map and are not influenced by
this cartographic necessity.

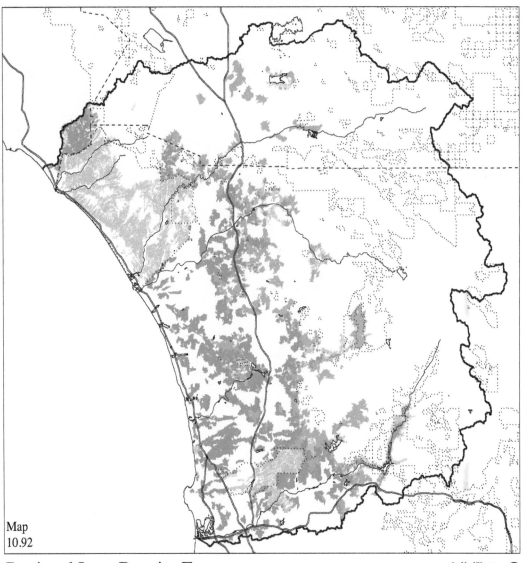

Map
10.92

Regional Low-Density Future
1,000,000 New Residents
California Gnatcatcher Potential Habitat Change

0   2.5   5 Kilometers
0   2.5   5 Miles

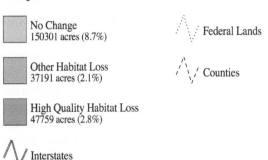

No Change
150301 acres (8.7%)

Other Habitat Loss
37191 acres (2.1%)

High Quality Habitat Loss
47759 acres (2.8%)

Interstates

Rivers

Federal Lands

Counties

NOTE: Some impacted areas are too
small to be depicted at the scale used in the
map above. In order to better show these
locations, they have been expanded by
the drawing of a uniform line around their
perimeters. Statistics reported in the legend
and in Table 10.9 are based on the data used
to create the map and are not influenced by
this cartographic necessity.

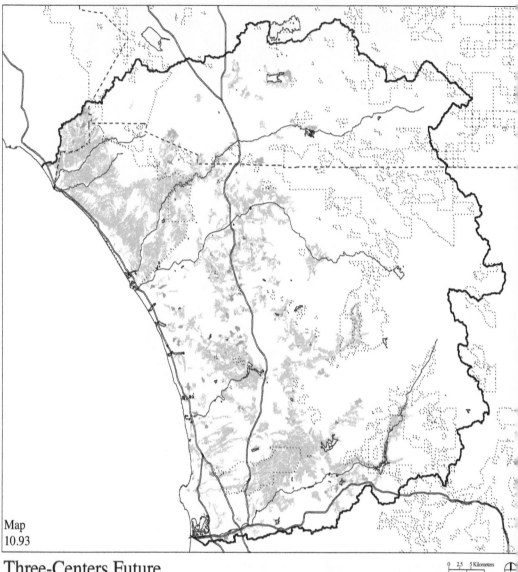

Map
10.93

Three-Centers Future
500,000 New Residents
California Gnatcatcher Potential Habitat Change

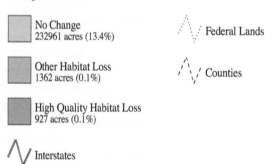

No Change
232961 acres (13.4%)

Other Habitat Loss
1362 acres (0.1%)

High Quality Habitat Loss
927 acres (0.1%)

Interstates

Rivers

Federal Lands

Counties

NOTE: Some impacted areas are too small to be depicted at the scale used in the map above. In order to better show these locations, they have been expanded by the drawing of a uniform line around their perimeters. Statistics reported in the legend and in Table 10.9 are based on the data used to create the map and are not influenced by this cartographic necessity.

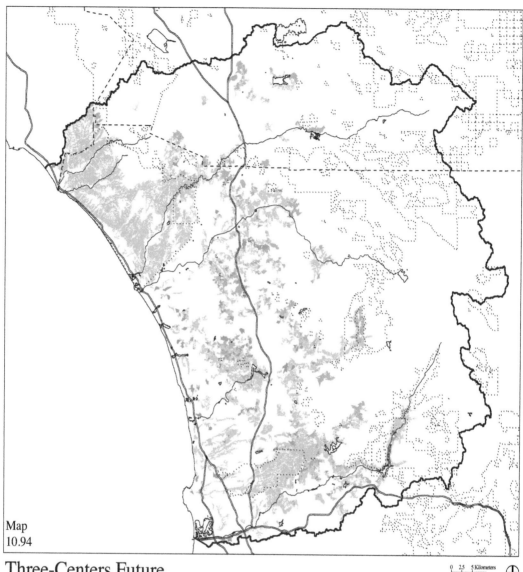

Map
10.94

Three-Centers Future
1,000,000 New Residents
California Gnatcatcher Potential Habitat Change

0 2.5 5 Kilometers
0 2.5 5 Miles

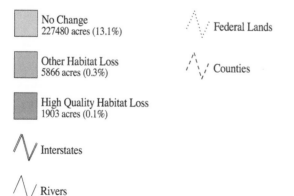

No Change
227480 acres (13.1%)

Other Habitat Loss
5866 acres (0.3%)

High Quality Habitat Loss
1903 acres (0.1%)

Interstates

Rivers

Federal Lands

Counties

NOTE: Some impacted areas are too small to be depicted at the scale used in the map above. In order to better show these locations, they have been expanded by the drawing of a uniform line around their perimeters. Statistics reported in the legend and in Table 10.9 are based on the data used to create the map and are not influenced by this cartographic necessity.

Western Snowy Plover

The western snowy plover (*Charadrius alexandrinus nivosus*) is a small shorebird listed as a threatened species under the U.S. Endangered Species Act. This species has a pointed black bill, long legs, and relatively short wings for its small body size. It weighs about 1.4 ounces and has a total length of about 6 inches.[62] Body coloration is tan to brown on the back with a white underbelly. A distinguishing black stripe may be seen just under the white portion of the neck during the breeding season.[63]

The western snowy plover breeds along the Pacific Coast of the United States and northern Mexico. A population of snowy plovers winters along the Pacific Coast from southern Oregon to Mexico. Although the western snowy plover's range is rather extensive from north to south, its restriction to the coastal habitats creates a problem of human-induced habitat destruction and encroachment.[64] Many of the threats faced by this species are a result of human habitation obligate species.[65] Ravens and crows may be found in association with human landscapes due to the abundance of food and their scavenging nature. The association of ravens and crows with human habitation and the intense human use of beaches increase the likelihood that ravens and crows will encounter a nesting bird, thereby increasing the likelihood of nest predation. Domesticated animals, crushing of eggs by humans, and increased frequencies of disturbance also reduce the success rate of nesting western snowy plovers. These factors have all resulted in western snowy plover population declines.[66]

Habitat for the western snowy plover consists of beaches and barren areas located close to the coast. Breeding pairs in Southern California have been found on salt evaporators, alkali flats, spits, and beaches.[67] For reproduction, the snowy plover appears to require a sandy, gravelly, or friable soil substrate on which to build nests. Feeding occurs in the sand where they can glean insects and over salt ponds or lakes where they can obtain brine flies.[68]

The potential habitat model for the western snowy plover was created based on land cover and distance from impacts caused by human disturbance. Beach and barren areas located within 1,600 feet

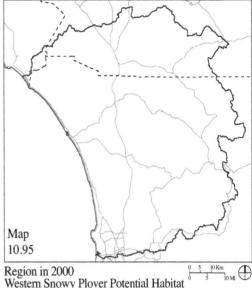

Map
10.95

Region in 2000
Western Snowy Plover Potential Habitat
Distance From Coastline

Within 1600 feet
15204 acres (0.9%)

of the coastline were identified as potential habitat (Map 10.95). Those areas greater than 300 feet from human development were classified as high value potential habitat. Thus, the resultant western snowy plover potential habitat map represents areas of beach or barren cover types located within 1,600 feet of the coastline with locations greater than 300 feet of human habitation identified as high value potential habitat. Although soil factors are important for nesting, soil information did not exist at an adequate resolution to be incorporated within the snowy plover's potential habitat model.

The potential habitat of western snowy plover creates a rather thin line of spatial habitat along the coast. The majority of the high quality habitat occurs within the boundaries of MCB Camp Pendleton. The remaining potential habitat is scattered along the coastline in a rather fragmented pattern. The relative low proportion of high value habitat in relation to the total amount of potential habitat reflects how the proximity to human infrastructure and associated human activities lessens the quality of snowy plover habitat.

Existing conditions and impacts of the futures are summarized in Table 10.10 and shown in Maps 10.96-10.104. The alternative futures vary in their degree of impact on western snowy plover potential habitat. The Three-Centers Future results in protection of the western snowy plover and as such does not impact potential habitat. The other three futures all decrease the amount of potential habitat available for the plover. Most of the reduction in habitat occurs just inland from the coast between the cities of Carlsbad and Encinitas. The high quality habitat remains roughly the same and is an expected result given that the majority of it occurs on MCB Camp Pendleton where no additional development is planned. The Coastal Future reduces the amount

of potential habitat to the greatest degree. This result is expected given that this future concentrates development along the coast. The Northern and Regional Low-Density Futures reduce potential habitat more than the Three-Centers.

Where potential habitat loss occurs, it is the result of the direct removal of habitat by housing in barren areas adjacent to beaches, not by indirect effects of human occupation. The high value habitat loss is the result of indirect effects caused by the proximity of prime habitat locations to new urban structures.

Table 10.10

Western Snowy Plover Potential Habitat (in acres)

	2000	Coastal		Northern		Reg. Low-Density		Three-Centers	
		500k	1,000k	500k	1,000k	500k	1,000k	500k	1,000k
Hi-Qual. Potential Habitat	385	385	383	383	383	385	383	385	385
Change from 2000		0	-2	-2	-2	0	-2	0	0
% Change from 2000		**0.0**	**-0.5**	**-0.5**	**-0.5**	**0.0**	**-0.5**	**0.0**	**0.0**
Other Habitat	1,446	1,394	1,367	1,404	1,382	1,414	1,402	1,446	1,446
Change from 2000		-52	-79	-42	-64	-32	-44	0	0
% Change from 2000		**-3.6**	**-5.5**	**-2.9**	**-4.4**	**-2.2**	**-3.0**	**0.0**	**0.0**
Total Habitat	1,831	1,779	1,750	1,787	1,765	1,799	1,785	1,831	1,831
Change from 2000		-52	-81	-44	-66	-32	-46	0	0
% Change from 2000		**-2.8**	**-4.4**	**-2.4**	**-3.6**	**-1.7**	**-2.5**	**0.0**	**0.0**

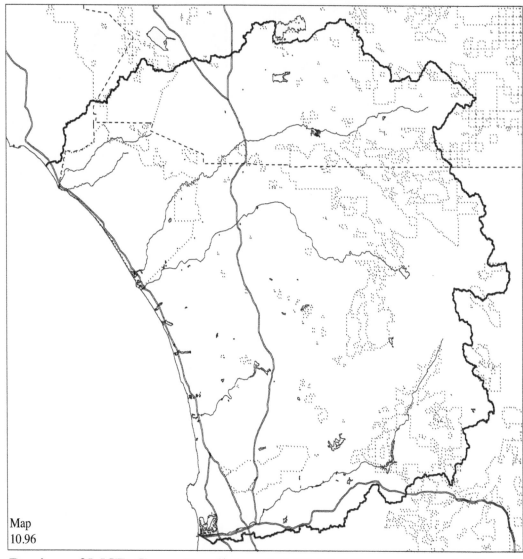

Map
10.96

Region of MCB Camp Pendleton & MCAS Miramar
Existing Conditions 2000
Western Snowy Plover Potential Habitat

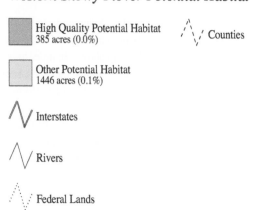

High Quality Potential Habitat
385 acres (0.0%)

Counties

Other Potential Habitat
1446 acres (0.1%)

Interstates

Rivers

Federal Lands

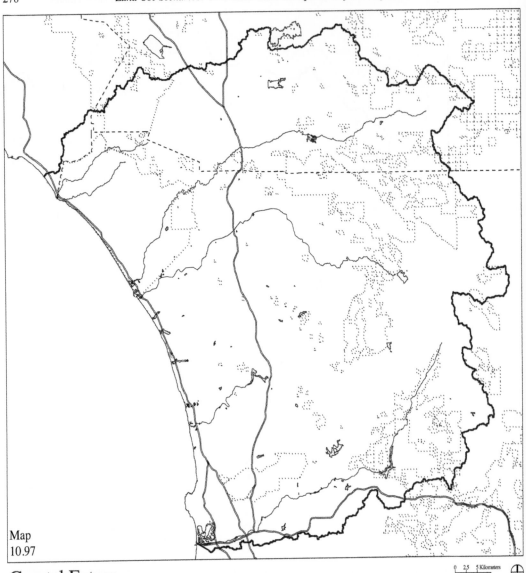

Map
10.97

Coastal Future
500,000 New Residents
Western Snowy Plover Potential Habitat Change

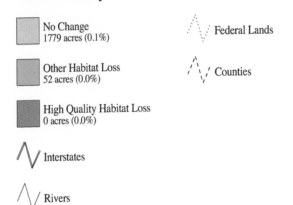

No Change
1779 acres (0.1%)

Other Habitat Loss
52 acres (0.0%)

High Quality Habitat Loss
0 acres (0.0%)

Interstates

Rivers

Federal Lands

Counties

NOTE: Some impacted areas are too small to be depicted at the scale used in the map above. In order to better show these locations, they have been expanded by the drawing of a uniform line around their perimeters. Statistics reported in the legend and in Table 10.10 are based on the data used to create the map and are not influenced by this cartographic necessity.

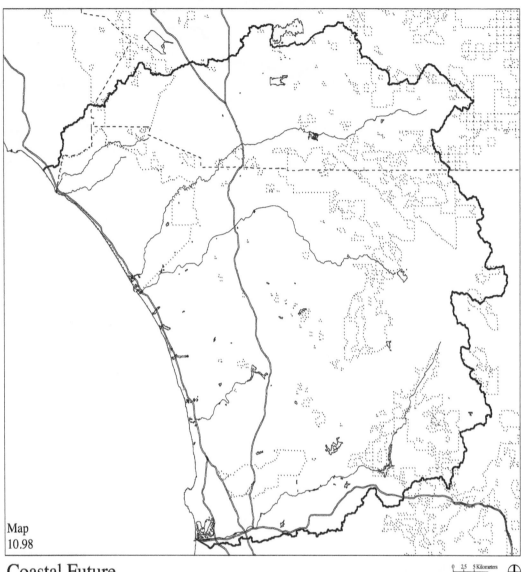

Map
10.98

Coastal Future
1,000,000 New Residents
Western Snowy Plover Potential Habitat Change

0   25   5 Kilometers
0   25   5 Miles

No Change
1752 acres (0.1%)

Other Habitat Loss
79 acres (0.0%)

High Quality Habitat Loss
2 acres (0.0%)

Interstates

Rivers

Federal Lands

Counties

NOTE: Some impacted areas are too small to be depicted at the scale used in the map above. In order to better show these locations, they have been expanded by the drawing of a uniform line around their perimeters. Statistics reported in the legend and in Table 10.10 are based on the data used to create the map and are not influenced by this cartographic necessity.

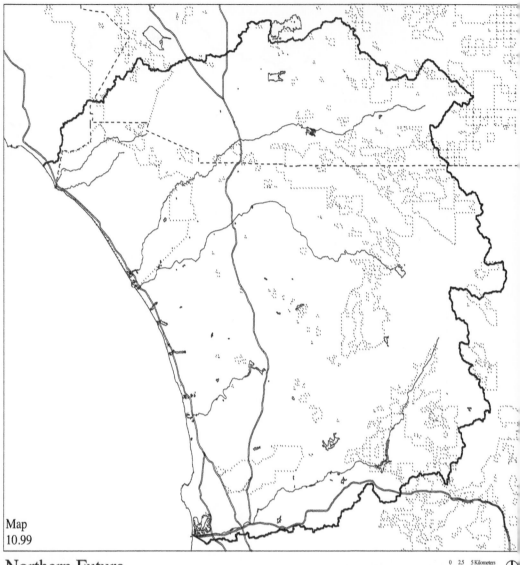

Map
10.99

Northern Future
500,000 New Residents
Western Snowy Plover Potential Habitat Change

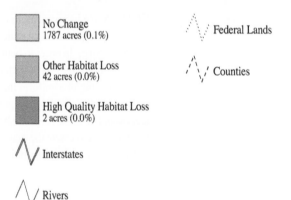

No Change
1787 acres (0.1%)

Other Habitat Loss
42 acres (0.0%)

High Quality Habitat Loss
2 acres (0.0%)

Interstates

Rivers

Federal Lands

Counties

NOTE: Some impacted areas are too small to be depicted at the scale used in the map above. In order to better show these locations, they have been expanded by the drawing of a uniform line around their perimeters. Statistics reported in the legend and in Table 10.10 are based on the data used to create the map and are not influenced by this cartographic necessity.

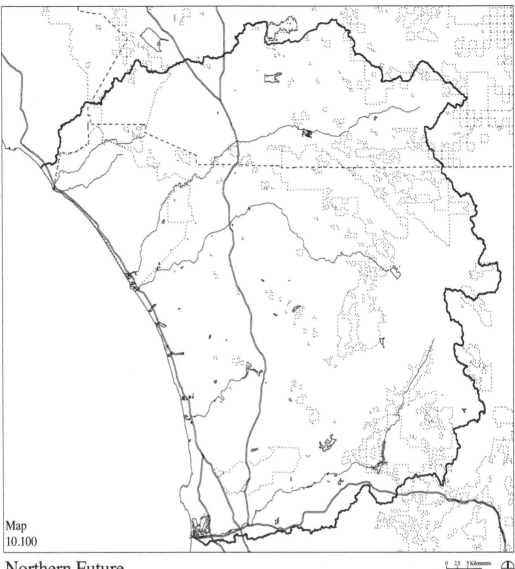

Map
10.100

Northern Future
1,000,000 New Residents
Western Snowy Plover Potential Habitat Change

0 2.5 5 Kilometers
0 2.5 5 Miles

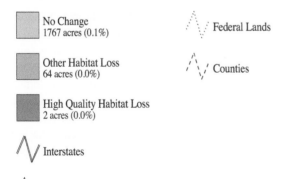

No Change
1767 acres (0.1%)

Other Habitat Loss
64 acres (0.0%)

High Quality Habitat Loss
2 acres (0.0%)

Interstates

Rivers

Federal Lands

Counties

NOTE: Some impacted areas are too small to be depicted at the scale used in the map above. In order to better show these locations, they have been expanded by the drawing of a uniform line around their perimeters. Statistics reported in the legend and in Table 10.10 are based on the data used to create the map and are not influenced by this cartographic necessity.

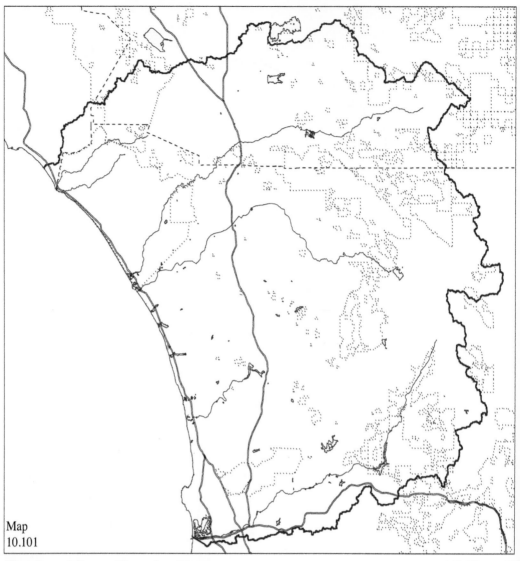

Map
10.101

Regional Low-Density Future
500,000 New Residents
Western Snowy Plover Potential Habitat Change

0   2.5   5 Kilometers
0   2.5   5 Miles

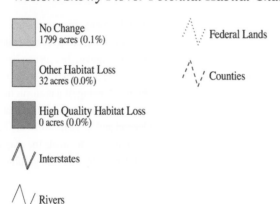

No Change
1799 acres (0.1%)

Other Habitat Loss
32 acres (0.0%)

High Quality Habitat Loss
0 acres (0.0%)

Interstates

Rivers

Federal Lands

Counties

NOTE: Some impacted areas are too small to be depicted at the scale used in the map above. In order to better show these locations, they have been expanded by the drawing of a uniform line around their perimeters. Statistics reported in the legend and in Table 10.10 are based on the data used to create the map and are not influenced by this cartographic necessity.

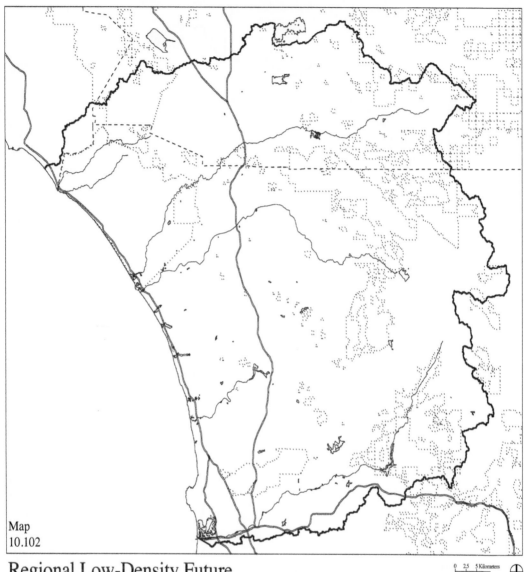

Regional Low-Density Future
1,000,000 New Residents
Western Snowy Plover Potential Habitat Change

Map
10.102

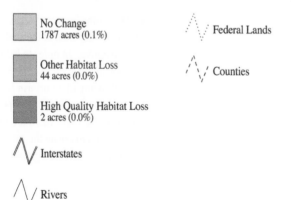

No Change
1787 acres (0.1%)

Other Habitat Loss
44 acres (0.0%)

High Quality Habitat Loss
2 acres (0.0%)

Interstates

Rivers

Federal Lands

Counties

NOTE: Some impacted areas are too small to be depicted at the scale used in the map above. In order to better show these locations, they have been expanded by the drawing of a uniform line around their perimeters. Statistics reported in the legend and in Table 10.10 are based on the data used to create the map and are not influenced by this cartographic necessity.

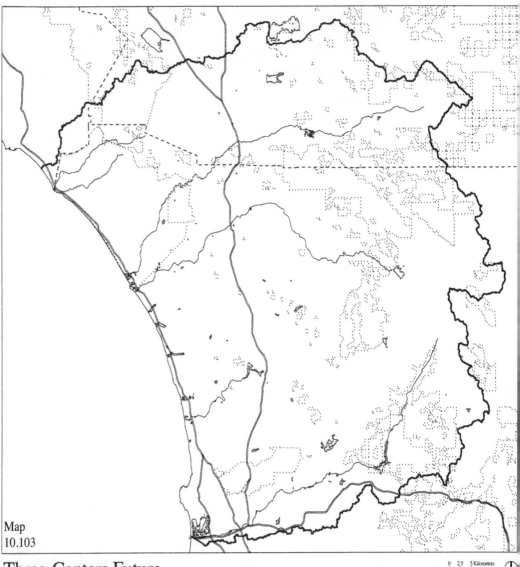

Map
10.103

Three-Centers Future
500,000 New Residents
Western Snowy Plover Potential Habitat Change

 No Change
1831 acres (0.1%)

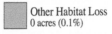 Other Habitat Loss
0 acres (0.1%)

NOTE: Some impacted areas are too small to be depicted at the scale used in the map above. In order to better show these locations, they have been expanded by the drawing of a uniform line around their perimeters. Statistics reported in the legend and in Table 10.10 are based on the data used to create the map and are not influenced by this cartographic necessity.

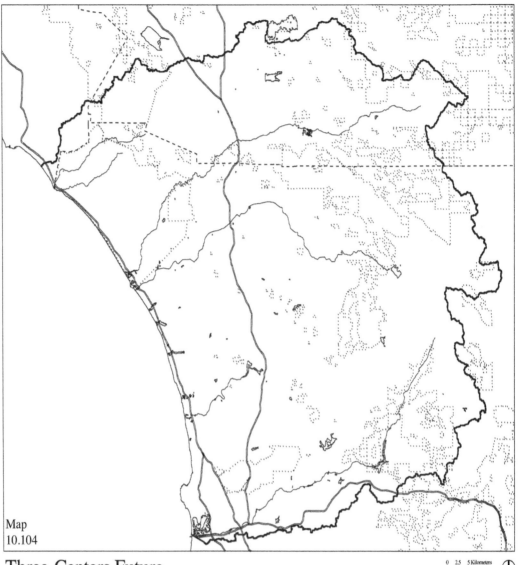

Map
10.104

Three-Centers Future
1,000,000 New Residents
Western Snowy Plover Potential Habitat Change

0 2.5 5 Kilometers
0 2.5 5 Miles

 No Change
1831 acres (0.1%)

 Other Habitat Loss
0 acres (0.1%)

NOTE: Some impacted areas are too small to be depicted at the scale used in the map above. In order to better show these locations, they have been expanded by the drawing of a uniform line around their perimeters. Statistics reported in the legend and in Table 10.10 are based on the data used to create the map and are not influenced by this cartographic necessity.

Miles

10 (ml)

Three Current Future
150,000 New Residents
Within Same Three-County Political Change

Southwestern Willow Flycatcher

The southwestern willow flycatcher (*Empidonax traillii extimus*) is a small neotropical migratory bird listed as an endangered species under the U.S. Endangered Species Act. It weighs approximately 0.47 ounces and attains a total length of 5.75 inches.[69] The species has a pale yellowish chest, grayish-olive back, dark wings, and a white throat. The southwestern willow flycatcher is almost indistinguishable from the alder flycatcher (*Empidonax alnorum*), but the voice provides a unique key for identification.[70] A subspecies of willow flycatcher, the southwestern willow flycatcher is slightly lighter in coloration than the other four subspecies.

The historic range of the southwestern willow flycatcher is similar to that of its current range with the reduction in quality and abundance of riparian habitat being the key for the endangered status.[71] Most of the loss or reduction in quality of riparian habitat is caused by alterations in hydrology.[72] Groundwater pumping and stream diversion have resulted in changes in the structure of riparian habitat resulting in declining flycatcher numbers.

Another threat is the invasion of the brown-headed cowbird (*Molothrus ater*).[73] Cowbirds are associated with anthropogenic features[74] and thus, riparian areas located next to human structures and agricultural practices are most at risk to nest parasitism by the cowbird. Since willow flycatchers are late nesters, re-nests may occur and these are infrequently parasitized because they occur after the main period of cowbird parasitic activity.[75]

The preferred habitat of the southwestern willow flycatcher consists of riparian areas with dense undergrowth. In most cases, the dense undergrowth tends to be within the first 10 to 13 feet of understory. Less dense riparian locations may be utilized as perches for feeding or as singing posts. The willow flycatcher is known to utilize both native and non-native vegetation for nesting. Common native species occupied include willow (*Salix* spp.), cottonwood (*Populus* spp.), box elder (*Acer negundo*), ash (*Fraxinus* spp.) and alder (*Alnus* spp.). Non-native species used as breeding sites include salt cedar (*Tamarix ramosissima*) and Russian olive (*Elaeagnus angustifolia*). Giant reed (*Arundo donax*) and tree of heaven (*Ailanthus altissima*) are not used by flycatchers for nesting and are of concern due to their invasive nature.[76] Preferred habitat also includes locations close to ponds, wet meadows, or other forms of standing water. The size of habitat patch occupied for nesting varies by location. The minimum riparian area identified as being inhabited by a nesting pair is 1.5 acres.[77] Although a minimum patch size area has been identified, it should be noted that flycatchers tend to cluster their territories into small portions of riparian sites.[78] Thus, major portions of riparian sites may be unoccupied at any given time.[79] Riparian sites occupied are typically greater than 30 feet wide.[80]

The determination of potential habitat for the southwestern willow flycatcher began by identifying riparian locations within the region. Riparian locations located within 100 feet of agricultural and urban lands were eliminated as potential habitat. These viable locations were then reduced to reflect the flycatcher's preference to be located close to water or moist soil by limiting potential habitat to locations within 150 feet of streams or lakes. Patches of habitat less than 1.5 acres in area were eliminated from consideration. Thus, the potential habitat for the southwestern willow flycatcher was identified as riparian areas located greater than 100 feet from agricultural areas, within 150 feet of streams or water bodies, and at least 1.5 acres in area. The relative width requirement of 30 feet for potential habitat sites could not be included due to the resolution of the land cover map being governed by a pixel size of approximately 100 feet square.

Potential habitat for the southwestern willow flycatcher is scattered throughout the study region. The largest contiguous patch of potential habitat exists along the lower reaches of the Santa Margarita River on MCB Camp Pendleton. Throughout the study area, potential habitat consists of small patches of riparian vegetation often separated by large distances. The dispersed nature of the habitat may explain the low sighting numbers of southwestern willow flycatchers in Southern California.

Existing conditions and the impacts of the futures are summarized in Table 10.11 and shown in Maps 10.105-10.113. The amount of available potential habitat decreases for all the alternative futures. The relatively large differences between the addition of 500,000 and 1,000,000 people is explained by the removal of suitably sized habitat patches. Within all the futures, potential habitat is removed along the edges, thereby reducing the area of each habitat patch.

The futures with a high amount of low-density residential development cause the greatest negative impact. Thus, the Coastal Future has the least amount of habitat loss and the Regional Low-Density Future causes the greatest loss. Based on the amount of low-density housing development occurring in the Northern Future, one would anticipate a loss in habitat similar to that of the Regional Low-Density Future. However, the Northern Future has substantially less habitat loss. This situation can be explained by the low

amount of flycatcher habitat in Western Riverside County. Future development within the study area emphasizing low-density development within San Diego County poses the greatest threat to the southwestern willow flycatcher.

Most of the habitat loss is the result of indirect impacts associated with residential development. Potential habitat is not directly removed; rather, with the increase in residential development the expectation is that brown-headed cowbirds will enlarge their distribution. Any increase in cowbird numbers around riparian vegetation, which provides habitat for the southwestern willow flycatcher, will reduce nesting success and render riparian vegetation habitat effectively unsuitable.

Table 10.11

Southwestern Willow Flycatcher Potential Habitat (in acres)

	2000	Coastal		Northern		Reg. Low-Density		Three-Centers	
		500k	1,000k	500k	1,000k	500k	1,000k	500k	1,000k
Potential Habitat	15,352	15,035	14,099	14,665	13,160	14,065	11,544	14,670	13,783
Change from 2000		-317	-1,253	-687	-2,192	-1,287	-3,808	-682	-1,569
% Change from 2000		-2.1	-8.2	-4.5	-14.3	-8.4	-24.8	-4.4	-10.2

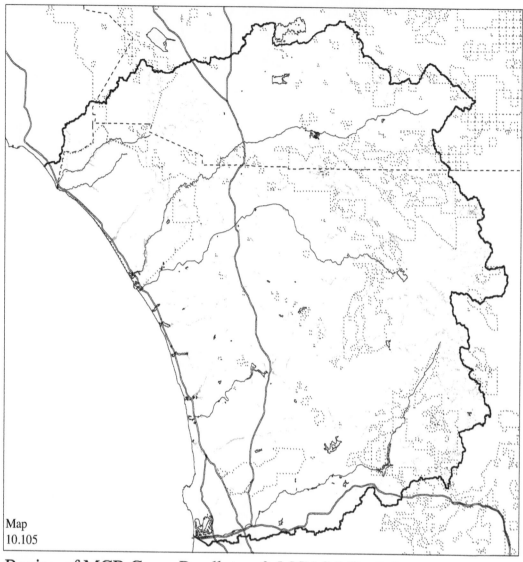

Map
10.105

Region of MCB Camp Pendleton & MCAS Miramar
Existing Conditions 2000
Southwestern Willow Flycatcher Potential Habitat

0 2.5 5 Kilometers
0 2.5 5 Miles

Potential Habitat
15352 acres (0.9%)

Interstates

Rivers

Federal Lands

Counties

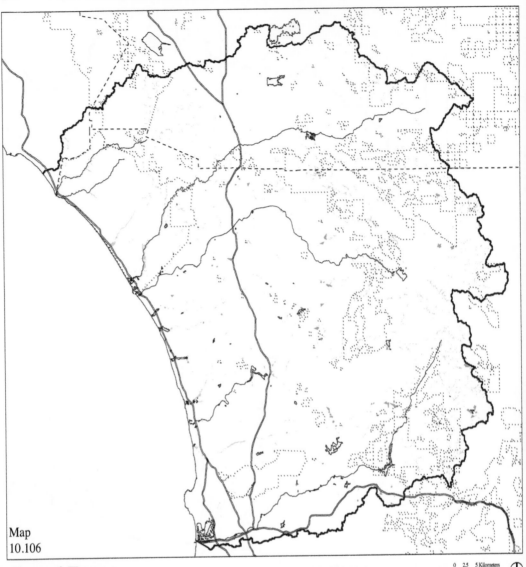

Map
10.106

Coastal Future
500,000 New Residents
Southwestern Willow Flycatcher Potential Habitat Change

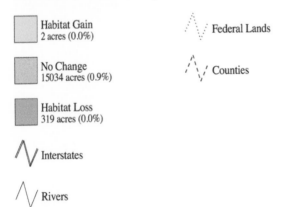

Habitat Gain
2 acres (0.0%)

No Change
15034 acres (0.9%)

Habitat Loss
319 acres (0.0%)

Interstates

Rivers

Federal Lands

Counties

NOTE: Some impacted areas are too
small to be depicted at the scale used in
the map above. In order to better show
these locations, they have been expanded
by the drawing of a uniform line around
their perimeters. Statistics reported in the
legend and in Table 10.11 are based on the
data used to create the map and are not
influenced by this cartographic necessity.

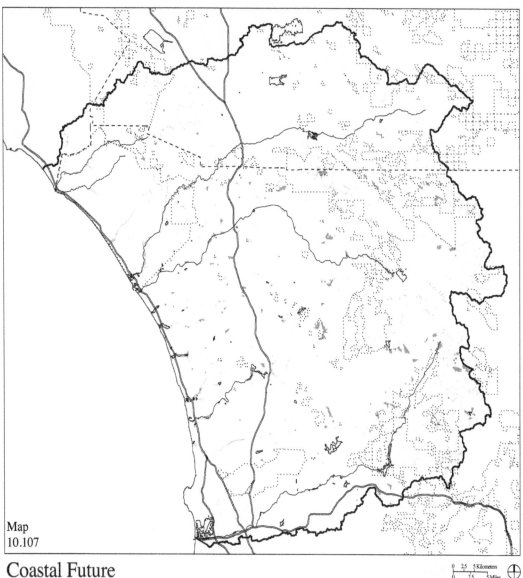

Map
10.107

Coastal Future
1,000,000 New Residents
Southwestern Willow Flycatcher Potential Habitat Change

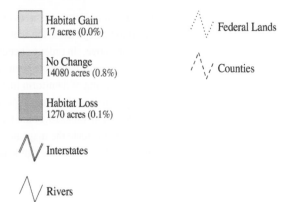

Habitat Gain
17 acres (0.0%)

No Change
14080 acres (0.8%)

Habitat Loss
1270 acres (0.1%)

Interstates

Rivers

Federal Lands

Counties

NOTE: Some impacted areas are too small to be depicted at the scale used in the map above. In order to better show these locations, they have been expanded by the drawing of a uniform line around their perimeters. Statistics reported in the legend and in Table 10.11 are based on the data used to create the map and are not influenced by this cartographic necessity.

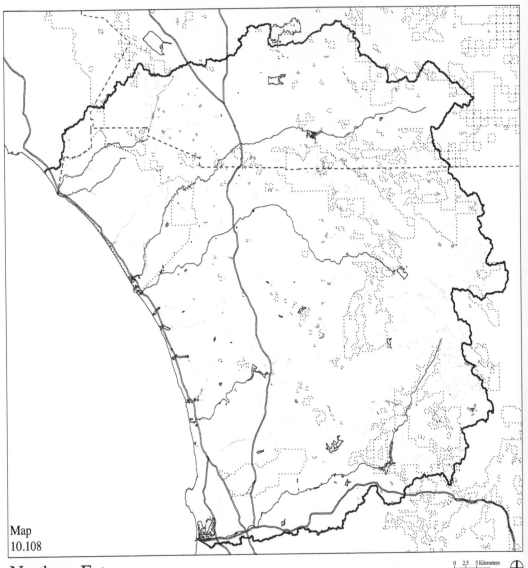

Map
10.108

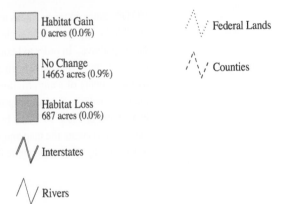

Northern Future
500,000 New Residents
Southwestern Willow Flycatcher Potential Habitat Change

Habitat Gain 0 acres (0.0%)	Federal Lands
No Change 14663 acres (0.9%)	Counties
Habitat Loss 687 acres (0.0%)	
Interstates	
Rivers	

NOTE: Some impacted areas are too small to be depicted at the scale used in the map above. In order to better show these locations, they have been expanded by the drawing of a uniform line around their perimeters. Statistics reported in the legend and in Table 10.11 are based on the data used to create the map and are not influenced by this cartographic necessity.

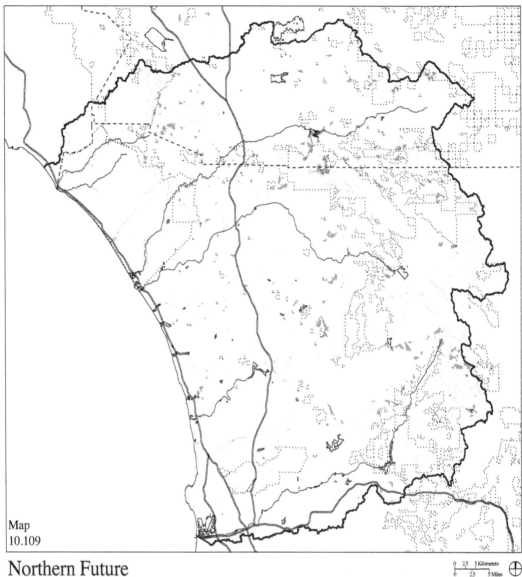

Map
10.109

Northern Future
1,000,000 New Residents
Southwestern Willow Flycatcher Potential Habitat Change

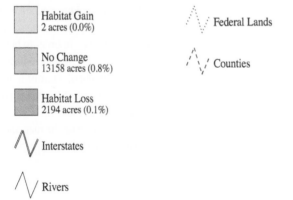

Habitat Gain
2 acres (0.0%)

No Change
13158 acres (0.8%)

Habitat Loss
2194 acres (0.1%)

Interstates

Rivers

Federal Lands

Counties

NOTE: Some impacted areas are too small to be depicted at the scale used in the map above. In order to better show these locations, they have been expanded by the drawing of a uniform line around their perimeters. Statistics reported in the legend and in Table 10.11 are based on the data used to create the map and are not influenced by this cartographic necessity.

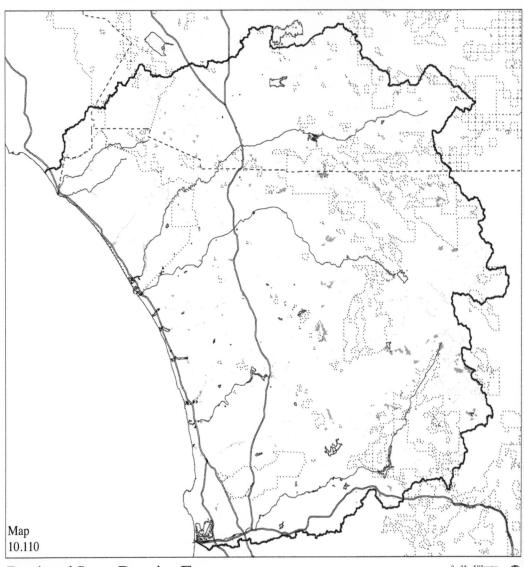

Map
10.110

Regional Low-Density Future
500,000 New Residents
Southwestern Willow Flycatcher Potential Habitat Change

0   2.5   5 Kilometers
0   2.5   3 Miles

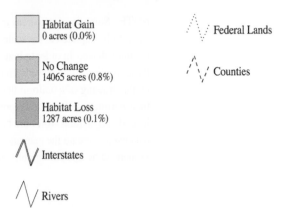

Habitat Gain
0 acres (0.0%)

No Change
14065 acres (0.8%)

Habitat Loss
1287 acres (0.1%)

Interstates

Rivers

Federal Lands

Counties

NOTE: Some impacted areas are too small to be depicted at the scale used in the map above. In order to better show these locations, they have been expanded by the drawing of a uniform line around their perimeters. Statistics reported in the legend and in Table 10.11 are based on the data used to create the map and are not influenced by this cartographic necessity..

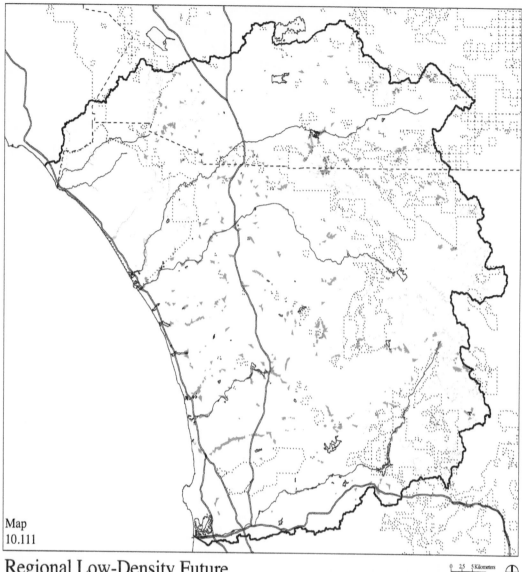

Map
10.111

Regional Low-Density Future
1,000,000 New Residents
Southwestern Willow Flycatcher Potential Habitat Change

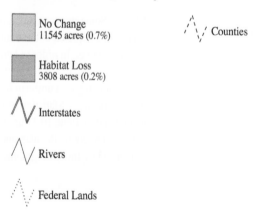

No Change
11545 acres (0.7%)

Habitat Loss
3808 acres (0.2%)

Interstates

Rivers

Federal Lands

Counties

NOTE: Some impacted areas are too
small to be depicted at the scale used in
the map above. In order to better show
these locations, they have been expanded
by the drawing of a uniform line around
their perimeters. Statistics reported in the
legend and in Table 10.11 are based on the
data used to create the map and are not
influenced by this cartographic necessity.

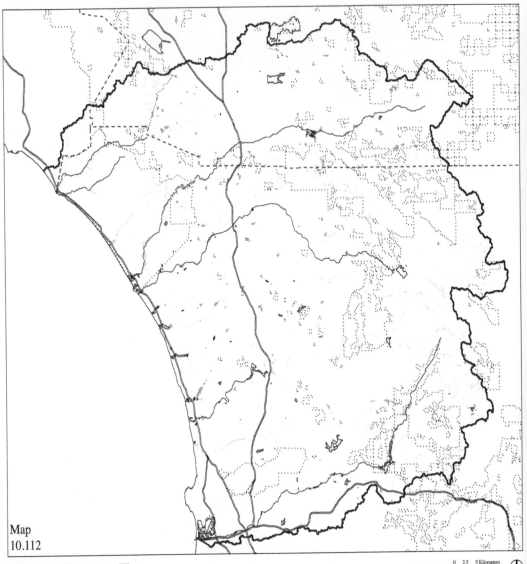

Map
10.112

Three-Centers Future
500,000 New Residents
Southwestern Willow Flycatcher Potential Habitat Change

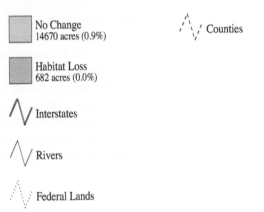

No Change
14670 acres (0.9%)

Habitat Loss
682 acres (0.0%)

Interstates

Rivers

Federal Lands

Counties

NOTE: Some impacted areas are too small to be depicted at the scale used in the map above. In order to better show these locations, they have been expanded by the drawing of a uniform line around their perimeters. Statistics reported in the legend and in Table 10.11 are based on the data used to create the map and are not influenced by this cartographic necessity.

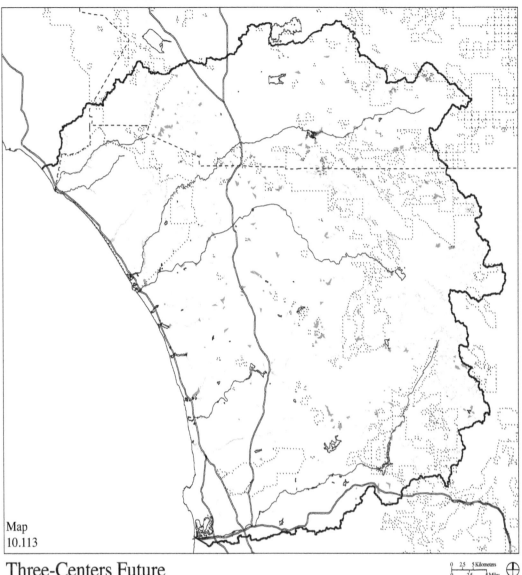

Map
10.113

Three-Centers Future
1,000,000 New Residents
Southwestern Willow Flycatcher Potential Habitat Change

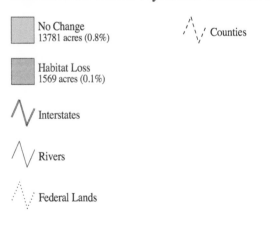

No Change
13781 acres (0.8%)

Habitat Loss
1569 acres (0.1%)

Interstates

Rivers

Federal Lands

Counties

NOTE: Some impacted areas are too small to be depicted at the scale used in the map above. In order to better show these locations, they have been expanded by the drawing of a uniform line around their perimeters. Statistics reported in the legend and in Table 10.11 are based on the data used to create the map and are not influenced by this cartographic necessity.

Least Bell's Vireo

The least Bell's vireo (*Vireo bellii pusillus*) is listed by the federal government as an endangered species. A subspecies of the Bell's vireo, it weighs 0.3 ounces and reaches a total length of 4 inches.[81] Back coloration varies for the least Bell's vireo from gray near the top of the back to a slight yellowish color as the tail is reached. Wings have barring and an eye ring is present. The bird's bill is small and the tail is rather long.[82]

The least Bell's vireo, endemic to California and Baja California, has been extirpated from all of its northern range and is now restricted in the United States to the eight southern-most counties of California. It is most abundant in San Diego County.[83] The reduction of the vireo's historic range results from the loss of adequate habitat. Habitat loss has been attributed to invasive exotic vegetation, water diversion, habitat fragmentation, livestock grazing, house pets, natural predators, parasitism by brown-headed cowbirds, and disturbance by humans.[84] Addressing these issues where populations of the vireos are in decline must be done to assist recovery efforts.

Preferred habitat of the vireo consists of high quality early successional riparian woodlands 5 to 10 years in age and dense willow riparian woodland with lush understory vegetation.[85] Willow species are most commonly used for nesting and most nest sites are located near the edges of thickets. A low dense shrub layer is essential for nesting. Riparian areas located adjacent to natural vegetation are preferred over those located in close proximity to agricultural and urban areas. Although the species is found in riparian vegetation no water requirements have been mentioned in the literature and the vireo has been known to nest in upland vegetation when riparian sites have been inundated by spring floodwaters.[86]

The male least Bell's vireo is a tenacious defender of his territory returning in successive years to nest in the same location.[87] The average territory size of 1.9 acres[88] is small, yet habitat fragmentation is frequently mentioned as a problem that could lead to extirpation of local subpopulations.[89]

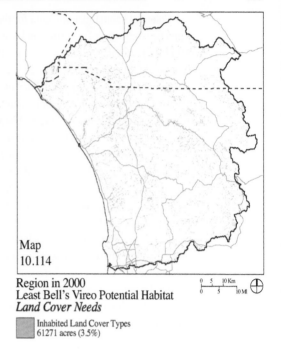

Map 10.114

Region in 2000
Least Bell's Vireo Potential Habitat
Land Cover Needs

Inhabited Land Cover Types
61271 acres (3.5%)

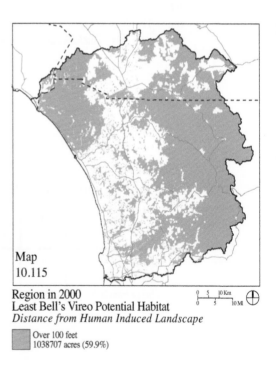

Map 10.115

Region in 2000
Least Bell's Vireo Potential Habitat
Distance from Human Induced Landscape

Over 100 feet
1038707 acres (59.9%)

Potential habitat identification for the least Bell's vireo began by identifying riparian locations (Map 10.114). Riparian areas located within 100 feet of agricultural and urban lands were eliminated as potential habitat (Map 10.115). Patches of habitat greater than 1.9 acres in area were maintained as potential habitat. Thus, the potential habitat for the least Bell's vireo was identified as riparian areas located greater than 100 feet from agricultural areas and at least 1.9 acres in area. Stand age estimates for riparian areas were not available and could not be readily modeled, thus, they are not included as part of the potential habitat model.

Least Bell's vireo potential habitat occurs throughout the study region with most of the large patches located along coastal streams. Very little potential habitat is present within Western Riverside County. As with the southwestern willow flycatcher, much of the vireo's habitat occurs as small isolated patches.

Existing conditions and impacts of the futures are summarized in Table 10.12 and shown in Mpas 10.116-10.124. All the futures result in a decrease in the amount of potential habitat for the least Bell's vireo. The Regional Low-Density Future creates the greatest loss. The loss can be attributed to the amount of low-density residential development placed within San Diego County and the indirect impacts associated with this development. The Northern Future has less loss than the Regional Low-Density Future and given the similarity in the amount of low-density residential development occurring in the two futures the difference in loss may be attributed to the lack of potential habitat in Western Riverside County. The Coastal and Three-Centers Futures result in similar low losses in habitat. These results are a function of the conservation techniques imposed under the Three-Centers Future and the compact nature of the Coastal Future's residential development.

As with the southwestern willow flycatcher, most of the least Bell's vireo potential habitat loss is the result of indirect impacts. Riparian vegetation and water areas are generally not compatible locations for residential development. Thus, most loss of potential habitat is attributed to the buffering of anthropogenic structures by 100 feet. This buffer was imposed to account for the nest parasitism of the brown headed cowbird and as such the habitat loss associated with the cowbird may be interpreted as an indirect effect on vireos by reducing their probability of successfully producing offspring.

Table 10.12

Least Bell's Vireo Potential Habitat (in acres)

	2000	Coastal		Northern		Reg. Low-Density		Three-Centers	
		500k	1,000k	500k	1,000k	500k	1,000k	500k	1,000k
Potential Habitat	28,061	27,279	25,400	26,756	24,030	25,415	21,253	26,868	25,363
Change from 2000		-783	-2,661	-1,305	-4,031	-2,646	-6,808	-1,193	-2,698
% Change from 2000		**-2.9**	**-9.5**	**-4.7**	**-14.4**	**-9.4**	**-24.3**	**-4.3**	**-9.6**

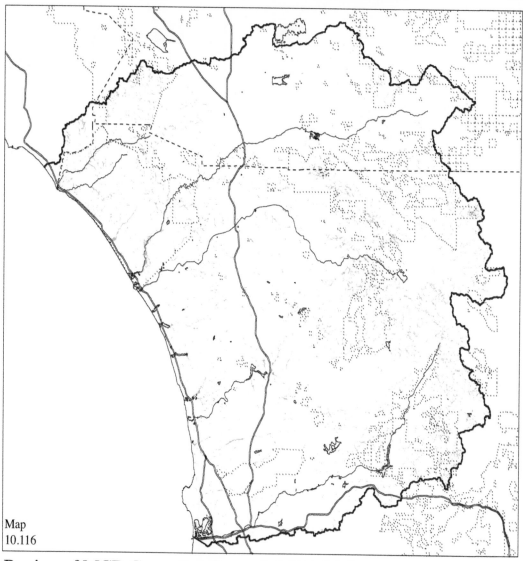

Map
10.116

Region of MCB Camp Pendleton & MCAS Miramar
Existing Conditions 2000
Least Bell's Vireo Potential Habitat

 Potential Habitat
28061 acres (1.6%)

 Interstates

Rivers

 Federal Lands

 Counties

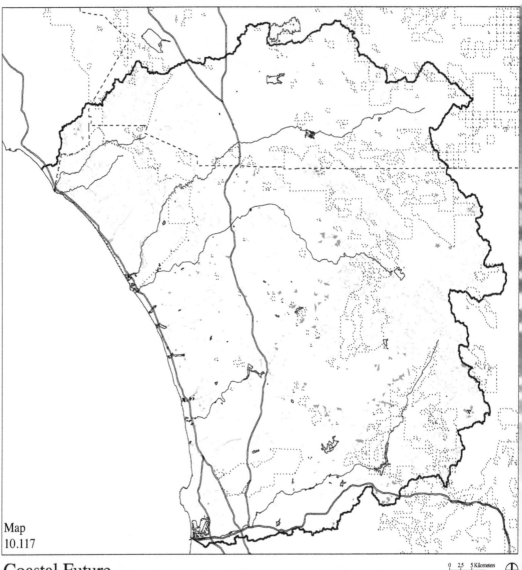

Map
10.117

Coastal Future
500,000 New Residents
Least Bell's Vireo Potential Habitat Change

0   2.5   5 Kilometers
0   2.5   5 Miles

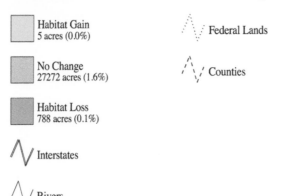

Habitat Gain
5 acres (0.0%)

No Change
27272 acres (1.6%)

Habitat Loss
788 acres (0.1%)

Interstates

Rivers

Federal Lands

Counties

NOTE: Some impacted areas are too small to be depicted at the scale used in the map above. In order to better show these locations, they have been expanded by the drawing of a uniform line around their perimeters. Statistics reported in the legend and in Table 10.12 are based on the data used to create the map and are not influenced by this cartographic necessity.

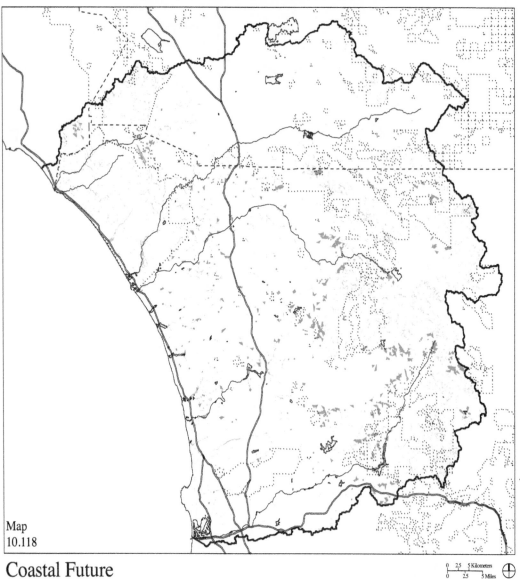

Map
10.118

Coastal Future
1,000,000 New Residents
Least Bell's Vireo Potential Habitat Change

0 2.5 5 Kilometers
0 2.5 5 Miles

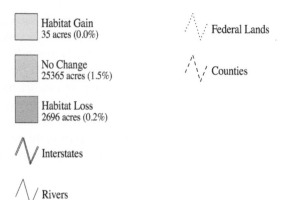

Habitat Gain
35 acres (0.0%)

No Change
25365 acres (1.5%)

Habitat Loss
2696 acres (0.2%)

Interstates

Rivers

Federal Lands

Counties

NOTE: Some impacted areas are too
small to be depicted at the scale used in
the map above. In order to better show
these locations, they have been expanded
by the drawing of a uniform line around
their perimeters. Statistics reported in the
legend and in Table 10.12 are based on
the data used to create the map and are not
influenced by this cartographic necessity.

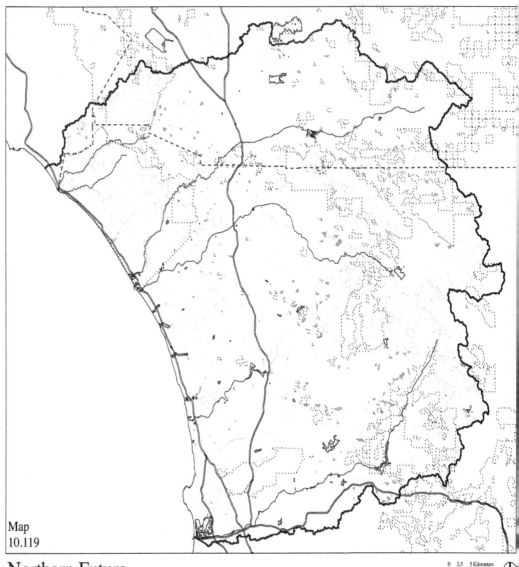

Map
10.119

Northern Future
500,000 New Residents
Least Bell's Vireo Potential Habitat Change

0   2.5   5 Kilometers
0   2.5   5 Miles

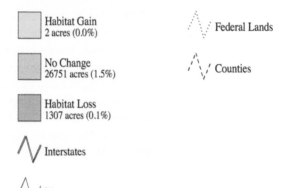

Habitat Gain
2 acres (0.0%)

No Change
26751 acres (1.5%)

Habitat Loss
1307 acres (0.1%)

Interstates

Rivers

Federal Lands

Counties

NOTE: Some impacted areas are too small to be depicted at the scale used in the map above. In order to better show these locations, they have been expanded by the drawing of a uniform line around their perimeters. Statistics reported in the legend and in Table 10.12 are based on the data used to create the map and are not influenced by this cartographic necessity.

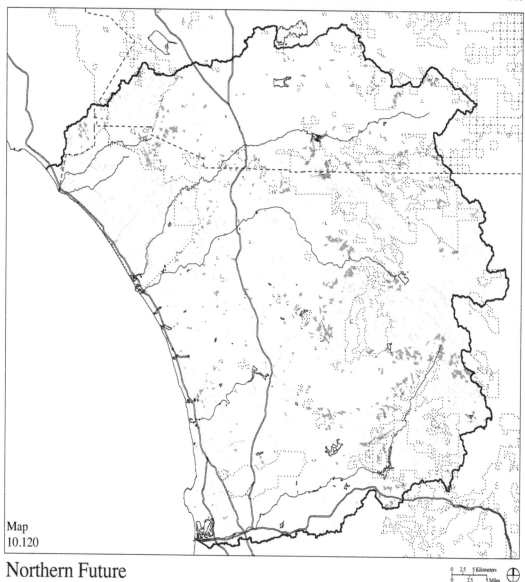

Map
10.120

Northern Future
1,000,000 New Residents
Least Bell's Vireo Potential Habitat Change

0   2.5   5 Kilometers
0   2.5   5 Miles

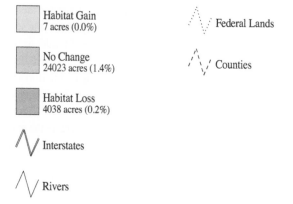

Habitat Gain
7 acres (0.0%)

No Change
24023 acres (1.4%)

Habitat Loss
4038 acres (0.2%)

Interstates

Rivers

Federal Lands

Counties

NOTE: Some impacted areas are too small to be depicted at the scale used in the map above. In order to better show these locations, they have been expanded by the drawing of a uniform line around their perimeters. Statistics reported in the legend and in Table 10.12 are based on the data used to create the map and are not influenced by this cartographic necessity.

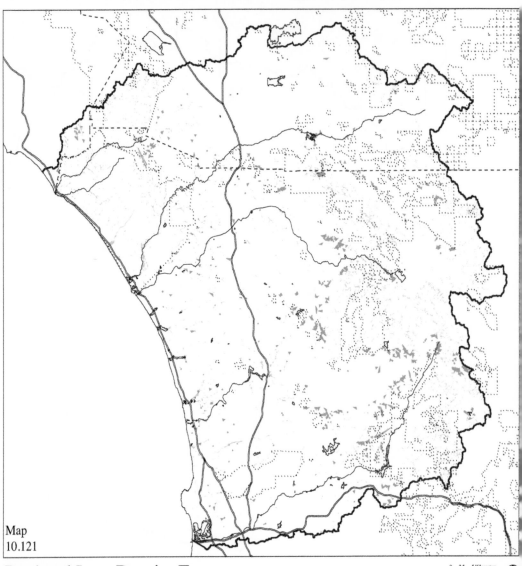

Map
10.121

Regional Low-Density Future
500,000 New Residents
Least Bell's Vireo Potential Habitat Change

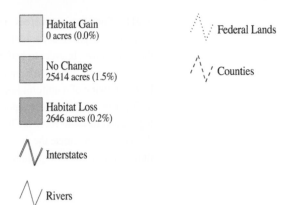

Habitat Gain
0 acres (0.0%)

No Change
25414 acres (1.5%)

Habitat Loss
2646 acres (0.2%)

Interstates

Rivers

Federal Lands

Counties

NOTE: Some impacted areas are too small to be depicted at the scale used in the map above. In order to better show these locations, they have been expanded by the drawing of a uniform line around their perimeters. Statistics reported in the legend and in Table 10.12 are based on the data used to create the map and are not influenced by this cartographic necessity.

0   2.5   5 Kilometers
0   2.5   5 Miles

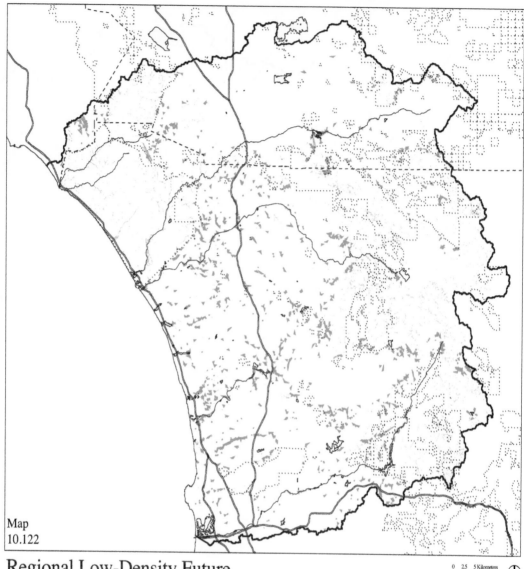

Map
10.122

Regional Low-Density Future
1,000,000 New Residents
Least Bell's Vireo Potential Habitat Change

Habitat Gain
0 acres (0.0%)

NOTE: Some impacted areas are too small to be depicted at the scale used in the map above. In order to better show these locations, they have been expanded by the drawing of a uniform line around their perimeters. Statistics reported in the legend and in Table 10.12 are based on the data used to create the map and are not influenced by this cartographic necessity.

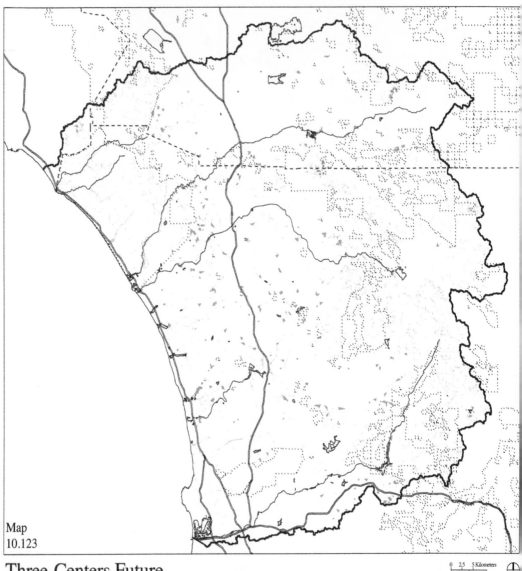

Map
10.123

Three-Centers Future
500,000 New Residents
Least Bell's Vireo Potential Habitat Change

0   2.5   5 Kilometers
0   2.5   5 Miles

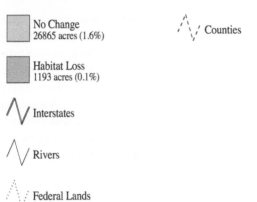

No Change
26865 acres (1.6%)

Habitat Loss
1193 acres (0.1%)

Interstates

Rivers

Federal Lands

Counties

NOTE: Some impacted areas are too small to be depicted at the scale used in the map above. In order to better show these locations, they have been expanded by the drawing of a uniform line around their perimeters. Statistics reported in the legend and in Table 10.12 are based on the data used to create the map and are not influenced by this cartographic necessity.

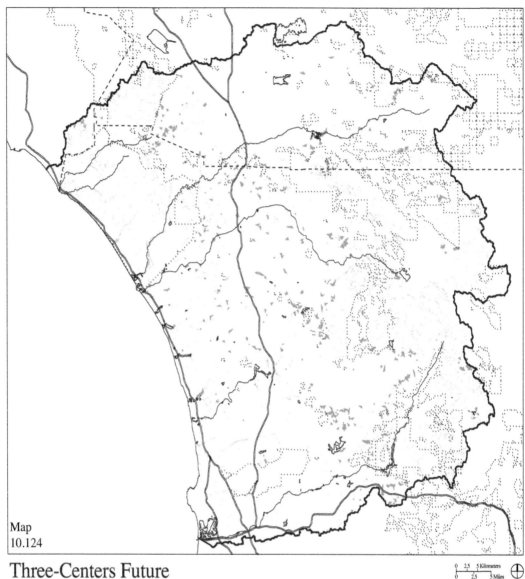

Map
10.124

Three-Centers Future
1,000,000 New Residents
Least Bell's Vireo Potential Habitat Change

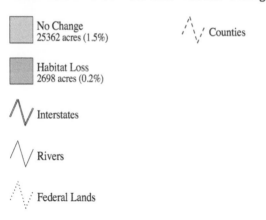

No Change
25362 acres (1.5%)

Habitat Loss
2698 acres (0.2%)

Interstates

Rivers

Federal Lands

Counties

NOTE: Some impacted areas are too small to be depicted at the scale used in the map above. In order to better show these locations, they have been expanded by the drawing of a uniform line around their perimeters. Statistics reported in the legend and in Table 10.12 are based on the data used to create the map and are not influenced by this cartographic necessity.

0 2.5 5 Kilometers
0 2.5 5 Miles

Western Meadowlark

The western meadowlark (*Sturnella neglecta*) was formerly considered to be a subspecies of the eastern meadowlark (*Sturnella magna*), but was granted species status in 1908. Both meadowlarks have short tails, long bills, and are heavy for their size of 3.4 ounces and length of 9 inches.[90] Both meadowlarks have yellow bellies and a pale gray-brown back. The western meadowlark has less white on the tail, whitish flanks, and more yellow located in the facial region than its eastern cousin.[91]

Breeding Bird Surveys (1968–1991) indicate that the breeding populations of the western meadowlark have declined slightly throughout the United States and Canada.[92] The declines have been attributed to habitat loss, declining habitat quality, and loss of disturbance regimes.[93] Other population reductions have been attributed to the type and timing of disturbances occurring on meadowlark inhabited land. Agricultural practices such as surface tillage in spring, early burning,[94] or the application of pesticides on fields[95] likely 1) destroy all nests, 2) damage incubating parents and flightless young, and 3) eradicate insects which serve as food.

Western meadowlarks are terrestrial songbirds. They inhabit grasslands and are most common in native grasslands and pastures, but occur in many other grassland habitats including alfalfa and hay fields, roadsides, and weedy margins of croplands.[96] The grasslands most frequently occupied have burned within the last 7 years.[97] The fire interval association and observations from other studies indicate that the meadowlark prefers areas with high quality grass and forbs.[98]

Territory size varies greatly throughout the range of the western meadowlark, with some estimates being as high as 32 acres[99] and others as low as 3 acres.[100] An in-depth study found that territories of 7 to 8 acres are common.[101] Information on the species territory dynamics and subsequent patch size effects show very little patch size effect indicating that the presence of meadowlarks is linked more closely to the quality of the habitat and less on the size.[102] Patch dynamic studies have also indicated little relationship between urban areas and the decrease or increase in species abundance.[103]

Potential habitat modeling for the western meadowlark began by identifying the inhabited land cover types (Map 10.125). All the agricultural fields dominated by row crops or pastures and the grasslands in the region were extracted. The potential habitat land cover types were combined into patches and those areas below 7.5 acres in area were eliminated. Thus, the potential habitat for the western meadowlark consists of the combination of agriculture and grasslands greater than 7.5 acres in area. Although the quality of meadowlark habitat is influenced by the species of grass or crop grown, no distinction was made in the potential habitat model to address this issue. These two factors limited the ability to assess the quality of meadowlark habitat.

Potential habitat for the western meadowlark occurs in four large patches. Three of the four large patches represent areas known to be of different vegetation composition. Row crops dominate the potential habitat locations within Western Riverside County and in the Ramona area of San Diego County. The large habitat patch near Lake Henshaw is composed of perennial grasses. The last habitat cluster is located on MCB Camp Pendleton and is comprised mainly of annual grasses. Each of these large clustered habitat locations would have a different disturbance frequency. This factor is not represented in the western meadowlark's potential habitat model, but is apparent from ground inspection. Intermingled around these large habitat clusters and dispersed throughout the study region are smaller patches of potential habitat.

Existing conditions and impacts associated with the futures are summarized in Table 10.13 and shown in Maps 10.126-10.134. All the alternative futures cause an increase in the potential habitat available for the western meadowlark. The differences between the additions of 500,000 and 1,000,000 new residents within each future may be attributed to the juxtaposition of the new housing lots in relationship to established development. This model result implies that with the addition of more people, the amount of habitat would roughly increase in a linear fashion with a like increase in population size.

The Coastal Future results in the least amount of habitat increase, the Regional Low-Density Future has the greatest increase, and the other two futures

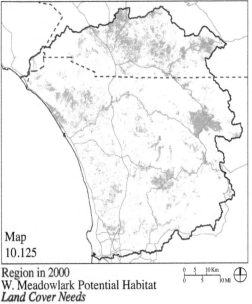

Map
10.125

Region in 2000
W. Meadowlark Potential Habitat
Land Cover Needs

0 5 10 Km
0 5 10 Mi

Inhabited Land Cover Types
231513 acres (13.4%)

produce intermediate increases. Virtually all the increase in potential habitat results from the removal of shrub vegetation surrounding low-density residential housing. With the removal of shrub vegetation, a transition to annual forbs and grasses is expected which the meadowlark will occupy. Even though the Northern Future has a relatively high amount of low-density residential development, the majority of the new development in Western Riverside County occurs on agricultural land that is already considered potential habitat.

These results indicate a relationship between housing density and amount of western meadowlark habitat. Low-density residential development results in more habitat and likewise high-density development has the potential to result in less habitat. Generally, housing development goes from low to high density as space becomes less available. Beyond the 1,000,000 new residents allocated in the futures examined, western meadowlark habitat may decrease if infill becomes the development norm. Thus, although this study shows an increase in habitat, readers should keep in mind the time frame of analysis and understand that at some point, as housing density increases, meadowlark habitat will decrease.

Table 10.13

Western Meadowlark Potential Habitat (in acres)

	2000	Coastal		Northern		Reg. Low-Density		Three-Centers	
		500k	1,000k	500k	1,000k	500k	1,000k	500k	1,000k
Potential Habitat	199,130	207,121	224,646	215,031	240,754	259,256	326,151	216,797	233,322
Change from 2000		7,991	25,516	15,901	41,623	60,126	127,021	17,687	34,192
% Change from 2000		4.0	12.8	8.0	20.9	30.2	63.8	8.9	17.2

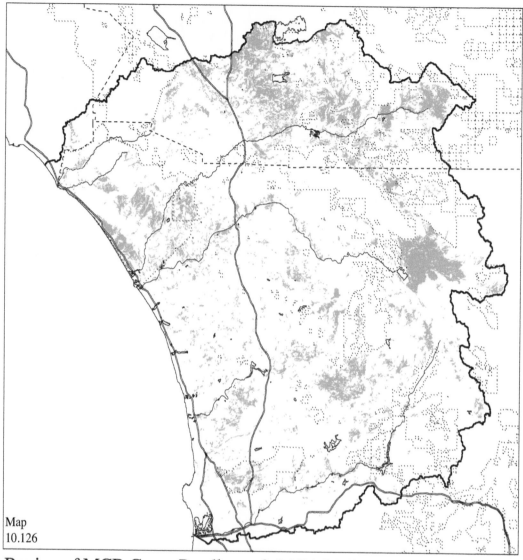

Map
10.126

Region of MCB Camp Pendleton & MCAS Miramar
Existing Conditions 2000
Western Meadowlark Potential Habitat

Potential Habitat
199130 acres (11.5%)

Interstates

Rivers

Federal Lands

Counties

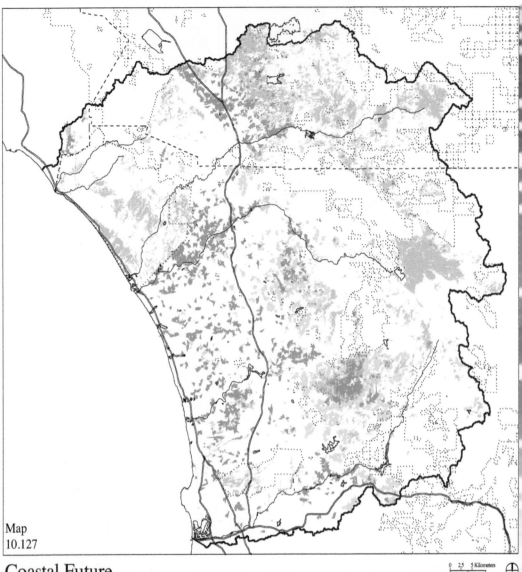

Map
10.127

Coastal Future
500,000 New Residents
Western Meadowlark Potential Habitat Change

0   2.5   5 Kilometers
0   2.5   5 Miles

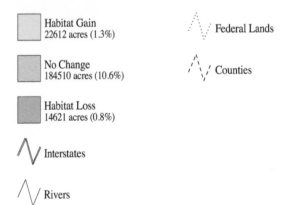

Habitat Gain
22612 acres (1.3%)

No Change
184510 acres (10.6%)

Habitat Loss
14621 acres (0.8%)

Interstates

Rivers

Federal Lands

Counties

NOTE: Some impacted areas are too small to be depicted at the scale used in the map above. In order to better show these locations, they have been expanded by the drawing of a uniform line around their perimeters. Statistics reported in the legend and in Table 10.13 are based on the data used to create the map and are not influenced by this cartographic necessity.

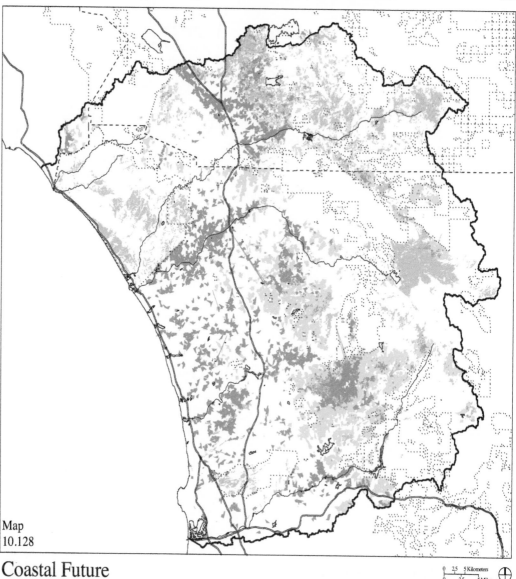

Map
10.128

Coastal Future
1,000,000 New Residents
Western Meadowlark Potential Habitat Change

0 25 5 Kilometers
0 25 5 Miles

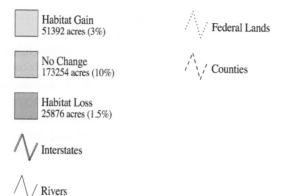

Habitat Gain
51392 acres (3%)

No Change
173254 acres (10%)

Habitat Loss
25876 acres (1.5%)

Interstates

Rivers

Federal Lands

Counties

NOTE: Some impacted areas are too small to be depicted at the scale used in the map above. In order to better show these locations, they have been expanded by the drawing of a uniform line around their perimeters. Statistics reported in the legend and in Table 10.13 are based on the data used to create the map and are not influenced by this cartographic necessity.

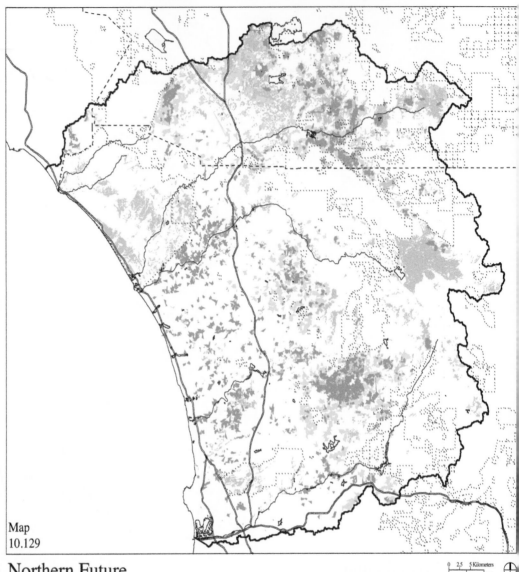

Map
10.129

Northern Future
500,000 New Residents
Western Meadowlark Potential Habitat Change

0   2.5   5 Kilometers
0   2.5   5 Miles

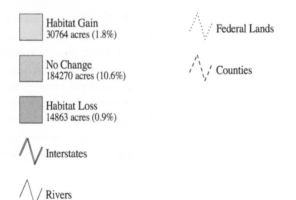

Habitat Gain
30764 acres (1.8%)

No Change
184270 acres (10.6%)

Habitat Loss
14863 acres (0.9%)

Interstates

Rivers

Federal Lands

Counties

NOTE: Some impacted areas are too
small to be depicted at the scale used in
the map above. In order to better show
these locations, they have been expanded
by the drawing of a uniform line around
their perimeters. Statistics reported in the
legend and in Table 10.13 are based on
the data used to create the map and are not
influenced by this cartographic necessity.

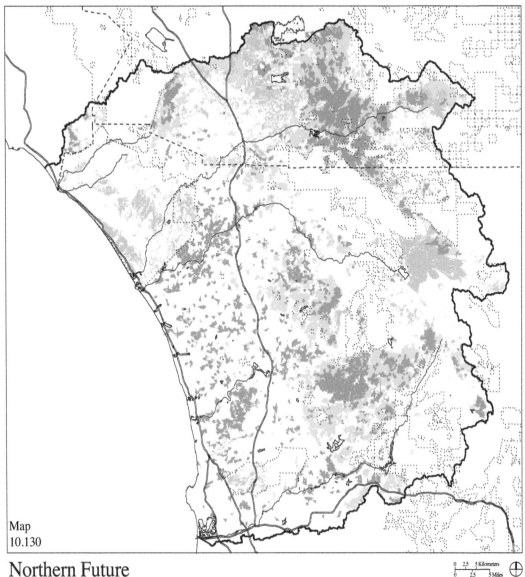

Map
10.130

Northern Future
1,000,000 New Residents
Western Meadowlark Potential Habitat Change

0 2.5 5 Kilometers
0 2.5 5 Miles

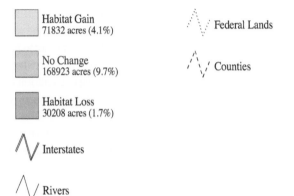

Habitat Gain
71832 acres (4.1%)

No Change
168923 acres (9.7%)

Habitat Loss
30208 acres (1.7%)

Interstates

Rivers

Federal Lands

Counties

NOTE: Some impacted areas are too small to be depicted at the scale used in the map above. In order to better show these locations, they have been expanded by the drawing of a uniform line around their perimeters. Statistics reported in the legend and in Table 10.13 are based on the data used to create the map and are not influenced by this cartographic necessity.

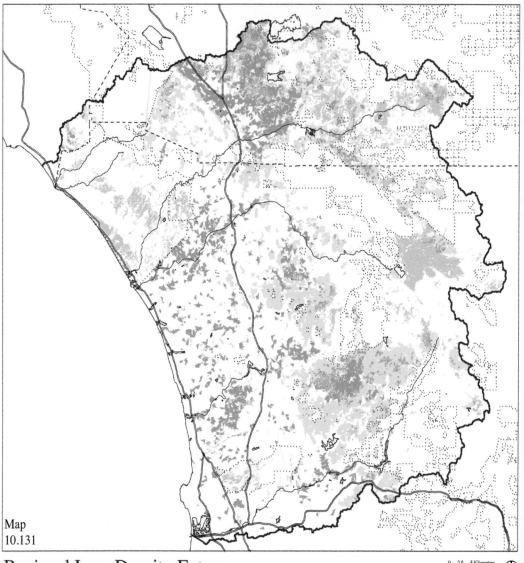

Map
10.131

Regional Low-Density Future
500,000 New Residents
Western Meadowlark Potential Habitat Change

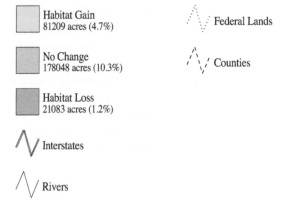

Habitat Gain
81209 acres (4.7%)

No Change
178048 acres (10.3%)

Habitat Loss
21083 acres (1.2%)

Interstates

Rivers

Federal Lands

Counties

NOTE: Some impacted areas are too small to be depicted at the scale used in the map above. In order to better show these locations, they have been expanded by the drawing of a uniform line around their perimeters. Statistics reported in the legend and in Table 10.13 are based on the data used to create the map and are not influenced by this cartographic necessity.

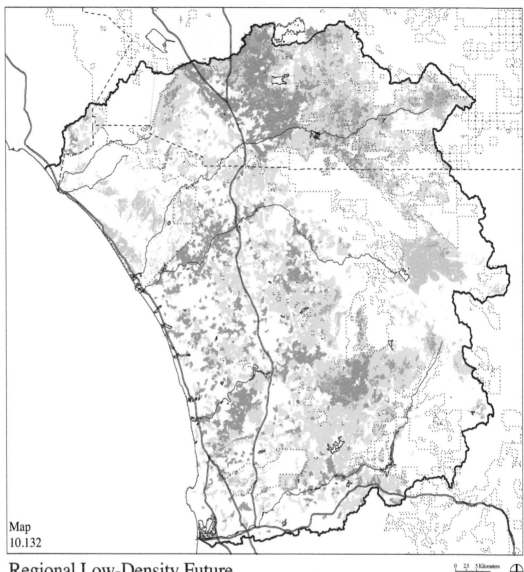

Map
10.132

Regional Low-Density Future
1,000,000 New Residents
Western Meadowlark Potential Habitat Change

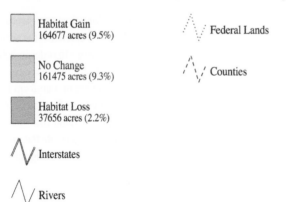

Habitat Gain
164677 acres (9.5%)

No Change
161475 acres (9.3%)

Habitat Loss
37656 acres (2.2%)

Interstates

Rivers

Federal Lands

Counties

NOTE: Some impacted areas are too small to be depicted at the scale used in the map above. In order to better show these locations, they have been expanded by the drawing of a uniform line around their perimeters. Statistics reported in the legend and in Table 10.13 are based on the data used to create the map and are not influenced by this cartographic necessity.

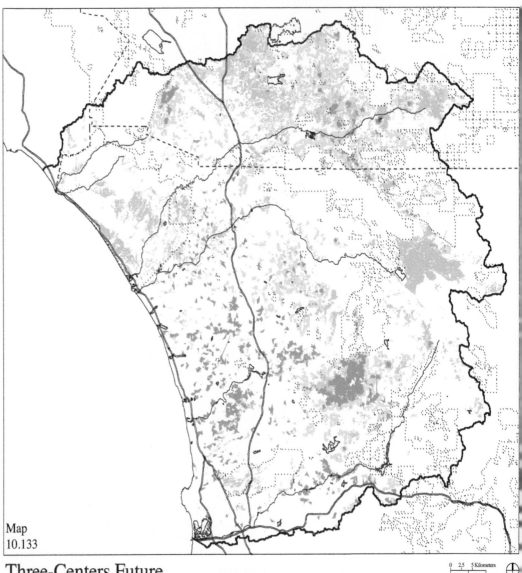

Map
10.133

Three-Centers Future
500,000 New Residents
Western Meadowlark Potential Habitat Change

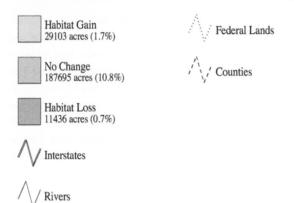

Habitat Gain
29103 acres (1.7%)

No Change
187695 acres (10.8%)

Habitat Loss
11436 acres (0.7%)

Interstates

Rivers

Federal Lands

Counties

NOTE: Some impacted areas are too
small to be depicted at the scale used in
the map above. In order to better show
these locations, they have been expanded
by the drawing of a uniform line around
their perimeters. Statistics reported in the
legend and in Table 10.13 are based on
the data used to create the map and are not
influenced by this cartographic necessity.

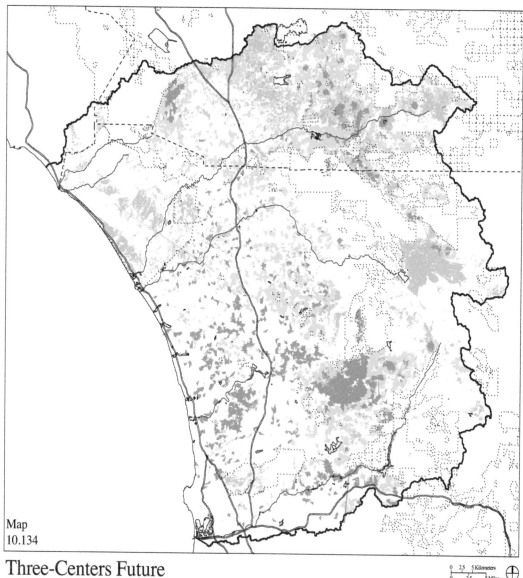

Map
10.134

Three-Centers Future
1,000,000 New Residents
Western Meadowlark Potential Habitat Change

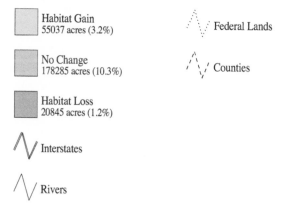

Habitat Gain
55037 acres (3.2%)

No Change
178285 acres (10.3%)

Habitat Loss
20845 acres (1.2%)

Interstates

Rivers

Federal Lands

Counties

0   2.5   5 Kilometers
0   2.5   5 Miles

NOTE: Some impacted areas are too
small to be depicted at the scale used in
the map above. In order to better show
these locations, they have been expanded
by the drawing of a uniform line around
their perimeters. Statistics reported in the
legend and in Table 10.13 are based on
the data used to create the map and are not
influenced by this cartographic necessity.

Great Horned Owl

The great horned owl (*Bubo virginianus*) is one of the largest owls weighing 3.1 pounds with a total length of 22 inches.[104] It has the widest prey base of all the owls and an extensive range from treeline in northern Alaska south to the most southerly latitudes of South America.[105] Colorations vary greatly by region over this large geographic range. The great horned owls of the Southwest are gray with a gray to rusty face. The bodies are broad with a large head. The most distinguishing features are the two stout ear-tufts giving the owl its distinctive "horned" name and a cat-like head shape.[106] The resident subspecies in Southern California is *B. v. pacificus*.[107]

The greatest concern for these species is the preservation of adequate nesting locations. Any loss of access to the big trees required to support this large owl could have localized effects on great horned owl subspecies. With 16 subspecies the threat of localized impacts is real within the land mosaic from which these owls operate.

The great horned owl is considered a habitat generalist, and inhabits a wide variety of conditions. This owl prefers open and secondary growth woodlands, agricultural areas, swamps, and other wetlands, as well as residential parks and urban areas. Preference is for fragmented habitats and areas of mature forest are avoided.[108] Avoidance of mature forest is due to the location of prey species for the owl. Most prey items are located in relatively open areas and the owl prefers to perch on trees, rock outcrops, telephone poles, and structures. As such, the most preferred hunting locations are located along the edge of open habitats with an abundance of tall features.[109]

Nest site selection is extremely variable and great horned owls have the widest range of nest sites for any bird in the Americas.[110] Most commonly, the owl relies on tree nests constructed by other species, but the owl will use cavities in trees and snags, cliff ledges, deserted buildings, and artificial platforms. In southeastern Los Angeles County, six of nine nests were located on narrow ledges of cliffs. The remaining three nests were located in eucalyptus trees (*Eucalyptus globulus*). These nests apparently were constructed by red-tailed hawks (*Buteo*

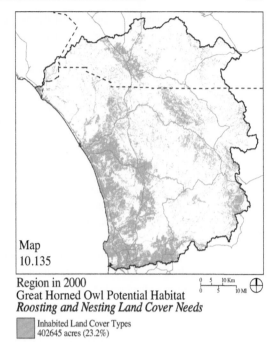

Map
10.135

Region in 2000
Great Horned Owl Potential Habitat
Roosting and Nesting Land Cover Needs

Inhabited Land Cover Types
402645 acres (23.2%)

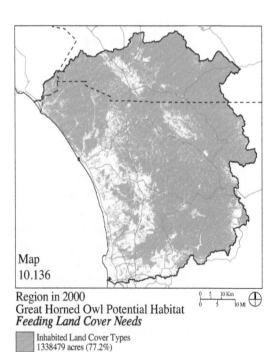

Map
10.136

Region in 2000
Great Horned Owl Potential Habitat
Feeding Land Cover Needs

Inhabited Land Cover Types
1338479 acres (77.2%)

jamancensis).[111] Once a nest site has been selected, eviction by other raptors is not possible.

Great horned owls are highly territorial and mated pairs occupy territories year-round. Territoriality appears to limit the number of breeding pairs and individuals that are prevented from establishing a territory exist as single "floaters" which concentrate on the boundaries of existing territories.[112] With year-round territories, the home range of a great horned owl is anticipated to be the same area as the territory. Home ranges vary from 0.5 to just over 1 square mile with the average home range area being a little more than 0.75 square miles.[113]

Great horned owl potential habitat occurs mainly along the edges of urban development and forested areas within the study region. Locations with large expanses of urban, shrub, agriculture, or grassland land covers do not contain much habitat. Thus, many expansive areas required for the survival of other species examined in this study, such as the gnatcatcher, are not as critical for the great horned owl.

Potential habitat for the great horned owl was modeled by analyzing the landscape matrix of the relationship between possible roosting/nesting sites (Map 10.135) and the location of feeding areas (Map 10.136). Potential roosting and nesting sites were identified from the land cover map extracting the woodland, tree, and urban cover types. Areas of greater than 75% slope, which approximated the location of cliff or steep rocky terrain, were added as potential roosting and nest sites. Feeding locations were identified from the land cover and included grassland, meadow, wetland, coastal sage scrub, chaparral, agriculture, and golf courses. Once potential roosting, nesting, and feeding locations were identified, a moving 0.75 square mile home range window was run across the entire extent of the study area. Those locations that contained a

minimum 10% roosting and nesting area and 10% feeding area were maintained as potential great horned owl habitat. Summarizing, great horned owl habitat was identified as locations with at least 10% roosting/nesting habitat and 10% feeding habitat within an owl's home range.

Existing contidtions and impacts assocated with the futures are summarized in Table 10.14 and shown in Maps 10.137-10.145. All four alternative futures result in an increase in great horned owl potential habitat. Habitat increase with the addition of 500,000 and 1,000,000 people follows a roughly linear trend with a doubling of residents resulting in a like increase in owl habitat. The Coastal Future has the lowest increase with the majority of the increase in habitat occurring in the low-density residential development housing lots. The Three-Centers Future has the largest increase in owl habitat. This result is a function of how housing lots must be positioned around proposed conservation areas and critical habitat locations causing low-density residential housing lots to disperse in the lower elevations. With each house lot representing a potential roosting or nesting location in a matrix of foraging land, dispersing housing across the landscape creates a habitat matrix ideal for owl survival. In the Northern and Regional Low-Density Futures, additional potential habitat is created in a similar fashion; however, the rural residential housing is more contiguous and does not fragment the foraging habitat to the degree of the Three-Centers Future.

The alternative future results illustrate how the great horned owl, which has a large home range and requires a unique edge habitat mosaic, can benefit from dispersed residential development. Future policies that emphasize the preservation of lowland vegetation patches and the creation of low-density residential development favor the creation of great horned owl habitat.

Table 10.14

Great Horned Owl Potential Habitat (in acres)

	2000	Coastal		Northern		Reg. Low-Density		Three-Centers	
		500k	1,000k	500k	1,000k	500k	1,000k	500k	1,000k
Potential Habitat	767,404	785,875	794,076	831,218	875,516	815,749	870,717	835,751	917,913
Change from 2000		18,471	26,672	63,814	108,112	48,345	103,313	68,347	150,509
% Change from 2000		2.4	3.5	8.3	14.1	6.3	13.5	8.9	19.6

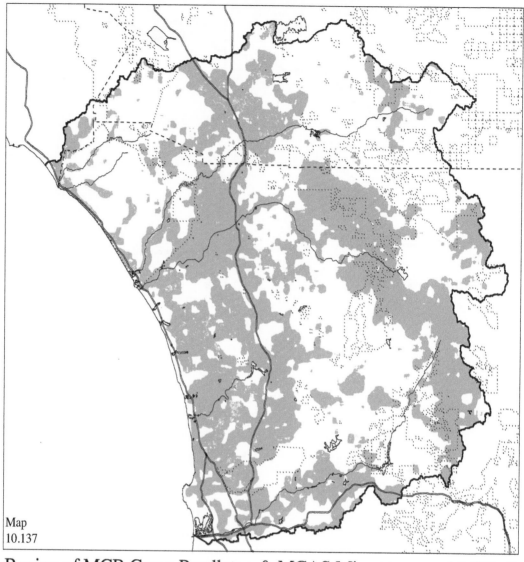

Map
10.137

Region of MCB Camp Pendleton & MCAS Miramar
Existing Conditions 2000
Great Horned Owl Potential Habitat

Potential Habitat
767404 acres (44.3%)

Interstates

Rivers

Federal Lands

Counties

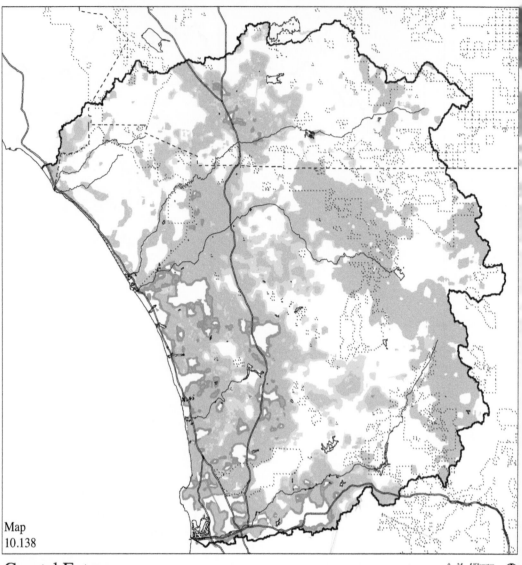

Map
10.138

Coastal Future
500,000 New Residents
Great Horned Owl Potential Habitat Change

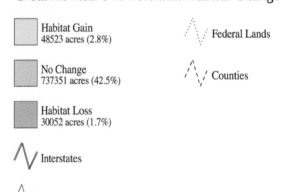

Habitat Gain
48523 acres (2.8%)

No Change
737351 acres (42.5%)

Habitat Loss
30052 acres (1.7%)

Interstates

Rivers

Federal Lands

Counties

0   2.5   5 Kilometers
0   2.5   5 Miles

NOTE: Some impacted areas are too small to be depicted at the scale used in the map above. In order to better show these locations, they have been expanded by the drawing of a uniform line around their perimeters. Statistics reported in the legend and in Table 10.14 are based on the data used to create the map and are not influenced by this cartographic necessity.

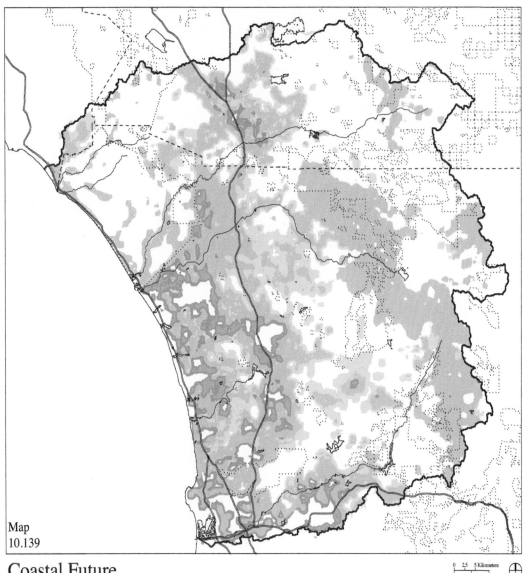

Map
10.139

Coastal Future
1,000,000 New Residents
Great Horned Owl Potential Habitat Change

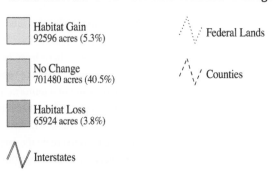

Habitat Gain
92596 acres (5.3%)

No Change
701480 acres (40.5%)

Habitat Loss
65924 acres (3.8%)

Interstates

Rivers

Federal Lands

Counties

NOTE: Some impacted areas are too small to be depicted at the scale used in the map above. In order to better show these locations, they have been expanded by the drawing of a uniform line around their perimeters. Statistics reported in the legend and in Table 10.14 are based on the data used to create the map and are not influenced by this cartographic necessity.

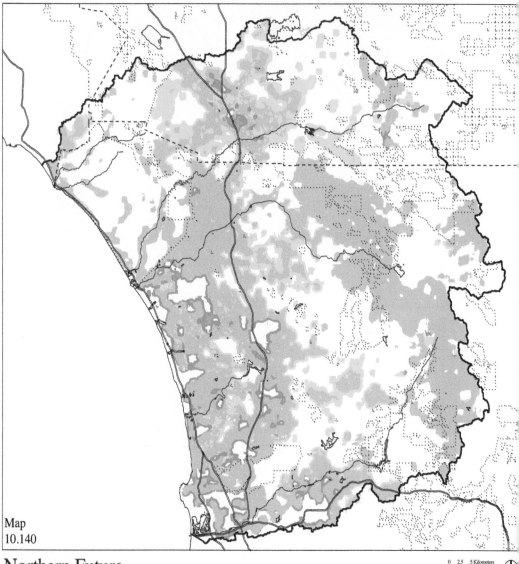

Map
10.140

Northern Future
500,000 New Residents
Great Horned Owl Potential Habitat Change

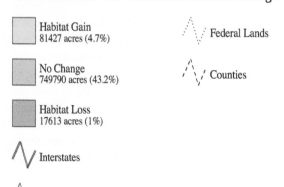

Habitat Gain
81427 acres (4.7%)

No Change
749790 acres (43.2%)

Habitat Loss
17613 acres (1%)

Interstates

Rivers

Federal Lands

Counties

NOTE: Some impacted areas are too
small to be depicted at the scale used in
the map above. In order to better show
these locations, they have been expanded
by the drawing of a uniform line around
their perimeters. Statistics reported in the
legend and in Table 10.14 are based on
the data used to create the map and are not
influenced by this cartographic necessity.

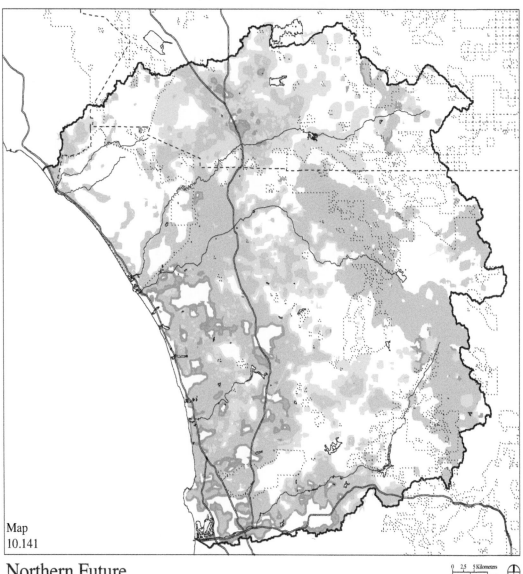

Map
10.141

Northern Future
1,000,000 New Residents
Great Horned Owl Potential Habitat Change

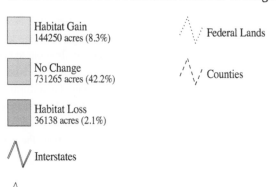

Habitat Gain
144250 acres (8.3%)

No Change
731265 acres (42.2%)

Habitat Loss
36138 acres (2.1%)

Interstates

Rivers

Federal Lands

Counties

NOTE: Some impacted areas are too
small to be depicted at the scale used in
the map above. In order to better show
these locations, they have been expanded
by the drawing of a uniform line around
their perimeters. Statistics reported in the
legend and in Table 10.14 are based on
the data used to create the map and are not
influenced by this cartographic necessity.

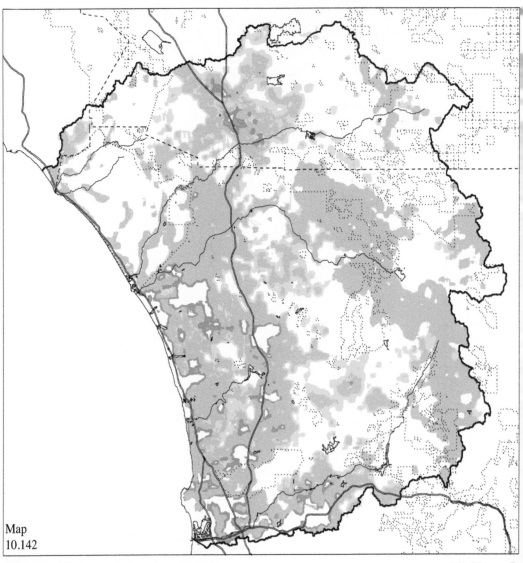

Map
10.142

Regional Low-Density Future
500,000 New Residents
Great Horned Owl Potential Habitat Change

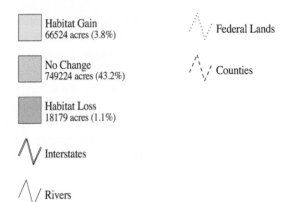

Habitat Gain
66524 acres (3.8%)

No Change
749224 acres (43.2%)

Habitat Loss
18179 acres (1.1%)

Interstates

Rivers

Federal Lands

Counties

NOTE: Some impacted areas are too
small to be depicted at the scale used in
the map above. In order to better show
these locations, they have been expanded
by the drawing of a uniform line around
their perimeters. Statistics reported in the
legend and in Table 10.14 are based on
the data used to create the map and are not
influenced by this cartographic necessity.

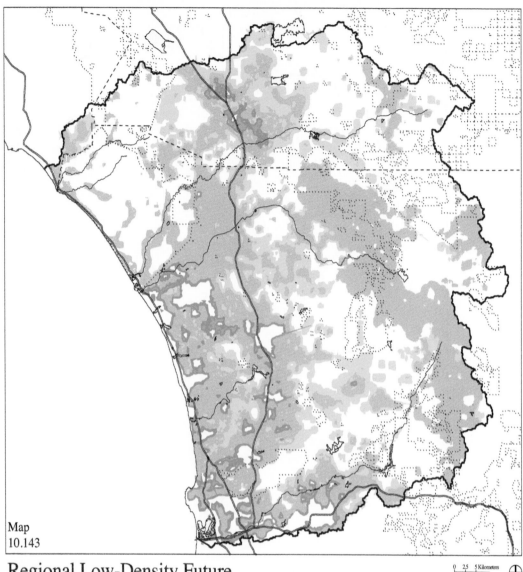

Map
10.143

Regional Low-Density Future
1,000,000 New Residents
Great Horned Owl Potential Habitat Change

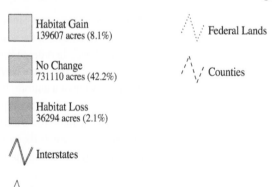

Habitat Gain
139607 acres (8.1%)

No Change
731110 acres (42.2%)

Habitat Loss
36294 acres (2.1%)

Interstates

Rivers

Federal Lands

Counties

NOTE: Some impacted areas are too small to be depicted at the scale used in the map above. In order to better show these locations, they have been expanded by the drawing of a uniform line around their perimeters. Statistics reported in the legend and in Table 10.14 are based on the data used to create the map and are not influenced by this cartographic necessity.

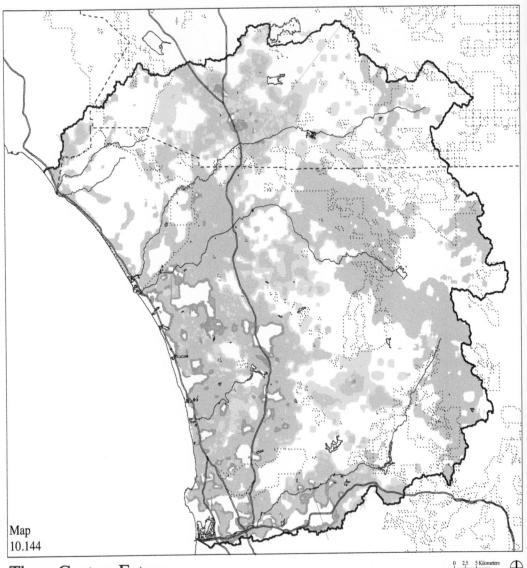

Map
10.144

Three-Centers Future
500,000 New Residents
Great Horned Owl Potential Habitat Change

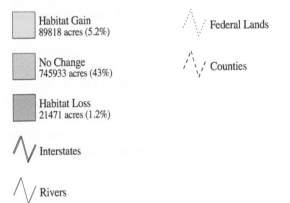

Habitat Gain
89818 acres (5.2%)

No Change
745933 acres (43%)

Habitat Loss
21471 acres (1.2%)

Interstates

Rivers

Federal Lands

Counties

0 2.5 5 Kilometers
0 2.5 5 Miles

NOTE: Some impacted areas are too small to be depicted at the scale used in the map above. In order to better show these locations, they have been expanded by the drawing of a uniform line around their perimeters. Statistics reported in the legend and in Table 10.14 are based on the data used to create the map and are not influenced by this cartographic necessity.

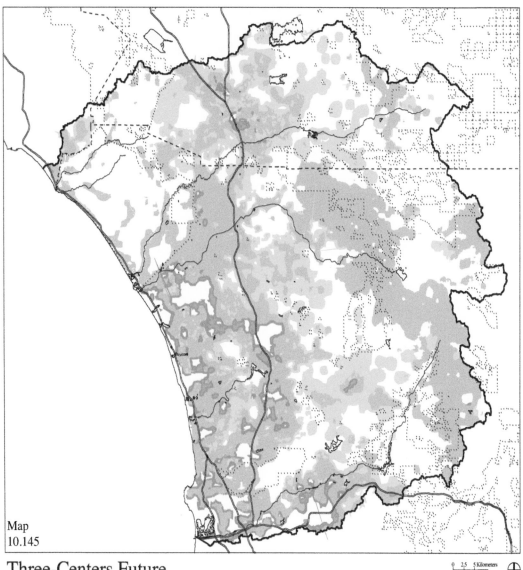

Map
10.145

Three-Centers Future
1,000,000 New Residents
Great Horned Owl Potential Habitat Change

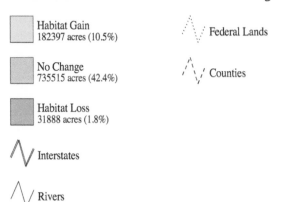

Habitat Gain
182397 acres (10.5%)

No Change
735515 acres (42.4%)

Habitat Loss
31888 acres (1.8%)

Interstates

Rivers

Federal Lands

Counties

NOTE: Some impacted areas are too
small to be depicted at the scale used in
the map above. In order to better show
these locations, they have been expanded
by the drawing of a uniform line around
their perimeters. Statistics reported in the
legend and in Table 10.14 are based on
the data used to create the map and are not
influenced by this cartographic necessity.

Cougar

The cougar (*Felis concolor californicus*) is a resident area sensitive species that requires large contiguous tracts of habitat. The cougar is a large cat with a weight ranging from 80 to 200 pounds. The fur is tawny to gray in coloration and the tip of the tail, ears, and sides of the nose are dark brown.[114]

In California, the cougar faces many threats including poaching, animal damage control activities, loss of habitat, habitat fragmentation, severed travel corridors, high road densities, declining mule deer populations, drought, inbreeding, and feline leukemia.[115] With the cougar ranging over large expanses of land, the need to maintain large contiguous patches of habitat is paramount for its survival.[116] Management options to maintain the genetic diversity and abundance have been proposed and consist of conservation of large tracts of land, maintaining mule deer (*Odocoileus hemionus californicus*), acquiring habitat corridors, and reducing livestock grazing.[117] Many of these proposed management objectives require a significant amount of planning and knowledge about the regional movement of deer and cougars.

Cougars are opportunistic predators; however, their principal food source in California is large ungulates such as black tailed deer and mule deer. Other food sources include rabbits, rodents, lizards, feral pigs, and occasionally livestock.[118] Cougars hunt by stalking, typically approaching to within 50 feet of the prey before attacking.[119] Stalking requires that adequate cover be present to conceal the cougar's approach. General features of good stalking habitat include rock outcrops, dense brush, and riparian areas. Since deer are a primary food source, there is a strong correlation between the overall habitat needs of mule deer and those of the cougar.

There are four classes of cougars within a population: resident males, resident females, dependent kittens, and transients. Home ranges include hunting areas, water resources, resting areas, lookouts, and denning sites.[120] Home ranges for males are larger than home ranges for females and are only maintained by resident animals. Male home ranges vary from 25 to 500 square miles and female home ranges vary from 8 to 400 square miles.[121]

Transient cougars will disperse over long distances to establish a territory.[122]

Mule deer inhabit a wide range of vegetation cover and most vegetation types in the region provide potential habitat. Thus, the cougar's potential habitat model focused on the need for stalking cover over their home range. Stalking cover was identified as a land cover dominated by shrubs, woodlands, or trees (Map 10.146). Stalking cover patches that did not meet the minimum area requirement for the home range of a single female were eliminated as potential habitat. Thus, the potential habitat for the cougar consisted of contiguous stalking habitat at least 8 square miles (or 5,120 acres).

The majority of potential cougar habitat exists within federal lands. The Cleveland National Forest contains a considerable amount of cougar habitat within its jurisdiction, an observation that underscores the significance of montane terrain and vegetation. MCB Camp Pendleton and MCAS Miramar provide low elevation cougar habitat close to the coast. These locations are the only large areas possessing these unique habitat qualities. Natural vegetation that is fragmented by anthropogenic land uses is generally avoided by cougars.

Existing conditions and impacts associated with the futures are summarized in Table 10.15 and shown in Maps 10.147-10.155. All the alternative futures reduce the amount of potential habitat for the cougar. The relative difference between the addition of 500,000 or 1,000,000 people is small; however, every future has slightly more than twice as much potential habitat loss with a doubling of additional people. The nonlinear loss is attributed to fragmentation and trimming of isolated patches of cougar habitat. With the addition of 500,000 people, some of the patches that are at least 8 square miles in area are trimmed along the edges and subsequently eliminated as potential habitat. This situation is exacerbated with the addition of 1,000,000 people.

The Coastal Future results in the least amount of habitat loss and the Regional Low-Density Future has the greatest loss. The Northern Future's habitat loss approximates the loss occurring under the Regional Low-Density Future, and the Three-Centers Future approximates the loss occurring under the Coastal Future. These results indicate that low density rural residential housing is a major

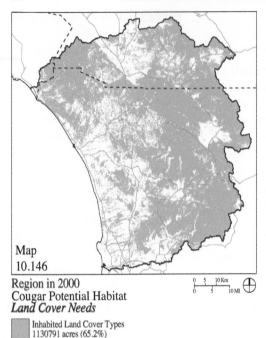

Map
10.146

Region in 2000
Cougar Potential Habitat
Land Cover Needs

0 5   10 Km
0 5   10 MI

Inhabited Land Cover Types
1130791 acres (65.2%)

threat to cougar habitat throughout the study region. Subtle differences in the location of loss occur for each of the futures. The Regional Low-Density Future results in more loss in San Diego County and increases the relative isolation of habitat occurring on MCAS Miramar. The Northern Future has more loss in Western Riverside County creating a potential problem of movement of cougars from the northern to southern locations in the study area. Movement just to the east of MCB Camp Pendleton would become more difficult for the cougar under the Northern Future. The Coastal and Three-Centers Futures isolate Miramar and reduce the potential north to south movement of cougars.

From the alternative future results, two broad situations become a concern. MCAS Miramar has the potential to become more isolated, and movement of cougars between the north and the south is threatened. Any future that results in development relatively close to the coast in the north threatens movement potential, and any development close to the coast in the south threatens to isolate MCAS Miramar. These factors should be considered when planning for cougar populations in Southern California.

Table 10.15

Cougar Potential Habitat (in acres)

	2000	Coastal		Northern		Reg. Low-Density		Three-Centers	
		500k	1,000k	500k	1,000k	500k	1,000k	500k	1,000k
Potential Habitat	980,933	940,750	894,228	888,964	763,720	872,508	738,674	923,144	854,205
Change from 2000		-40,183	-86,705	-91,969	-217,213	-108,425	-242,259	-57,789	-126,728
% Change from 2000		**-4.1**	**-8.8**	**-9.4**	**-22.1**	**-11.1**	**-24.7**	**-5.9**	**-12.9**

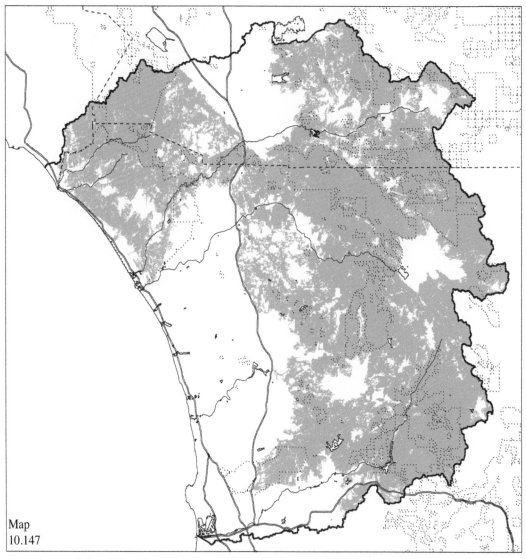

Map
10.147

Region of MCB Camp Pendleton & MCAS Miramar
Existing Conditions 2000
Cougar Potential Habitat

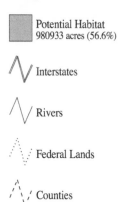

Potential Habitat
980933 acres (56.6%)

Interstates

Rivers

Federal Lands

Counties

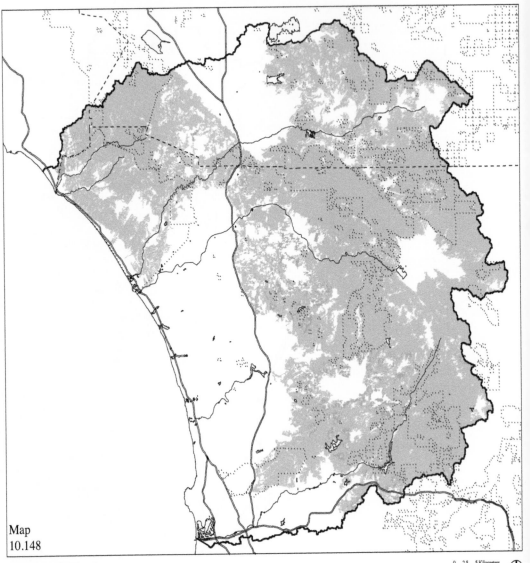

Map
10.148

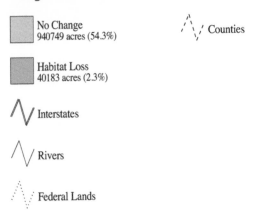

Coastal Future
500,000 New Residents
Cougar Potential Habitat Change

No Change
940749 acres (54.3%)

Habitat Loss
40183 acres (2.3%)

Interstates

Rivers

Federal Lands

Counties

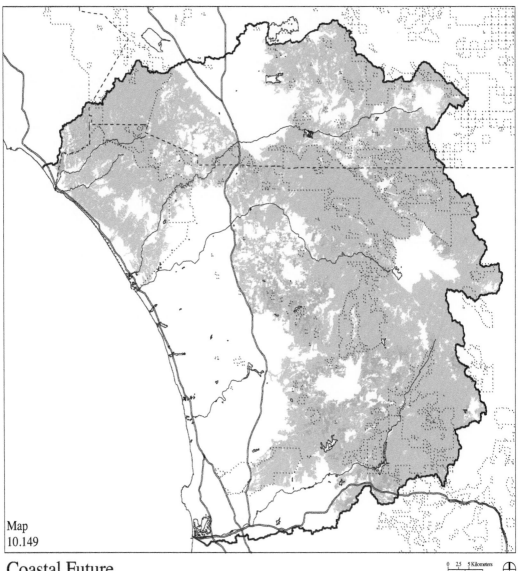

Map
10.149

Coastal Future
1,000,000 New Residents
Cougar Potential Habitat Change

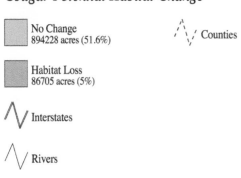

No Change
894228 acres (51.6%)

Habitat Loss
86705 acres (5%)

Interstates

Rivers

Federal Lands

Counties

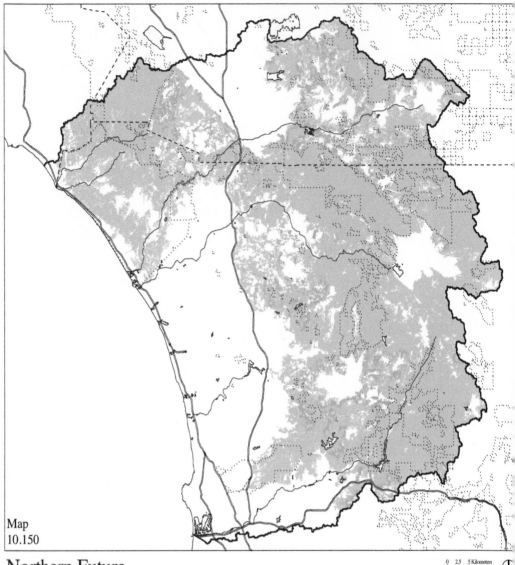

Map
10.150

Northern Future
500,000 New Residents
Cougar Potential Habitat Change

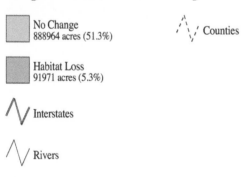

No Change
888964 acres (51.3%)

Habitat Loss
91971 acres (5.3%)

Interstates

Rivers

Federal Lands

Counties

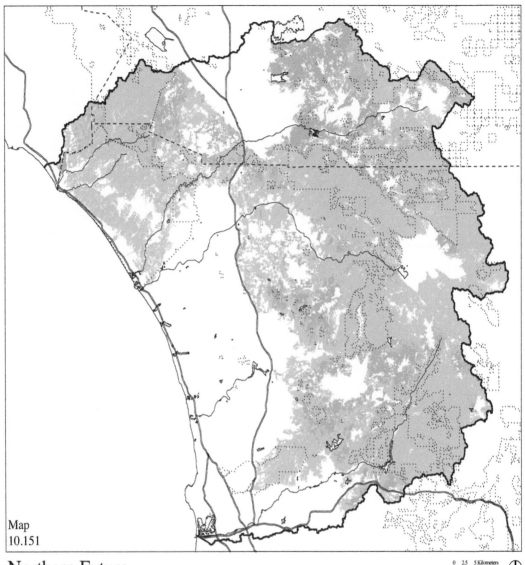

Map
10.151

Northern Future
1,000,000 New Residents
Cougar Potential Habitat Change

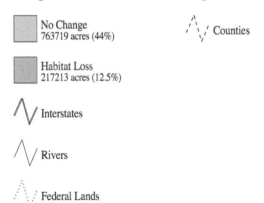

No Change
763719 acres (44%)

Habitat Loss
217213 acres (12.5%)

Interstates

Rivers

Federal Lands

Counties

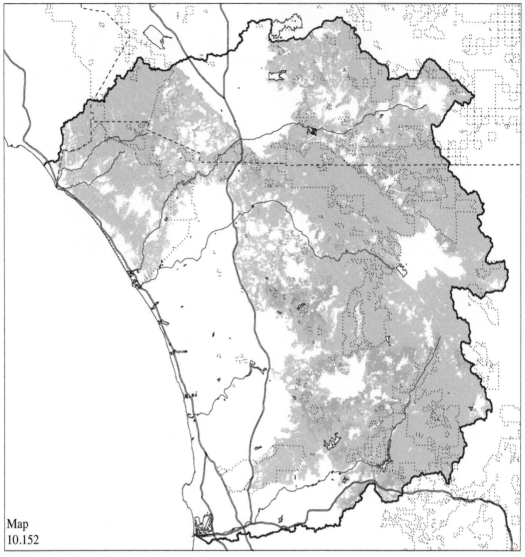

Map
10.152

Regional Low-Density Future
500,000 New Residents
Cougar Potential Habitat Change

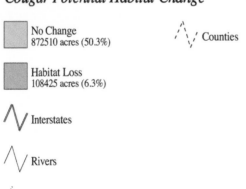

No Change
872510 acres (50.3%)

Habitat Loss
108425 acres (6.3%)

Interstates

Rivers

Federal Lands

Counties

0   2.5   5 Kilometers
0   2.5   5 Miles

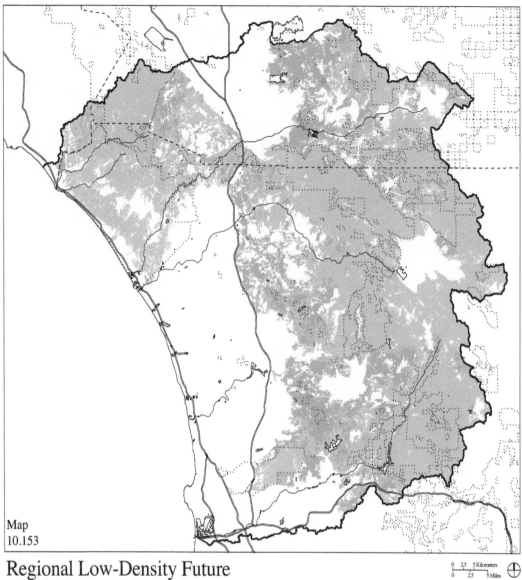

Map
10.153

Regional Low-Density Future
1,000,000 New Residents
Cougar Potential Habitat Change

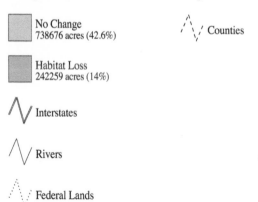

No Change
738676 acres (42.6%)

Habitat Loss
242259 acres (14%)

Interstates

Rivers

Federal Lands

Counties

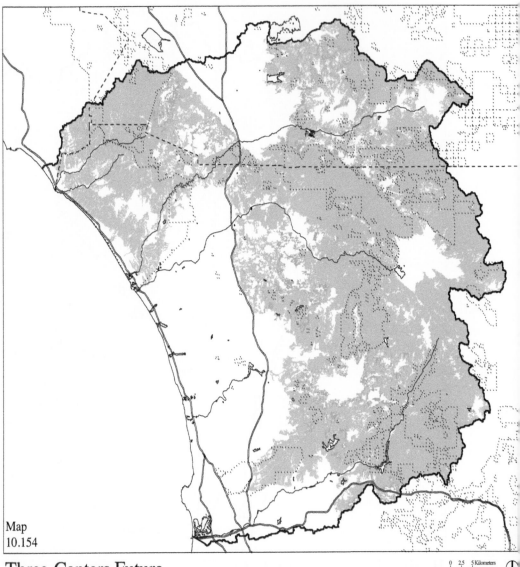

Map
10.154

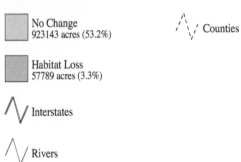

Three-Centers Future
500,000 New Residents
Cougar Potential Habitat Change

No Change
923143 acres (53.2%)

Habitat Loss
57789 acres (3.3%)

Counties

Interstates

Rivers

Federal Lands

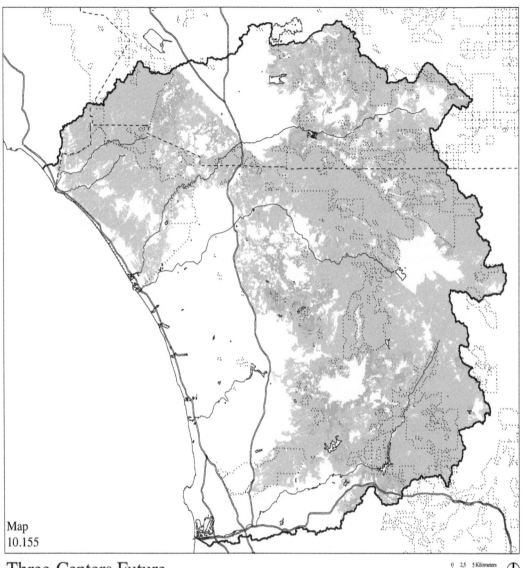

Map
10.155

Three-Centers Future
1,000,000 New Residents
Cougar Potential Habitat Change

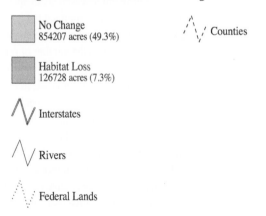

No Change
854207 acres (49.3%)

Habitat Loss
126728 acres (7.3%)

Interstates

Rivers

Federal Lands

Counties

Biological Consequences Associated with the Alternative Futures: Species Richness

Scott D. Bassett

Species richness assessments examine the distribution of a set of species to determine the total overall number of individual species residing at any location.[123] Generally, the assessment technique does not model in-depth the habitat characteristics of any individual species, but rather attempts to make use of generalized information to adequately assess the biodiversity of the region.[124] The approach carries the assumption that the overall condition of a region's biodiversity may be determined through an accounting of the total number of individual species inhabiting an area. Decreases or increases in richness levels indicate trends in biodiversity health.

Species richness assessments of terrestrial vertebrates rely on simple spatial information that is linked to wildlife habitat relationship (WHR) models to determine the potential distribution of each individual species on a species-by-species basis.[125] Linkages between spatial information and habitat requirements may be as elementary as the presence or absence of species for each land cover present in a region. More complex assessments utilize a combination of land cover and other ancillary spatial information to derive potential species distributions.[126] Ancillary spatial information may include elevation, slope, aspect, distance to water, type of water (salt or fresh), distance to urban structures, and a host of other factors. The inclusion or exclusion of ancillary spatial information is dependent on its availability and may also be limited by available computing resources. These limitations dictate that information may be ignored to allow for the mapping of potential species distributions within an adequate time frame for decision-making.

In this study, potential distribution maps used for calculating species richness were created using a combination of land cover, elevation, and salt water input layers. Land cover emphasized the simple presence or absence of a species within each individual land cover type. Elevation ranges were applied to species that inhabited specific elevation zones. Ocean water attributes were applied in two different ways. Some species potential distribution maps were generated utilizing a simple presence or absence in ocean water. Other species showed an affinity to inhabit terrestrial land cover near the ocean. For example, a species potential distribution could include the land cover types: beach, coastal sage scrub, and barren within 1 mile of the ocean and below 1,000 feet. The potential distribution maps were species specific and inclusion or exclusion of any factor was done based on the habitat requirements of the individual species.

A total of 410 species potential distributions were modeled for the study region. This figure includes 16 amphibians, 43 reptiles, 286 birds, and 65 mammals. Of the total 410 species, 23 consisted of species living exclusively in the ocean or on islands off the California coast. Thus, only 387 species were utilized for calculations of species richness. Of these 387 species, 16 were non-native.

Two different assessments of species richness were made. The first summed the number of native species predicted to inhabit a given piece of land. The second summed the number of non-native species predicted to inhabit a given piece of land. This approach allowed for an accurate accounting of the non-native influence in species richness calculations. Although the number of non-native species is small, their inclusion in a total species richness map would show positive elevated effects in biodiversity for urban regions even though many natural resources managers would support eradication of non-native species.

The native species richness for the study area ranges from a low of 36 species to a high of 253 species. The lowest species richness occurs on areas of barren land mostly devoid of vegetation. Urban locations have a species richness value of 57. The highest value occurs within riparian vegetation relatively close to coastal areas. Many of the locations with high richness values occur in chaparral vegetation and riparian areas.

The introduced species richness for the study area ranges from a low of 1 species to a high of 14 species. The low occurs in areas of barren land and the high occurs in riparian areas. Urban locations have a total introduced richness value of 9 species. The urban locations have a relatively high number of non-native species as compared with native species.

Table 10.16 summarizes native species richness change. Table 10.17 presents the number of acres that gain or lose native and non-native species richness. Maps 10.156-10.173 show species richness distribution for current conditions and in each of the futures. The alternative futures result in native species richness declines in a number of areas and in increases in non-native species richness. The Northern Future results in the largest area decrease in native species richness. The Northern and Regional Low-Density Futures cause the greatest increase in area of non-native species richness indicating that differences in vegetation can alter the overall terrestrial vertebrate species assemblages. Also the results show the effect low-density housing has on the expansion of non-native species.

Table 10.16

Native Species Richness Change (in acres)

	2000	Coastal		Northern		Reg. Low-Density		Three-Centers	
		500k	1,000k	500k	1,000k	500k	1,000k	500k	1,000k
36 to 60 Species	279473	322,292	364,953	322,140	364,784	321,849	363,958	322,440	364,947
Change from 2000		42,811	85,472	42,659	85,303	42,368	84,477	42,959	85,466
% Change from 2000		**15.3**	**30.6**	**15.3**	**30.5**	**15.2**	**30.2**	**15.4**	**30.6**
61 to 160 Species	144,151	130,625	118,678	163,403	188,347	132,498	124,456	139,298	137,625
Change from 2000		-13,529	-25,476	19,249	44,193	-11,656	-19,698	-4,856	-6,529
% Change from 2000		**-9.4**	**-17.7**	**13.4**	**30.7**	**-8.1**	**-13.7**	**-3.4**	**-4.5**
161 to 190 Species	471,837	468,126	468,647	476,049	491,391	519,645	539,496	476,849	483,786
Change from 2000		-3,725	-3,203	4,198	19,541	47,794	67,645	4,998	11,936
% Change from 2000		**-0.8**	**-0.7**	**0.9**	**4.1**	**10.1**	**14.3**	**1.1**	**2.5**
191 to 210 Species	746,109	720,587	689,387	680,239	598,263	667,689	613,826	703,095	655,408
Change from 2000		-25,542	-56,742	-65,890	-147,866	-78,441	-132,303	-43,034	-90,721
% Change from 2000		**-3.4**	**-7.6**	**-8.8**	**-19.8**	**-10.5**	**-17.7**	**-5.8**	**-12.2**
211 to 253 Species	92,477	92,464	92,428	92,263	91,309	92,413	92,358	92,413	92,328
Change from 2000		-15	-51	-216	-1,170	-66	-121	-66	-151
% Change from 2000		**>-0.1**	**-0.1**	**-0.2**	**-1.3**	**-0.1**	**-0.1**	**-0.1**	**-0.2**

Table 10.17

Net Species Richness Change (in acres)

	Coastal		Northern		Reg. Low-Density		Three-Centers	
	500k	1,000k	500k	1,000k	500k	1,000k	500k	1,000k
Area Gain in Native Species Richness From 2000	4,782	9,839	8,682	14,304	15,482	59,536	5,432	9,352
Area Loss in Native Species Richness From 2000	-63,902	-132,542	-108,281	-229,478	-112,869	-200,071	-70,758	-137,909
Area Gain in Non-native Species Richness From 2000	58,276	121,966	110,354	231,783	119,727	244,447	76,191	147,261
Area Loss in Non-native Species Richness From 2000	-10,427	-20,416	-6,445	-11,807	-8,623	-15,161	0	0

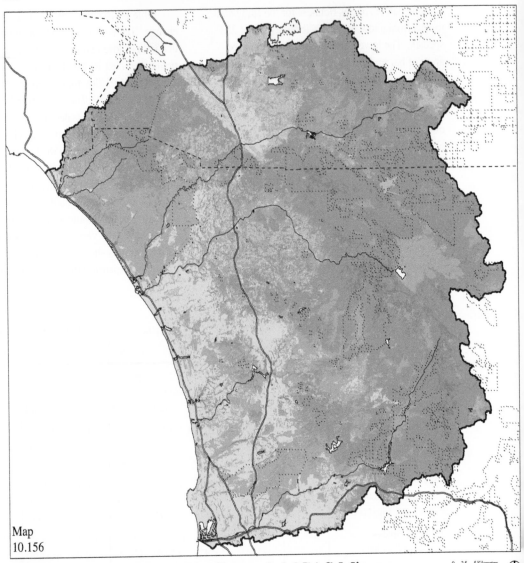

Map
10.156

Region of MCB Camp Pendleton & MCAS Miramar
Existing Conditions 2000
Native Species Richness

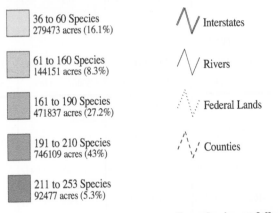

36 to 60 Species
279473 acres (16.1%)

61 to 160 Species
144151 acres (8.3%)

161 to 190 Species
471837 acres (27.2%)

191 to 210 Species
746109 acres (43%)

211 to 253 Species
92477 acres (5.3%)

Interstates

Rivers

Federal Lands

Counties

— See color insert following page 204 —

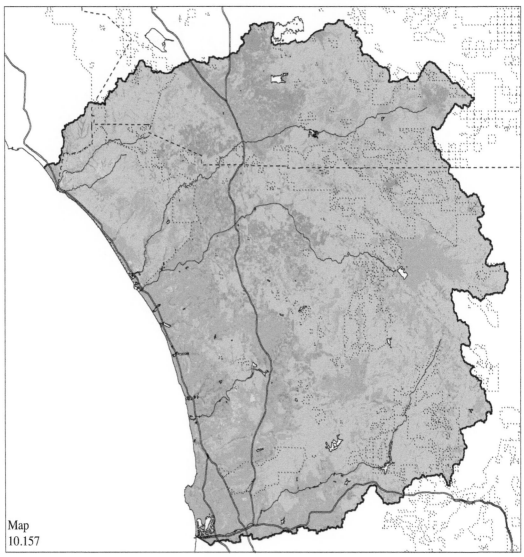

Map
10.157

Region of MCB Camp Pendleton & MCAS Miramar
Existing Conditions 2000
Non-Native Species Richness

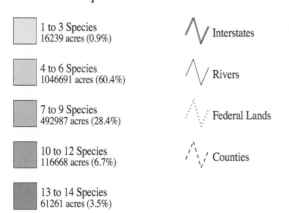

1 to 3 Species
16239 acres (0.9%)

4 to 6 Species
1046691 acres (60.4%)

7 to 9 Species
492987 acres (28.4%)

10 to 12 Species
116668 acres (6.7%)

13 to 14 Species
61261 acres (3.5%)

Interstates

Rivers

Federal Lands

Counties

0 2.5 5 Kilometers
0 2.5 5 Miles

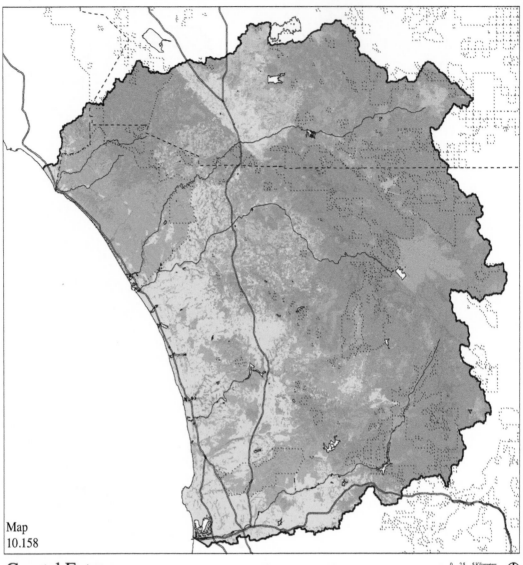

Map
10.158

Coastal Future
500,000 New Residents
Native Species Richness

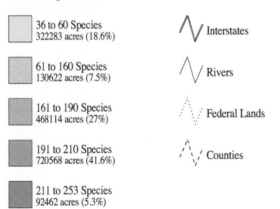

36 to 60 Species
322283 acres (18.6%)

61 to 160 Species
130622 acres (7.5%)

161 to 190 Species
468114 acres (27%)

191 to 210 Species
720568 acres (41.6%)

211 to 253 Species
92462 acres (5.3%)

Interstates

Rivers

Federal Lands

Counties

0   2.5   5 Kilometers
0   2.5   5 Miles

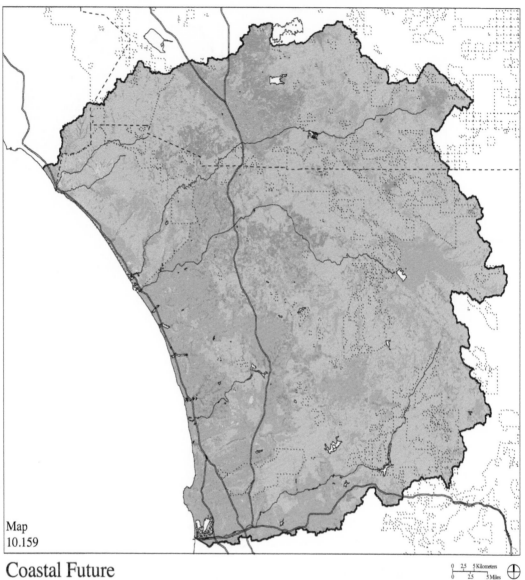

Map
10.159

Coastal Future
500,000 New Residents
Non-Native Species Richness

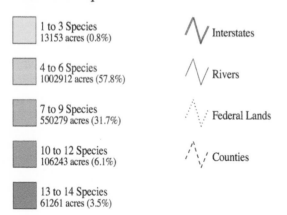

1 to 3 Species
13153 acres (0.8%)

4 to 6 Species
1002912 acres (57.8%)

7 to 9 Species
550279 acres (31.7%)

10 to 12 Species
106243 acres (6.1%)

13 to 14 Species
61261 acres (3.5%)

Interstates

Rivers

Federal Lands

Counties

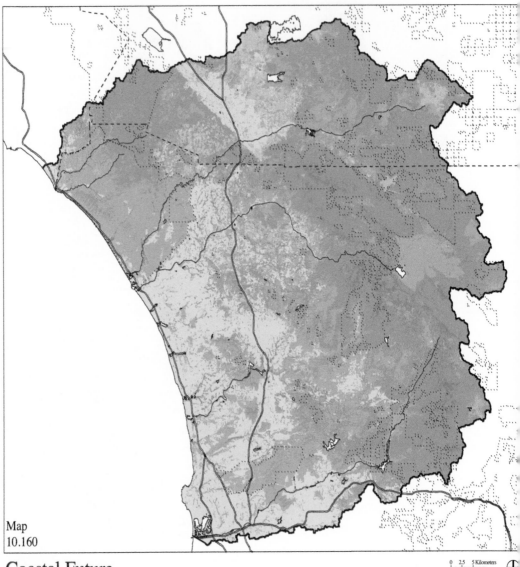

Map
10.160

Coastal Future
1,000,000 New Residents
Native Species Richness

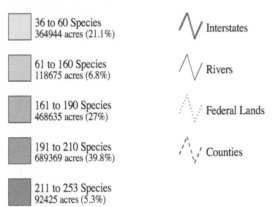

0   2.5   5 Kilometers
0   2.5   5 Miles

36 to 60 Species
364944 acres (21.1%)

61 to 160 Species
118675 acres (6.8%)

161 to 190 Species
468635 acres (27%)

191 to 210 Species
689369 acres (39.8%)

211 to 253 Species
92425 acres (5.3%)

Interstates

Rivers

Federal Lands

Counties

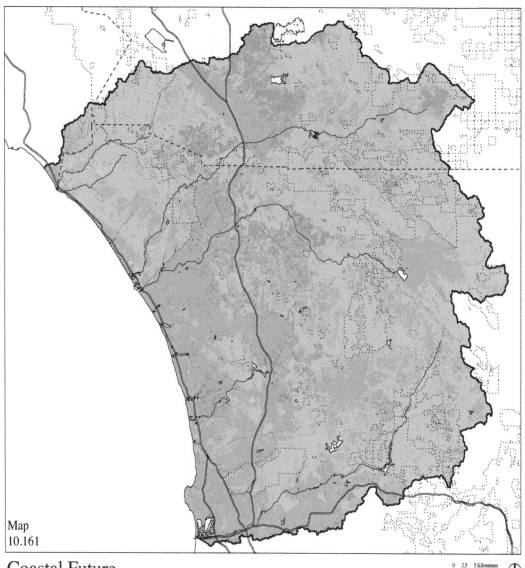

Map
10.161

Coastal Future
1,000,000 New Residents
Non-Native Species Richness

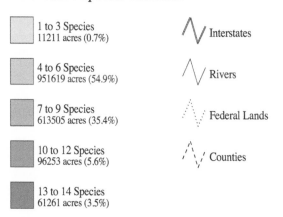

1 to 3 Species
11211 acres (0.7%)

4 to 6 Species
951619 acres (54.9%)

7 to 9 Species
613505 acres (35.4%)

10 to 12 Species
96253 acres (5.6%)

13 to 14 Species
61261 acres (3.5%)

Interstates

Rivers

Federal Lands

Counties

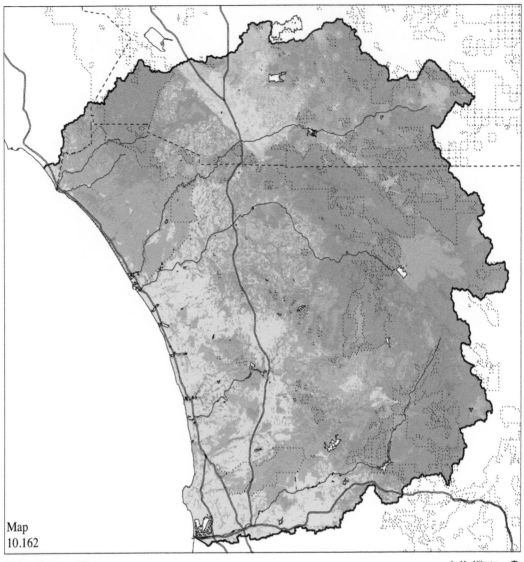

Map
10.162

Northern Future
500,000 New Residents
Native Species Richness

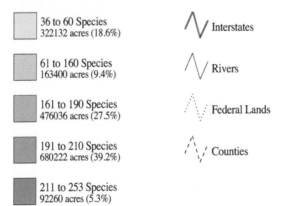

36 to 60 Species 322132 acres (18.6%)		Interstates
61 to 160 Species 163400 acres (9.4%)		Rivers
161 to 190 Species 476036 acres (27.5%)		Federal Lands
191 to 210 Species 680222 acres (39.2%)		Counties
211 to 253 Species 92260 acres (5.3%)		

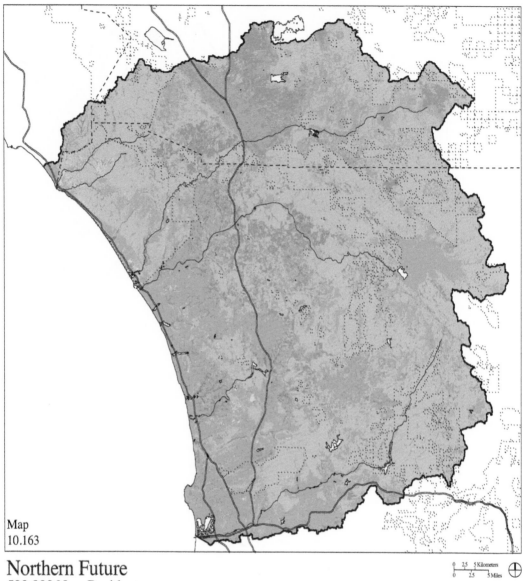

Map
10.163

Northern Future
500,000 New Residents
Non-Native Species Richness

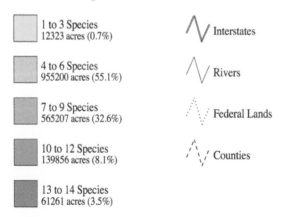

1 to 3 Species
12323 acres (0.7%)

4 to 6 Species
955200 acres (55.1%)

7 to 9 Species
565207 acres (32.6%)

10 to 12 Species
139856 acres (8.1%)

13 to 14 Species
61261 acres (3.5%)

Interstates

Rivers

Federal Lands

Counties

0   2.5   5 Kilometers
0   2.5   5 Miles

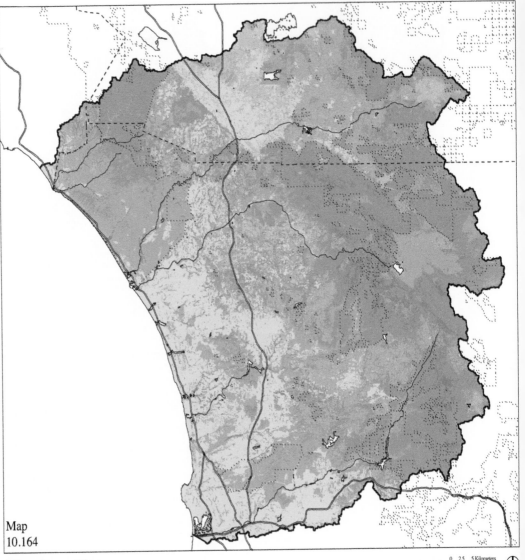

Map
10.164

Northern Future
1,000,000 New Residents
Native Species Richness

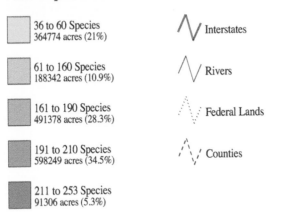

36 to 60 Species
364774 acres (21%)

61 to 160 Species
188342 acres (10.9%)

161 to 190 Species
491378 acres (28.3%)

191 to 210 Species
598249 acres (34.5%)

211 to 253 Species
91306 acres (5.3%)

Interstates

Rivers

Federal Lands

Counties

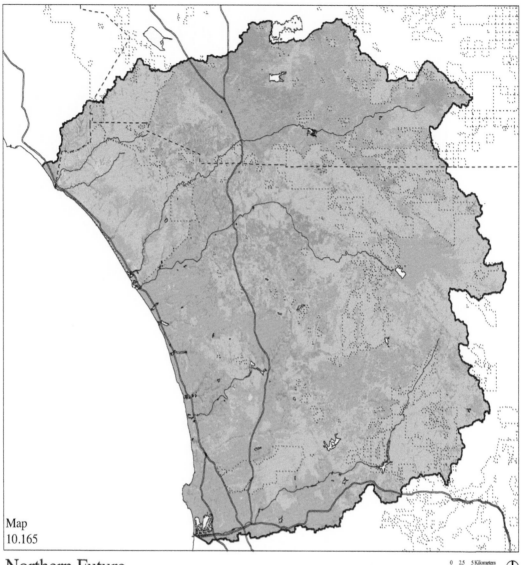

Map
10.165

Northern Future
1,000,000 New Residents
Non-Native Species Richness

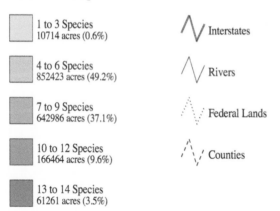

1 to 3 Species
10714 acres (0.6%)

4 to 6 Species
852423 acres (49.2%)

7 to 9 Species
642986 acres (37.1%)

10 to 12 Species
166464 acres (9.6%)

13 to 14 Species
61261 acres (3.5%)

Interstates

Rivers

Federal Lands

Counties

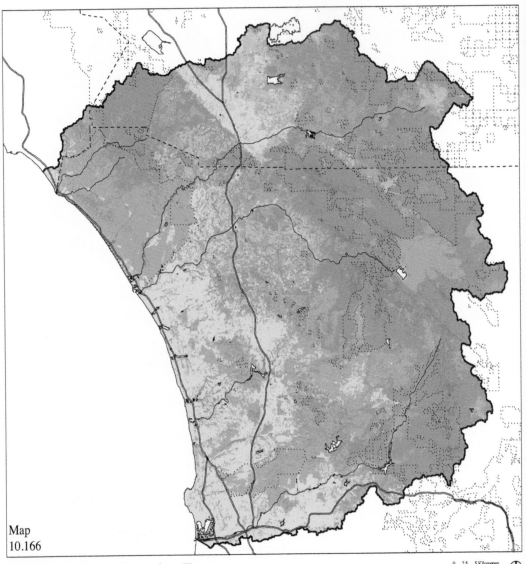

Map
10.166

Regional Low-Density Future
500,000 New Residents
Native Species Richness

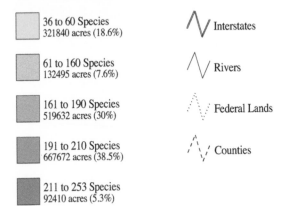

36 to 60 Species
321840 acres (18.6%)

61 to 160 Species
132495 acres (7.6%)

161 to 190 Species
519632 acres (30%)

191 to 210 Species
667672 acres (38.5%)

211 to 253 Species
92410 acres (5.3%)

Interstates

Rivers

Federal Lands

Counties

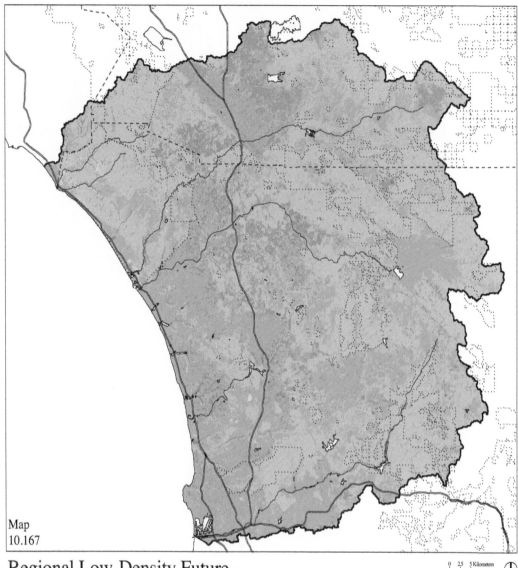

Map
10.167

Regional Low-Density Future
500,000 New Residents
Non-Native Species Richness

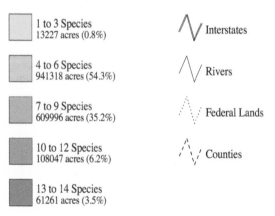

1 to 3 Species
13227 acres (0.8%)

4 to 6 Species
941318 acres (54.3%)

7 to 9 Species
609996 acres (35.2%)

10 to 12 Species
108047 acres (6.2%)

13 to 14 Species
61261 acres (3.5%)

Interstates

Rivers

Federal Lands

Counties

0 2.5 5 Kilometers
0 2.5 5 Miles

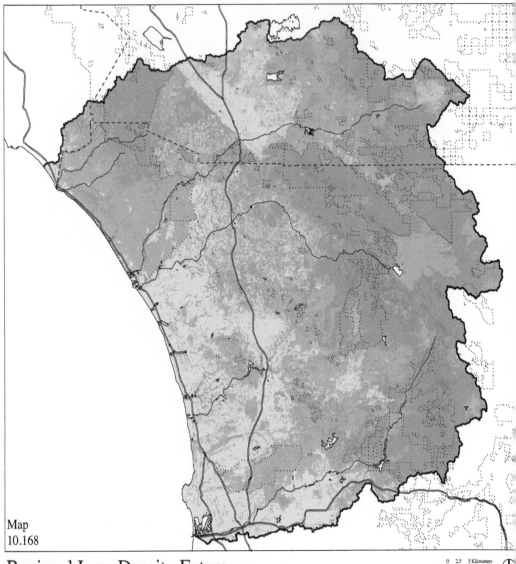

Map
10.168

Regional Low-Density Future
1,000,000 New Residents
Native Species Richness

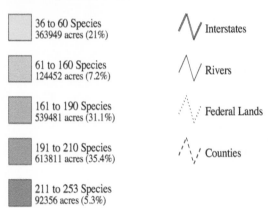

36 to 60 Species
363949 acres (21%)

61 to 160 Species
124452 acres (7.2%)

161 to 190 Species
539481 acres (31.1%)

191 to 210 Species
613811 acres (35.4%)

211 to 253 Species
92356 acres (5.3%)

Interstates

Rivers

Federal Lands

Counties

0   2.5   5 Kilometers
0   2.5   5 Miles

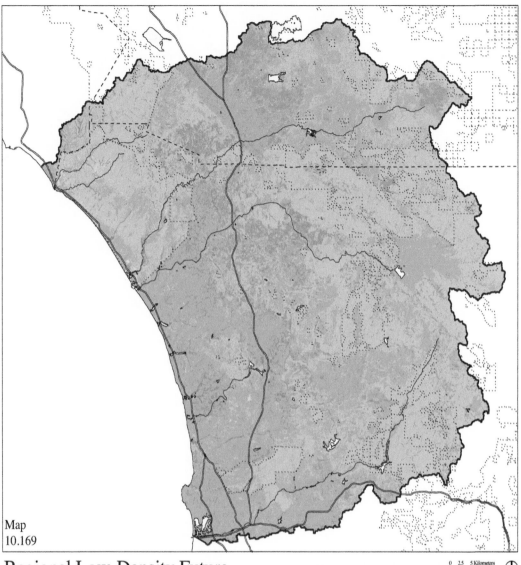

Map
10.169

Regional Low-Density Future
1,000,000 New Residents
Non-Native Species Richness

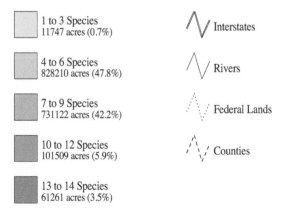

0   2.5   5 Kilometers
0   2.5   5 Miles

1 to 3 Species
11747 acres (0.7%)

4 to 6 Species
828210 acres (47.8%)

7 to 9 Species
731122 acres (42.2%)

10 to 12 Species
101509 acres (5.9%)

13 to 14 Species
61261 acres (3.5%)

Interstates

Rivers

Federal Lands

Counties

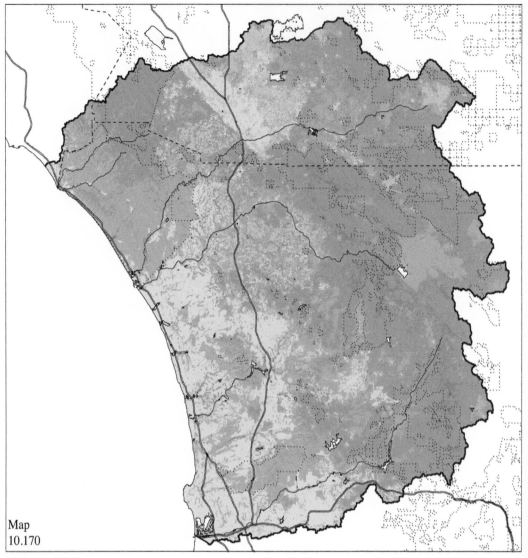

Map
10.170

Three-Centers Future
500,000 New Residents
Native Species Richness

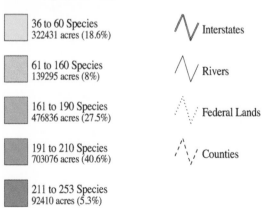

36 to 60 Species
322431 acres (18.6%)

61 to 160 Species
139295 acres (8%)

161 to 190 Species
476836 acres (27.5%)

191 to 210 Species
703076 acres (40.6%)

211 to 253 Species
92410 acres (5.3%)

Interstates

Rivers

Federal Lands

Counties

0   2.5   5 Kilometers
0   2.5   5 Miles

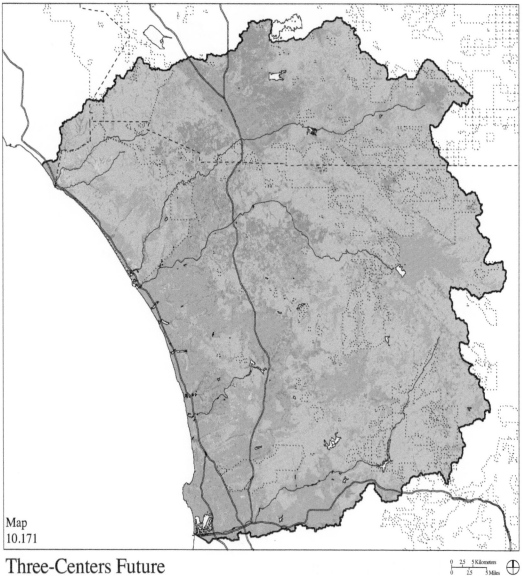

Map
10.171

Three-Centers Future
500,000 New Residents
Non-Native Species Richness

0   2.5   5 Kilometers
0   2.5   5 Miles

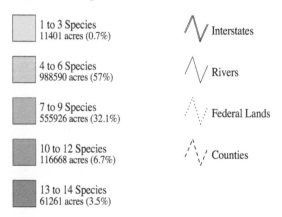

1 to 3 Species
11401 acres (0.7%)

4 to 6 Species
988590 acres (57%)

7 to 9 Species
555926 acres (32.1%)

10 to 12 Species
116668 acres (6.7%)

13 to 14 Species
61261 acres (3.5%)

Interstates

Rivers

Federal Lands

Counties

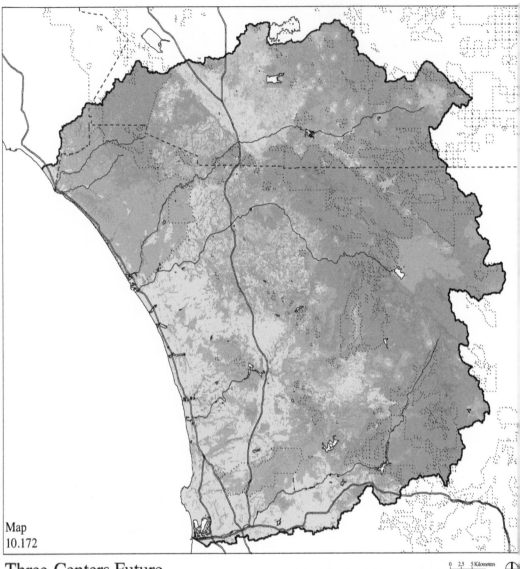

Map
10.172

Three-Centers Future
1,000,000 New Residents
Native Species Richness

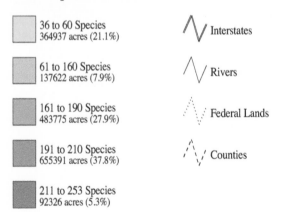

36 to 60 Species 364937 acres (21.1%)	Interstates
61 to 160 Species 137622 acres (7.9%)	Rivers
161 to 190 Species 483775 acres (27.9%)	Federal Lands
191 to 210 Species 655391 acres (37.8%)	Counties
211 to 253 Species 92326 acres (5.3%)	

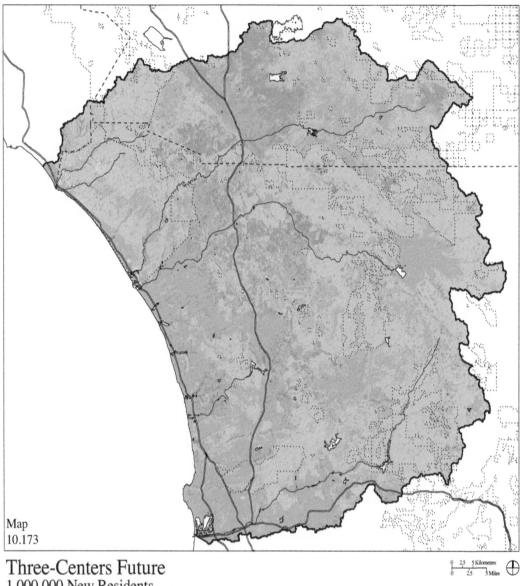

Map
10.173

Three-Centers Future
1,000,000 New Residents
Non-Native Species Richness

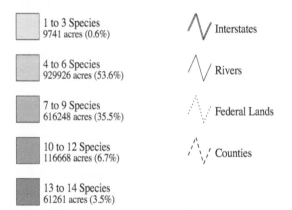

1 to 3 Species
9741 acres (0.6%)

4 to 6 Species
929926 acres (53.6%)

7 to 9 Species
616248 acres (35.5%)

10 to 12 Species
116668 acres (6.7%)

13 to 14 Species
61261 acres (3.5%)

Interstates

Rivers

Federal Lands

Counties

0   2.5   5 Kilometers
0   2.5   5 Miles

NOTES:

[1] William L. Halvorson, "Changes in Landscape Values and Expectations: What Do We Want and How Do We Measure It?" in R. Gerald Wright, ed., *National Parks and Protected Areas* (Cambridge, Massachusetts: Blackwell Science, Inc., 1996), pp. 15–17.

[2] Reed F. Noss and Allen Y. Cooperrider, *Saving Nature's Legacy: Protecting and Restoring Biodiversity* (Washington, D.C.: Island Press, 1994), pp. 14–17.

[3] Matthew Alan Cahn, *Environmental Deceptions: The Tension between Liberalism and Environmental Policymaking in the United States* (Albany, New York: State University of New York Press, 1995), pp. 126–127.

[4] Herman E. Daly and John B. Cobb, Jr., *For the Common Good: Redirecting the Economy toward Community, the Environment, and a Sustainable Future*, 2nd edition (Boston, Massachusetts: Beacon Press, 1994), pp. 252–267.

[5] Frank R. Burden, *Environmental Monitoring Handbook* (New York, New York: McGraw–Hill, 2002), pp. 1.3–27.16.

[6] The supporting information cited here denotes the importance of future planning based on analysis of the existing conditions from an ethics perspective, noting the importance this type of analysis has for human communities. Larry L. Rasmussen, *Earth Community Earth Ethics* (Maryknoll, New York: Orbis Books, 1996), pp. 322–343.

[7] Thomas C. Edwards, Jr., Collin G. Homer, Scott D. Bassett, Allan Falconer, R. Douglas Ramsey, and Doug W. Wight, *Utah Gap Analysis: An Environmental Information System*, Final Project Report 95–1 (Logan, Utah: Utah Cooperative Fish and Wildlife Research Unit, Utah State University, 1995), pp. 5.1–5.3.

[8] H. Resit Akcakaya and Jonathan L. Atwood, "A Habitat–based Metapopulation Model of the California Gnatcatcher," *Conservation Biology* 11:2 (April 1997), pp. 422–434.

[9] Richard T. T. Forman and Michel Godron, *Landscape Ecology* (New York, New York: Wiley, 1986).

[10] Richard T. T. Forman, *Land Mosaics: The Ecology of Landscapes and Regions* (Cambridge, England: Cambridge University Press, 1995), pp. 3–38.

[11] Wenche E. Dramstad, James D. Olson, and Richard T.T. Forman, *Landscape Ecology Principles in Landscape Architecture and Land–use Planning* (Washington, D.C.: Island Press, 1996), p. 14.

[12] U.S. Fish and Wildlife Service, *Standards for the Development of Habitat Suitability Index Models*, 103, ESM (Washington, D.C.: U.S. Fish and Wildlife Service, Division of Ecological Services, 1981) pp. 3.1–3.54.

[13] Data mining classification techniques can be used when an abundance of spatial information is available for specific point locations. For classification techniques using binary regression trees see Leo Beiman, Jerome Friedman, Charles J. Stone, and R. A. Olshen, *Classification and Regression Trees* (Boca Raton, Florida: CRC Press, 1984). For classification techniques using multi–node classification trees see Jerome H. Friedman, "Multivariate Adaptive Regression Splines," *The Annals of Statistics* 19:1 (March 1991), pp. 1–141.

[14] James C. Hickman, *The Jepson Manual: Higher Plants of California* (Berkeley, California: University of California Press, 1993), p. 721.

[15] Gerald A. Scheid, "Habitat Characteristics of Willowy Monardella in San Diego County: Site Selection for Transplants" in Thomas S. Elias, ed., *Conservation and Management of Rare and Endangered Plants* (Sacramento, California: California Native Plant Society, 1987), pp. 501–506.

[16] James S. Fralish and Scott B. Franklin, *Taxonomy and Ecology of Woody Plants in North American Forests* (New York: John Wiley & Sons, Inc., 2002), p. 201.

[17] Edwin F. Katibeh, "A Brief History of Riparian Forests in the Central Valley of California" in Richard E. Warner and Kathleen M. Hendrix, eds., *California Riparian Ecosystems: Ecology, Conservation, and Productive Management* (Berkeley, California: University of California Press, 1984), pp. 23–29.

[18] Timothy S. Brothers, *Riparian Species Distributions in Relation to Stream Dynamics, San Gabriel River, California*, Ph.D. Dissertation (Los Angeles, California: University of California, 1985).

[19] Katibeh (1984), p. 28.

[20] Philip N. Hooge, Mark T. Stanback, and Walter D. Koeing, "Nest–site Selection in the Acorn

Woodpecker," *The Auk* 116:1 (1999), p. 51.

[21] Edward J. Tarbuck and Frederick K. Lutgens, *Earth Science*, 7th edition (Englewood Cliffs, New York: Macmillan Publishing Company, 1994), p. 96.

[22] Detailed information may be required to avoid confusion of these two species with other fairy shrimp species, see Ellen Bauder, D. Ann Kreager, and Scott C. McMillan, *Vernal Pools of Southern California: Recovery Plan* (Portland, Oregon: U.S. Department of the Interior Fish and Wildlife Service, 1998), pp. 16–20.

[23] A probable reason for the ability of the two species to co–occur appears to be related to their development rates from the cyst. For a more detailed description see Stacie A. Hathaway and Marie A. Simovich, "Factors Affecting the Distribution and Co–occurrence of Two Southern Californian Anostracans (*Branchiopoda*), *Branchinecta sandiegonensis* and *Streptocephalus woottoni*," *Journal of Crustacean Biology* 16:4 (1996), p. 675.

[24] Stacie A. Hathaway and Marie A. Simovich (1996), p. 670.

[25] Richard J. Gonzalez, Jeff Drazen, Stacie Hathaway, Brent Bauer, and Marie Simovich, "Physiological Correlates of Water Chemistry Requirements in Fairy Shrimps (Anostraca) from Southern California," *Journal of Crustacean Biology* 16:2 (1996), pp. 315–322.

[26] Charles Black and Paul H. Zedler, "An Overview of 15 years of Vernal Pool Restoration and Construction Activities in San Diego County, California" in C. W. Wiltham, E. T. Bauder, D. Belk, W. R. Ferren Jr., and R. Ornduff, eds., *Ecology, Conservation, and Management of Vernal Pool Ecosystems*, Proceedings from a 1996 Conference, (Sacramento, California: California Native Plant Society, 1998), pp. 195–205.

[27] Edith A. Purer, "Ecological Study of Vernal Pools, San Diego County," *Ecology* 20:2 (June 1937), p. 217.

[28] Wayne R. Ferren, Jr., David M. Hubbard, Anuja K. Parikh, and Nathan Gale, "Review of Ten Years of Vernal Pool Restoration and Creation in Santa Barbara, California" in C. W. Wiltham, E. T. Bauder, D. Belk, W. R. Ferren Jr., and R. Ornduff, eds., *Ecology, Conservation, and Management of Vernal Pool Ecosystems*, Proceedings from a 1996 Conference (Sacramento, California: California

Native Plant Society, 1998), p. 206.

[29] Bauder, et al. (1998), p. 22.

[30] Some of the best physical descriptions are available by way of the world–wide–web from various pest management commercial and institutional sources. Descriptions of the Argentine ant come from: University of Arizona, "Argentine Ants," Urban Integrated Pest Management, available on-line at <http://ag.arizona.edu/urbanipm/insects/ants/argentineants.html>; Forest and Bird, "Argentine Ants – Lets Stop the Invasion," (2003) available on-line at <http://www.forestandbird.org.nz/biosecurity/argentinants.asp>; and Western Exterminator Company, "Argentine ant," available on-line at <http://www.west–ext.com/argentine_ant.html>.

[31] Andrew V. Suarez, David A. Holway, and Ted J. Case, "Patterns of Spread in Biological Invasions Dominated by Long–distance Jump Dispersal: Insights from Argentine Ants," *Proceedings of the National Academy of Sciences* 98:3 (January 2001), p. 1095.

[32] The ants appear most successful in Mediterranean and subtropical climates, but do not survive more arid environments. For a more detailed literature based accounting of the occurrence of Argentine ants see Suarez, et al. (2001), pp. 1095–1096.

[33] David A. Holway, "Distribution of the Argentine Ant (*Linepithema humile*) in Northern California," *Conservation Biology* 9:6 (December 1995), pp. 1634–1637.

[34] For a discussion on nest movement behavior, see Bert Hölldobler and Edward O. Wilson, *The Ants* (Cambridge, Massachusetts: Harvard University Press, 1990), pp. 212–213.

[35] David A. Holway and Ted J. Case, "Mechanisms of Dispersed Central–place Foraging in Polydomous Colonies of the Argentine Ant," *Animal Behaviour* 59 (2000), pp. 433–441.

[36] An explanation of the economic importance for the spread of Argentine ants is weak, but an ecological examination of ant impacts is discussed in detail, see, Nathan J. Sanders, Kasey E. Barton, and Deborah M. Gordon, "Long–term Dynamics of the Distribution of the Invasive Argentine ant, *Linepithema humile*, and Native Ant Taxa in Northern California," *Oecologia* 127 (January 2001), pp. 128–129.

[37] Competition is noted as a major cause, see Kathleen G. Human and Deborah M. Gordon, "Effects of Argentine Ants on Invertebrate Biodiversity in Northern California," *Conservation Biology* 11:5 (October 1997), pp. 1242–1248. Although predation as another ant reduction mechanism is not clear some have speculated about it being a factor, see F. Russell Cole, Arthur C. Medeiros, Lloyd L. Loope, and William W. Zuehlke, "Effects of the Argentine Ant on Arthropod Fauna of Hawaiian High–elevation Shrubland," *Ecology* 73:4 (1992), pp. 1313–1322.

[38] Andrew V. Suarez, Jon Q. Richmond, and Ted J. Case, "Prey Selection in Horned Lizards Following the Invasion of Argentine Ants in Southern California," *Ecological Applications* 10:3 (2000), pp. 711–725.

[39] Stream flow and subsequently soil moisture were found as a statistically significant factor for invasion in riparian systems, see David A. Holway, "Factors Governing Rate of Invasion: A Natural Experiment Using Argentine Ants," *Oecologia* 115 (1998), pp. 206–212. For temperature influences, see Xim Cerda, Javier Retana, and Sebastian Cros, "Thermal Disruption of Transitive Hierarchies in Mediterranean Ant Communities," *Journal of Animal Ecology* 66 (1997), pp. 363–374. For preferred soil characteristics, see M. J. Way, M. E. Cammell, M. R. Paiva, and C. A. Collingwood, "Distribution and Dynamics of the Argentine Ant *Linepithema* (*Iridomyrmex*) *humile* (Mayr) in Relation to Vegetation, Soil Conditions, Topography, and Native Competitor Ants in Portugal," *Insectes Sociaux* 44 (1997), pp. 425–428.

[40] Andrew V. Suarez, Douglas T. Bolger, and Ted J. Case, "Effects of Fragmentation and Invasion on Native Ant Communities in Coastal Southern California," *Ecology* 79 (1998), p. 2053.

[41] Suarez, et al. (1998), p. 2048.

[42] Suarez, et al. (1998), p. 2050.

[43] Robert C. Stebbins, *Western Reptiles and Amphibians*, 2nd edition (New York: Houghton Mifflin Company, 1985), pp. 139–140.

[44] Stephen R. Goldberg, "Reproduction of the Coast Horned Lizard, *Phrynosoma coronatum*, in Southern California," *The Southwestern Naturalist* 28:4 (1983), pp. 478–479.

[45] Eric R. Pianka and William S. Parker, "Ecology of Horned Lizards: A Review with Special Reference to *Phrynosoma platyrhinos*," *Copeia* 1975:1 (1975), pp. 141–162.

[46] David A. Holway, "Effect of Argentine Ant Invastions on Ground–dwelling Arthropods in Northern California Riparian Woodlands," *Oecologia* 116 (1998), pp. 254–255.

[47] Robert N. Fisher, Andrew V. Suarez, and Ted J. Case, "Spatial Patterns in the Abundance of the Coastal Horned Lizard," *Conservation Biology* 16:1 (2002), pp. 211–214.

[48] Preference for chaparral and sandy substrates is backed by sound statistics in Fisher, et al. (2002), p. 211. Other descriptions of the vegetation requirements for the coast horned lizard may be found in the life history section of S. Morey, "R029, Coast Horned Lizard, *Phrynosoma coronatum*," *California Wildlife Habitat Relationships System*, Version 7.0, (Sacramento, California: California Department of Fish and Game, 2001), spp. R029.

[49] Morey (2001), spp. R029.

[50] The most detailed physical description for the rufous–crowned sparrow is found in Roger T. Peterson, *A Field Guide to Western Birds*, 3rd edition (New York: Houghton Mifflin Company, 1990), p. 320. For bird call information and a brief physical description of the rufous–crowned sparrow see John Farrand, Jr., *National Audubon Society Field Guide to North American Birds: Western Region*, 2nd edition (New York: Alfred A. Knopf, Inc., 1994), pp. 725–726.

[51] Paul W. Collins, "Rufous–crowned Sparrow," *The Birds of North America* 472 (1999), pp. 1–27.

[52] D. Dobkin, "B487 Rufous–crowned Sparrow, *Aimophila ruficeps*," *California Wildlife Habitat Relationships System*, Version 7.0, (Sacramento, California: California Department of Fish and Game, 2001), spp. B487.

[53] Douglas T. Bolger, Thomas A. Scott, and John T. Rotenberry, "Breeding Bird Abundance in an Urbanizing Landscape in Coastal Southern California," *Conservation Biology* 11:2 (April 1997), pp. 406–421.

[54] Scott A. Morrison and Douglas T. Bolger, "Lack of an Urban Edge Effect on Reproduction in a Fragmentation–sensitive Sparrow," *Ecological Applications* 12:2 (April 2002), pp. 398 – 411.

[55] The best physical description emphasized by an artist rendition of the California gnatcatcher is found in David A. Sibley, *The Sibley Guide to Birds*

(New York: Alfred A. Knopf, Inc., 2000), p. 396. A further description is found in Peterson (1990), pp. 268–269.

[56] Robert M. Zink, George F. Barrowclough, Jonathan L. Atwood, and Rachelle C. Blackwell–Rago, "Genetics, Taxonomy, and Conservation of the Threatened California Gnatcatcher," *Conservation Biology* 14:5 (October 2000), pp. 1394–1405.

[57] Daniel Rubinoff, "Evaluating the California Gnatcatcher as an Umbrella Species for Conservation of Southern California Coastal Sage Scrub," *Conservation Biology* 15:5 (October 2001), pp. 1374–1383.

[58] Gerald T. Braden, Robert L. McKernan and Shawn M Powell, "Effects of Nest Parasitism by the Brown–headed Cowbird on Nesting Success of the California Gnatcatcher," *Condor* 99:4 (November 1997), p. 863.

[59] T. Kucera, "B553 California Gnatcatcher, *Polioptila californica*," *California Wildlife Habitat Relationships System*, Version 7.0, (Sacramento, California: California Department of Fish and Game, 2001), spp. B553.

[60] Jonathan L. Atwood and Jeffrey S. Bolsinger, "Elevational Distribution of California Gnatcatchers in the United States," *Journal of Field Ornithology* 63:2 (Spring 1992), pp. 159–168.

[61] Jonathan L. Atwood, "California Gnatcatchers and Coastal Sage Scrub: The Biological Basis for Endangered Species Listing" in J. E. Keeley, ed. *Interface between Ecology and Land Development in California* (Los Angeles, California: Southern California Academy of Sciences, 1993), pp. 149–169.

[62] Sibley (2000), p. 165.

[63] Peterson (1990), pp. 124–125.

[64] Kevin D. Lafferty, "Disturbance to Wintering Western Snowy Plovers," *Biological Conservation* 101 (2001), pp. 315–325.

[65] Gary W. Page, Lynne E. Stenzel and Christine A. Ribic, "Nest Site Selection and Clutch Predation in the Snowy Plover," *The Auk* 102 (April 1985), pp. 347–353.

[66] Abby Powell, "Western Snowy Plovers and California Least Terns," available on-line at <http:biology.usgs.gov/s+t/SNT/nofram/ca168.htm>.

[67] Gary W. Page and Lynne E. Stenzel, "The Breeding Status of the Snowy Plover in California," *Western Birds* 12:1 (1981), pp. 1–40.

[68] M. Rigney, "B154 Snowy Plover, *Charadrius alexandrinus*," *California Wildlife Habitat Relationships System*, Version 7.0, (Sacramento, California: California Department of Fish and Game, 2001), spp. B154.

[69] Sibley (2000), p. 326.

[70] Peterson (1990), pp. 238–239.

[71] U.S. Fish and Wildlife Service, *Southwestern Willow Flycatcher Draft Recovery Plan* (Albuquerque, New Mexico: Fish and Wildlife Service Reference Service, 2001), p. iv.

[72] Robert M. Marshall and Scott H. Stoleson, "Threats" in Deborah M. Finch and Scott H. Stoleson, eds, *Status, Ecology, and Conservation of the Southwestern Willow Flycatcher*, General Technical Report RMRS–GTR–60 (Ogden, Utah: U.S. Department of Agriculture, Forest Service, Rocky Mountain Research Station, 2000), pp. 14–16.

[73] Bryan T. Brown, "Rates of Brood Parasitism by Brown–headed Cowbirds on Riparian Passerines in Arizona," *Journal of Field Ornithology* 65:2 (1994), pp. 160–168.

[74] J. Edward Gates and Daniel R. Evans, "Cowbirds Breeding in the Central Appalachians: Spatial and Temporal Patterns and Habitat Selection," *Ecological Applications* 8:1 (February 1998), pp. 27–40.

[75] James A. Sedgwick and Fritz L. Knopf, "A High Incidence of Brown–headed Cowbird parasitism of Willow Flycatchers," *Condor* 90 (1988), p. 254.

[76] Mark K. Sogge, "Breeding Season Ecology," in *Status, Ecology, and Conservation of the Southwestern Willow Flycatcher*, General Technical Report RMRS–GTR–60, Deborah M. Finch and Scott H. Stoleson, eds, (Ogden, Utah: U.S. Department of Agriculture, Forest Service, Rocky Mountain Research Station, 2000), pp. 62–65.

[77] Mark K. Sogge, Timothy J. Tibbitts, and Jim Petterson, "Status and Ecology of the Southwestern Willow Flycatcher in the Grand Canyon," *Western Birds* 28 (1997), p. 146.

[78] Sogge (2000), p. 61.

[79] U.S. Fish and Wildlife Service, *Southwestern Willow Flycatcher Draft Recovery Plan* (Albuquerque, New Mexico: Fish and Wildlife Service Reference Service, 2001), pp. 11–12.

[80] Sogge et al. (1997), p. 149.

[81] Sibley (2000), p. 345.

[82] Peterson (1990), pp. 284–285.

[83] U.S. Fish and Wildlife Service, *Least Bell's Vireo Draft Recovery Plan* (Portland, Oregon: Fish and Wildlife Service Reference Service, 1998).

[84] K.E. Franzreb, *Ecology and Conservation of the Endangered Least Bell's Vireo* (Sacramento, California: U.S. Fish and Wildlife Service Endangered Species Office, 1989), pp. 1–17.

[85] U.S. Marine Corps, *BO–ESA Consultation on Riparian and Estuarine Programmatic Conservation Plan* (Camp Pendleton (San Diego County), California: U.S. Marine Corps, October 30, 1995).

[86] Franzreb (1989), pp. 1–17.

[87] L. Salata, *Status of the Least Bell's Vireo on Camp Pendleton, California* (Laguna Niguel, California: U.S. Fish and Wildlife Service, unpublished report), pp. 1–63.

[88] Regional Environmental Consultants, *Draft Comprehensive Species Management Plan for the Least Bell's Vireo, Vireo bellii pusillus* (San Diego, California: SANDAG, 1988), pp. 1–22.

[89] Michael E. Soule, "Land Use Planning and Wildlife Maintenance," *Journal of the American Planning Association* 57:3 (1991), pp. 313–323.

[90] Sibley (2000), p. 508.

[91] Peterson (1990), pp. 310–311.

[92] Wesley E. Lanyon, "*Sturnella neglecta*," *Birds of North America* 104 (1994), p. 14.

[93] Douglas H. Johnson and Lawrence D. Igl, "Area Requirements of Grassland Birds: A Regional Perspective," *The Auk* 118:1 (2001), pp. 24–34.

[94] Lanyon (1994), pp. 14–15.

[95] T. Luke George, Lowell C. McEwen, and Brett E. Peterson, "Effects of Grasshopper Control programs on Rangeland Breeding Bird Populations," *Journal of Range Management* 48 (1995), pp. 336–342.

[96] Lanyon (1994), p. 4.

[97] Elizabeth M. Madden, Andrew J. Hansen, and Robert K. Murphy, "Influence of Prescribed Fire History on Habitat and Abundance of Passerine Birds in Northern Mixed–grass Prairie," *The Canadian Field–Naturalist* 113 (1999), pp. 627–640.

[98] J. A. Wiens and J. T. Rotenberry, "Habitat Associations and Community Structure in Shrubsteppe Environments," *Ecological Monographs* 51 (1981), pp. 21–41.

[99] S. Charles Kendeight, "Birds of a Prairie Community," *Condor* 43 (1941), p. 169.

[100] Wesley E. Lanyon, "The Comparative Biology of the Meadowlarks (*Sturnella*) in Wisconsin," *Nuttal Ornithological Club* 1 (1957), p. 28.

[101] Lanyon (1957), p. 28.

[102] Steven T. Knick and John T. Rotenberry, "Landscape Characteristics of Fragmented Shrubsteppe Habitats and Breeding Passerine Birds", *Conservation Biology* 9:5 (1995), pp. 1059–1071.

[103] S. L. Haire, C. E. Bock, B. S. Cade, and B. C. Bennett, "The Role of Landscape and Habitat Characteristics in Limiting Abundance of Grassland Nesting Songbirds in an Urban Open Space," *Landscape and Urban Planning* 48 (2000), pp. 65–82.

[104] Sibley (2000), p. 274.

[105] American Ornithologists' Union, *Check–list of North American Birds*, 6th edition (Washington, D.C.: American Ornithologists' Union, 1983), pp. 277–279.

[106] Peterson (1990), pp. 200–201.

[107] C. Stuart Houston, Dwight G. Smith, and Christoph Rohner, "Great Horned Owl: *Bubo virginianus*," *The Birds of North America* 372 (1998), p 3.

[108] Thomas E. Morrell and Richard H. Yahner, "Habitat Characteristics of Great Horned Owls in South–central Pennsylvania," *Journal of Raptor Research* 28:3 (1994), pp. 164–170.

[109] Christoph Rohner and Charles J. Krebs, "Owl Predation on Snowshoe Hares: Consequences of Antipredator Behavior," *Oecologia* 108 (1996), pp. 303–310.

[110] Houston et al. (1998), p. 13.

[111] J. C. Turner, Jr., *Physiological Ecology of the Great Horned Owl, Bubo virginianus pacificus*, M.S. Thesis (Fullerton, California: California State College at Fullerton, 1969)

[112] Christoph Rohner, "Non–territorial "Floaters" in Great Horned Owls: Space Use During a Cyclic Peak of Snowshoe Hares," *Animal Behavior* 53 (1997), pp. 901–912.

[113] C. Polite, "B265 Great Horned Owl, *Bubo virginianus*," *California Wildlife Habitat Relationships System*, Version 7.0, (Sacramento, California: California Department of Fish and Game, 2001), spp. B265.

[114] William H. Burt and Richard P.

Grossenheider, *A Field Guide to the Mammals*, 3rd edition (New York: Houghton Mifflin Company, 1990), p. 77.

[115] Kevin Hansen, *Cougar, the American Lion* (Flagstaff, Arizona: Northland Publishing, 1992), pp. 55 – 85; see also Kenneth R. Russell, "Mountain Lion" in John L. Schmidt and Douglas L. Gilbert, eds. *Big Game of North America*, (Harrisburg, Pennsylvania: Stockpole Books, 1978), pp. 207–225.

[116] Paul Beier, "Determining Minimum Habitat Areas and Habitat Corridors for Cougars," *Conservation Biology* 7:1 (1993), pp. 94–108.

[117] Hansen (1992), pp. 43–48; see also Russell (1978) pp. 207–225.

[118] Hansen (1992), pp. 43–48.

[119] Russell (1978), p. 213.

[120] Hansen (1992), pp. 24–27.

[121] Hansen (1992), pp. 25–26.

[122] Paul Beier, "Dispersal of Juvenile Cougars in Fragmented Habitats," *Journal of Wildlife Management* 59:2 (1995), pp. 228–237.

[123] For a simple definition, see Michael Begon, John L. Harper, and Colin R. Townsend, *Ecology: Individuals, Populations, and Communities*, 2nd edition, (Boston, Massachusetts: Blackwell Scientific Publications, 1990), p. 862. For an example of an application using species richness, see Jeremy T. Kerr, "Species Richness, Endemism, and the Choice of Areas for Conservation," *Conservation Biology* 11:5 (October 1997), pp. 1094–1100.

[124] Carl Steinitz, Michael Binford, Paul Cote, Thomas Edwards, Jr., Stephen Ervin, Richard T. T. Forman, Craig Johnson, Ross Kiester, David Mouat, Douglas Olson, Allan Shearer, Richard Toth, and Robin Wills, *Biodiversity and Landscape Planning: Alternative Futures for the Region of Camp Pendleton, California* (Cambridge, Massachusetts: Harvard University Graduate School of Design, 1996), pp. 78–81.

[125] J. Michael Scott, Blair Csuti, James D. Jacobi, and Jack E. Estes, "Species Richness: A Geographic Approach to Protecting Future Biological Diversity," *BioScience* 37 (1987), pp. 782–788.

[126] Edwards, et al. (1995), pp. 4.1–4.5.

Chapter 11

Discussion and Conclusions

Allan W. Shearer, Scott D. Bassett, David A. Mouat, and Michael W. Binford

Each day at all levels of jurisdiction steps are taken to meet the needs of society. Within the spectrum of land planning and natural resources management, these decisions include where, when, and how to implement construction, conservation, and restoration initiatives. To some degree, the ultimate success of these actions will be determined by conditions that are both precisely defined and easily contained within the fence line of a given property. But also to some degree, and often to a very large degree, the success of these same actions will be determined by broader trends that are both more difficult to delineate and beyond immediate control. This study identifies some of the larger forces of change that could influence regional development around MCB Camp Pendleton, neighboring NWS Fallbrook, and MCAS Miramar, and explores some of the potential consequences that could result if these forces dominate the shaping of the future.

In reviewing the findings, it should be underscored that the futures considered for this analysis are possibilities, not eventualities. Further, while the scenarios represent a range of potential futures, they do not capture the full spectrum of all possible futures. Nevertheless, these scenarios can be of use to better understand uncertainties that may influence the conservation of regional natural resources.

Using the Scenarios and Alternative Futures as Aids in a Decision-Making Process

In this investigation, four futures are considered for the region, each named for the primary location of new construction. In the Coastal Future, high density development fills open areas between Interstate-15 and the Pacific Ocean. In the Northern Future, rural residential and suburban growth dominates southwestern Riverside County and northern San Diego County. In the Regional Low-Density Future, residential homes on large lots are distributed throughout the study region. And in the Three-Centers Future, urban and suburban growth is focused around Temecula, Valley Center, and Ramona.

Each of the futures is portrayed in two related ways: 1) as a scenario, which describes the evolution of the new conditions, and 2) as a pair of alternative futures, which capture specific moments in time. Both are useful for coming to terms with the significance of what might occur, but they are helpful in different ways. Referring to the analogy given in Chapter 1, a scenario is like a movie projected in the theater. The film strip speeds by at a rate that gives the illusion of continuous motion and allows us an understanding of the reasons for change over time. An alternative future is like a single frame of the film. By stopping the projector, the representation of evolving conditions is lost, but the opportunity for a close analysis of a single instant is gained.

In applying the scenarios and alternative futures to current decision-making needs, it is important to remember several points. First, each scenario is a story that is intended to adequately capture a sense of how and why conditions change. But like all stories, none of the scenarios provides an exhaustive accounting of a possible future. Readers should supplement the material that is provided by speculating on its implications as it might pertain to specific concerns. Doing so can contribute to a more complete consideration of possible future conditions and, in turn, to better decisions. For example, both the Forest Service and the Department of Defense are federal agencies that share statutory responsibilities; however, each organization uses its respective lands differently and each property contains unique geological and biological features. As such, the same futures may impact each land holding differently. For example, the Regional Low-Density Scenario describes a future in which undirected regional growth leaves the existing public lands as the only option for the conservation of natural resources. Subsequently, in order to protect these remaining open places from being over used, land use restrictions are put in place. The details of these restrictions are not specified in the scenario. As a policy exercise, it could be valuable for current environmental managers to consider the ways in which this future might impact their lands and stated missions. Will it become necessary to have all forest visitors stay on specific trails, rather than roam freely? Will it become necessary to have all training exercises stay within established movement corridors? Could closing some or all of the grounds for part of the year be required in order to provide adequate breeding conditions for some species? And if so, will that impact other mission needs? Will fire management be stopped near borders with adjacent homes? And if so, can a substitute management practice achieve the same results? If not, will the activities that currently take place on these lands change as the vegetation transitions from one community to another?

Second, it is also useful to consider variations on the scenario plots that could produce similar, but not necessarily identical, results. For example, the Coastal Scenario was generated, in part, over concerns about having an adequate water supply and the environmental impacts of growth. Desalinization is one potential source of water and has been considered in Southern California, but the costs of production are perceived as prohibitively high.[1] As written, the scenario calls for the construction of wind farms off the coast of Baja California to supply inexpensive and emission-free power to Southern California. Interest in renewable energy sources is growing throughout the world, and Mexico is certainly no exception. On a relatively small scale, the state government of Baja California is currently investigating several options. It has commissioned a feasibility study for a 300-megawatt

wind farm in Sierra de la Rumorosa to service the city of Mexicali[2] and there are also plans to build a 25-megawatt solar plant within Mexicali.[3] To the best of the researchers' knowledge, there are no plans for a large scale coastal wind farm as described in the scenario, but the future consideration of such a facility is conceivable given the government's interest in emerging technologies and the public's demand for electricity. Closer to fruition, there are plans for up to six fossil fuel power plants in Baja California to serve, in large part, the growing population and power needs of Southern California.[4] Once operational, these plants may begin to influence electricity prices in the study region. On the one hand, like the speculated wind farm, these conventional generating facilities could help to make desalinization plants (more) viable in the study region. But on the other hand, pollutants from fossil fuels may contribute to other environmental problems that are not addressed in the scenario.

Third, while adjacent frames on a movie reel are similar to each other, they are not identical and the choice of one frame vs. another will influence the results of an analysis. In this study, the two "frames" chosen for close inspection as alternative futures are based on allocations of 500,000 and 1,000,000 new residents. These anticipated increases are easily remembered and referenced, and are sufficiently distinct that meaningful differences can be measured and compared. But readers should remember that beyond providing conceptual clarity and operational expedience, there is nothing inherently significant in the choice of these projected moments. The impacts of growth are cumulative and incremental, and each new development contributes some additional strain on natural systems.

Development within the Alternative Futures

All of the futures accommodate the same numbers of new residents, but each scenario follows different rules of housing allocation. Table 11.1: "Land Used for Housing" summarizes the acreage of land dedicated to new residential development in each of the alternatives. Note that while all of the scenarios call for Ex-Urban housing to account for 1% of total growth, the Coastal and Three-Centers alternatives consume more land in this development

Table 11.1

Land Used for Housing (in acres)

	2000	Coastal		Northern		Reg. Low-Density		Three-Centers	
		500k	1,000k	500k	1,000k	500k	1,000k	500k	1,000k
Ex-Urban	86,132	102,509	128,287	87,765	94,746	87,518	104,555	119,680	123,592
Change from 2000		16,377	42,155	1,633	8,614	1,386	18,423	33,548	37,460
% Change from 2000		**19.0**	**48.9**	**1.9**	**10.0**	**1.6**	**21.4**	**38.9**	**43.5**
Rural Residential	207,870	222,175	239,087	283,140	378,036	326,864	442,321	213,455	248,232
Change from 2000		14,305	31,217	75,270	170,166	118,994	234,451	5,585	40,362
% Change from 2000		**6.9**	**15.0**	**36.2**	**81.9**	**57.2**	**112.8**	**2.7**	**19.4**
Suburban & Urban	198,925	237,732	279,478	235,909	275,383	241,340	290,523	244,046	292,193
Change from 2000		38,807	80,553	36,984	76,458	42,415	91,598	45,121	93,268
% Change from 2000		**19.5**	**40.5**	**18.6**	**38.4**	**21.3**	**46.0**	**22.7**	**46.9**
Total Housing	492,927	562,416	646,852	606,814	748,165	655,722	837,399	577,181	664,017
Change from 2000		69,489	153,925	113,887	255,238	162,795	344,472	84,254	171,090
% Change from 2000		**14.1**	**31.2**	**23.1**	**51.8**	**33.0**	**69.9**	**17.1**	**34.7**

category than the Northern and Regional Low-Density. This result is due to the fact that the two former scenarios have a higher proportion of housing at higher densities that allow for slightly larger Ex-Urban plots. In general, Ex-Urban homes in the Coastal and Three-Centers Futures are built on 10 to 15 acre parcels, while many of the Ex-Urban homes in the Northern and Regional Low-Density scenarios are built on 5 to 10 acre parcels. Overall, the Coastal 1,000k Future, which has a housing mixture weighted toward higher densities of development, transforms almost 154,000 acres of unbuilt land to housing, the lowest amount of the four futures. The Three-Centers Future, which concentrates new growth in and around Temecula, Valley Center, and Ramona, transforms the second least amount of land into housing with slightly over 171,000 acres changed. The Northern 1,000k Future, which emphasizes growth in southwestern Riverside and northern San Diego Counties, adds over 255,000 acres to the four types of housing. It includes an approximately 82% increase in Rural Residential housing and, with it, the largest increase in orchards and vineyards, 85% (see Table 10.1: "Vegetation"). Finally, the Regional Low-Density 1,000k Future, which distributes growth throughout the region, transforms over 344,000 acres to housing, a figure that more than doubles the amount of land converted to housing in the Coastal Future. Of particular note in the Regional Low-Density Future is the more

than 112% increase in Rural Residential acreage. However, unlike rural development in the Northern Future, which includes avocado and orange trees, this future increases grasslands by more than 85% (see Table 7.1: "Land Cover" and Table 10.1: "Vegetation").

General Issues Raised by the Analysis

The hydrological analysis shows that all four possible patterns of development would alter the Santa Margarita River's flow following 25- and 100-year storms—that is, precipitation events with a statistical likelihood of occurring once every 25 and once every 100 years, respectively. In general terms, all of the futures increased the amount of surface water runoff and the velocity of that runoff. The relative differences among the futures are a function of changes in ground cover. Most easily noticed are the amounts and locations of new impervious surfaces (i.e. roads and rooftops). But it should also be remembered that changes in vegetation can also contribute to increased surface water runoff. Grasslands, which accompany development, allow for faster runoff than shrub and tree plant communities. While the other rivers in the study area were not examined, the same physical processes are at work and similar outcomes can be expected. These potential consequences will present challenges throughout the region and could result in significant costs to both society and the natural environment. Imagine the teams of engineers who are responsible for the design of bridges, dams, levies, channels, and other elements of infrastructure that are in or near rivers. As a hypothetical example, say they design their structures to accommodate the effects of a 100-year storm (which would provide a high degree of confidence). As discussed in the chapter on hydrology, changing patterns of development may exacerbate the effects of such a storm and could allow, perhaps, the flood peak of the future 100-year storm to be similar to the current 110- or 120-year storm. Next, imagine the home- or business-owner who invests in property near the downstream stretch of a river. Future up-stream development could place these holdings at a greater risk of flood damage. Finally, imagine the land and natural resources managers responsible for maintaining riparian habitat along a river. Increased flood surges may rip away native vegetation and allow the opportunity for exotic species, such as Giant Reed (*Arundo donax*), to take root. Once such exotics become established, it can be very difficult and expensive to restore the native plant communities. Equally problematic, changes in river flow may not provide some native animal species with adequate conditions to successfully forage for food or to breed. Natural resources managers for areas up-stream may also face water-related difficulties as rainwater that is lost to the sea by rapid overland flow will not be available to seep into the root zones of terrestrial plant communities.

The biological analysis is based on changes in the amount of vegetation and its spatial distribution. When people think about the loss of natural environments, they often envision the construction of homes and gardens. But while these direct impacts to native ecosystems can be significant, they are not the only sources of potential habitat loss. Indirect impacts occur as the natural landscape is fragmented and species are blocked from migratory paths or isolated from other colonies that are needed to maintain genetic diversity and long-term survival. Other indirect impacts on vegetation that often accompany development are related to the natural process of fire. Periodic and varied fire events are necessary for the propagation of many of the region's native plant communities. Woody

plants and trees typically require infrequent and high-intensity fires, while grasses and forbs require more frequent and less intense fires. Development-driven changes to the natural fire regimes can occur at opposite extremes. First, development (including the construction of roads through remote areas) increases the likelihood of an ignition. A series of frequent small fires will likely produce or maintain grasslands at the expense of scrub, chaparral, and forest communities. Second, development usually, and understandably, brings fire suppression practices that are intended to protect homes and structures from damage. Active suppression can result in the transformation of grasslands to scrublands, and scrublands to forests. It should be noted, of course, that such transformations are not guaranteed as a variety of other factors such as soil, available water, available direct sunlight, and micro-climate are also important. Nevertheless, fire suppression can significantly contribute to vegetation change. Also of importance to both homeowners and natural resource managers is fuel load. As fires are suppressed, successive seasons of dead plant material accumulate and produce an increasingly greater threat of a larger and hotter fire. In addition to posing a greater risk to homes and other structures, these more intense fires can irreparably damage native plant communities.

The effects of high intensity fires and severe storm events may also combine to produce particularly devastating consequences. Immediately after a fire, vegetation root structures may not provide a sufficient mechanism to hold soils in place. Subsequently, soils are susceptible to erosion and heavy rains can result in mud slides that can cause immediate property damage and change the long-term conditions that would allow native plant communities to re-establish.

Three kinds of biodiversity analysis were undertaken in this study: landscape ecological pattern, single species potential habitat, and species richness. Each contributes to an understanding of the consequences of possible development, but no single approach provides a comprehensive assessment.

Given limits of time, financial resources, societal patience, and scientific knowledge, some advocate the protection of ecosystems at a landscape or regional scale by maintaining the

Table 11.2
Summary of Potential Habitat (in acres)

	2000	Coastal		Northern		Reg. Low-Density		Three-Centers	
		500k	1,000k	500k	1,000k	500k	1,000k	500k	1,000k
Willowy Monardella	8,401	7,704	7,026	7,815	7,128	7,591	6,908	8,401	8,401
Change from 2000		-697	-1,376	-586	-1,273	-810	-1,493	0	0
% Change from 2000		**-8.3**	**-16.4**	**-7.0**	**-15.2**	**-9.7**	**-17.8**	**0**	**0**
California Sycamore	52,872	51,328	49,734	49,583	46,025	51,157	49,400	50,752	49,734
Change from 2000		-1,544	-3,138	-3,289	-6,847	-1,715	-3,472	-2,120	-3,138
% Change from 2000		**-2.9**	**-5.9**	**-6.2**	**-13.0**	**-3.2**	**-6.6**	**-4.0**	**-5.9**
Fairy Shrimp	11,633	10,635	9,991	10,842	10,264	10,588	9,809	11,633	11,633
Change from 2000		-988	-1,732	-791	-1,369	-1,045	-1,824	0	0
% Change from 2000		**-8.6**	**-14.9**	**-6.8**	**-11.8**	**-9.0**	**-15.7**	**0**	**0**
Argentine Ant	530,766	620,016	719,358	712,696	884,448	746,882	930,485	649,594	782,136
Change from 2000		89,250	188,592	181,930	353,682	216,116	399,719	118,828	251,370
% Change from 2000		**16.8**	**35.5**	**34.3**	**66.6**	**40.7**	**75.3**	**22.4**	**47.4**
SD Coast Horned Lizard	721,305	651,479	577,957	580,554	462,700	554,663	410,209	626,933	522,011
Change from 2000		-69,826	-143,348	-140,751	-258,605	-166,642	-311,096	-94,372	-199,294
% Change from 2000		**-9.7**	**-19.9**	**-19.5**	**-35.9**	**-23.1**	**-43.1**	**-13.1**	**-27.6**
Rufous-crowned Sparrow	343,882	340,403	334,191	334,964	317,807	331,511	310,183	339,924	331,315
Change from 2000		-3,479	-9,691	-8,918	-26,076	-12,372	-33,699	-3,959	-12,568
% Change from 2000		**-1.0**	**-2.8**	**-2.6**	**-7.6**	**-3.6**	**-9.8**	**-1.2**	**-3.7**
California Gnatcatcher *	235,252	219,554	202,101	216,139	204,567	213,883	150,302	232,963	227,483
Change from 2000		-15,698	-33,151	-19,113	-30,685	-21,369	-84,950	-2,289	-7,769
% Change from 2000		**-6.7**	**-14.1**	**-8.1**	**-13.0**	**-9.1**	**-36.1**	**-1.0**	**-3.3**
Western Snowy Plover *	1,831	1,779	1,750	1,787	1,765	1,799	1,785	1,831	1,831
Change from 2000		-52	-81	-44	-66	-32	-46	0	0
% Change from 2000		**-2.8**	**-4.4**	**-2.4**	**-3.6**	**-1.7**	**-2.5**	**0.0**	**0.0**
SW Willow Flycatcher	15,352	15,035	14,099	14,665	13,160	14,065	11,544	14,670	13,783
Change from 2000		-317	-1,253	-687	-2,192	-1,287	-3,808	-682	-1,569
% Change from 2000		**-2.1**	**-8.2**	**-4.5**	**-14.3**	**-8.4**	**-24.8**	**-4.4**	**-10.2**
Least Bell's Vireo	28,061	27,279	25,400	26,756	24,030	25,415	21,253	26,868	25,363
Change from 2000		-783	-2,661	-1,305	-4,031	-2,646	-6,808	-1,193	-2,698
% Change from 2000		**-2.9**	**-9.5**	**-4.7**	**-14.4**	**-9.4**	**-24.3**	**-4.3**	**-9.6**
Western Meadowlark	199,130	207,121	224,646	215,031	240,754	259,256	326,151	216,797	233,322
Change from 2000		7,991	25,516	15,901	41,623	60,126	127,021	17,687	34,192
% Change from 2000		**4.0**	**12.8**	**8.0**	**20.9**	**30.2**	**63.8**	**8.9**	**17.2**
Great Horned Owl	767,404	785,875	794,076	831,218	875,516	815,749	870,717	835,751	917,913
Change from 2000		18,471	26,672	63,814	108,112	48,345	103,313	68,347	150,509
% Change from 2000		**2.4**	**3.5**	**8.3**	**14.1**	**6.3**	**13.5**	**8.9**	**19.6**
Cougar	980,933	940,750	894,228	888,964	763,720	872,508	738,674	923,144	854,205
Change from 2000		-40,183	-86,705	-91,969	-217,213	-108,425	-242,259	-57,789	-126,728
% Change from 2000		**-4.1**	**-8.8**	**-9.4**	**-22.1**	**-11.1**	**-24.7**	**-5.9**	**-12.9**

* Includes potential habitat of varying quality.

underlying conditions that support a variety of life, including not only plants and animals, but also invertebrates, fungi, and bacteria that also contribute to biodiversity.[5] This position is supported by findings from the fields of landscape ecology and conservation biology that suggest spatial pattern is an integral feature of the natural processes.[6] Further, it is hypothesized that maintaining structural integrity of the natural patch-corridor matrix will conserve the majority of—but not necessarily all—natural processes. The results of landscape ecological pattern analysis are not easily reduced to tables. The transformation of a given parcel not only results in the loss or gain in a specific category of analysis, but also alters the structure of the larger landscape matrix of patches and corridors. All of the futures result in a decrease in contiguous vegetation and stream corridors. Notably, when each of the futures is allocated with 1,000,000 new residents, the thin northwest–southeast corridor south of Temecula is lost. The Northern and Regional Low-Density Futures, which have the greatest amount of low density development, produce the greatest impacts on the structure of the regional ecosystem.

While conservation based on the principles of landscape ecology may provide a robust approach to maintaining biodiversity, some species warrant special consideration. The federal Endangered Species Act (ESA) and equivalent state laws require the protection of plants and animals that are judged to be at risk of extinction. Even without formal legal standing, a species, sub-species, or population can become the focus of management or preservation initiatives because of local interest. Table 11.2 summarizes change in potential habitat in each of the alternative futures. Readers should note that the acreage reported for California gnatcatcher and the western snowy plover represents a combination of potential habitat of varying qualities. The tables throughout Chapter 10 on biological impacts provide more precise distinctions. Overall the entries show a loss of potential habitat for many native species and a gain of potential habitat for some non-native species.

Single species management efforts can be effective in addressing specific conservation objectives, but the species-by-species approach has been criticized as ineffective, and even shortsighted, given the known large number of species and

unknown number of species that have yet to be identified.[7] A more generalized analysis of potential habitat is provided by the investigation of species richness. While the models used for this analysis are not as precise as those used in the single species modeling effort, they do provide a useful general indication of biological conditions. Similar to the findings above, all four scenarios lead to decreases of native species and increases in non-native species throughout the region. From the perspective of trying to maintain areas of high native species richness, any future concern over the impact of development should take into account the amount of rural residential housing and its proximity to the coast.

Specific Consequences of the Futures

Coastal Future

In the Coastal Future, high-density urban residential development is concentrated between the Pacific Ocean and Interstate-15, and there is a relatively small amount of new low-density housing in inland rural locations. Since much of the new development is placed outside the Santa Margarita River Basin, this future results in the least severe hydrologic impacts on this river. Virtually all of the biological impacts with 1,000,000 new residents are roughly twice those that occur with 500,000 new residents. A notable exception to this generalization comes from riparian species that exhibit sensitivity to being located close to residential development. By concentrating new development in a dense pattern, the Coastal Future places residential houses closer to riparian habitat. As a result, doubling the allocation of new residents from 500k to 1,000k triples the impact on sensitive riparian species.

Species relying on beach and other unbuilt lands in relative close proximity to the coastline are most likely to decline in the Coastal Futures. Both the rare coastal habitats as well as the coastal sage scrub will be deleteriously affected by this intensive coastal development pattern. Species requiring isolated beaches or estuaries away from human disturbance would be at particular risk.

However, a pattern of urban development close to the coast would afford some protection to locations east of the Coast Range from the indirect impacts of sprawl: the spread of certain invasive species would slow; removal of natural vegetation to protect residences from wildfire would be minimized; landscape fragmentation is less; and the overall human residence footprint would be smaller. These results demonstrate that a pattern of compact growth can mitigate ecological impacts.

Northern Future

The Northern Future concentrates suburban and rural residential development in southwestern Riverside County and the northern part of San Diego County. Its mixture of high- and low-density housing within this sub-region of the study area provides for limited control of urban sprawl. The majority of housing is placed in subdivisions that are relatively close to incorporated towns and associated infrastructure; however, a number of low-density rural lots are also placed throughout the region.

The concentration of development in the northern part of the study area produces the most dramatic increases in flood flow on the Santa Margarita River. Relative to current conditions, the 1,000k allocation produces a more than 71% increase in peak discharge for the 25-year storm event and a more than 41% increase for the 100-year storm event.

As with the other scenarios the impacts on riparian species are severe and result from the indirect effects associated with development. Low-density development in the Northern Future also results in the loss of some areas currently dominated by shrub land cover types (primarily coastal sage scrub and chaparral). Beyond the vegetation lost due to the placement of new structures, there is also loss associated with the creation of fire breaks. Typically, the vegetation in these fire breaks is dominated by annual grasses and forbs that have limited habitat value. Thus, with an increase in low-density development within shrub dominated land cover comes a concomitant increase in the amount of degraded grasslands. With the increase in grasslands comes an increase in terrestrial faunal species that prefer small grassland patches or landscapes represented by a mosaic of shrubs and grasslands. The Northern Future also removes the greatest amount of California Sycamore.

Regional Low-Density Future

The Regional Low-Density Future spreads development across the study area and the majority of new construction is suburban. Unlike the Northern Future which places a large number of new rural homes on abandoned agricultural fields, the Regional Low-Density Future concentrates rural housing in areas currently dominated by native shrubs. Much of this development occurs in San Diego County.

Overall, many of the modeled impacts were substantially larger for this scenario than the others. There are significant impacts on the coastal sage scrub vegetation and the fauna associated with it. Both the 500k and 1,000k allocations produce the greatest loss of potential habitat for the willowy monardella, fairy shrimp, San Diego coast horned lizard, rufous-crowned sparrow, California gnatcatcher, southwestern willow flycatcher, least Bell's vireo, and cougar. This future also results in the greatest increase in Argentine ant and western meadowlark potential habitat. Additionally, the landscape ecological pattern is fragmented to the greatest degree. As such, this future poses the greatest risk to the biodiversity conservation.

Three-Centers Future

The Three-Centers Future concentrates development and takes steps toward the conservation of some flora and fauna. Much of the emphasis falls on the potential habitat of the federally listed threatened and endangered species in the region, but this focus is not exclusive. The future incorporates the publicly available habitat conservation plans that were available as of 2000. (NOTE: Many of these plans were under development during the course of this study. It is very likely that the versions used in this investigation have been recently updated or otherwise amended by local and regional agencies.) The emphasis on biological conservation directs much of the future housing development to the fringes of existing residential development and current agricultural fields. While farming contributes to the local economy, crops are relatively of little value as nesting sites or forage grounds for native animal species. The Three-Centers pattern of development lessens the amount of rural development sprawl throughout the south-central and southeastern parts of the study region; however, new rural residential housing does spread out in the north. Compact development in the south assists in conserving much of the coastal sage scrub and chaparral located in the lower elevations and thereby provides some protection to species relying on these types of vegetation for habitat. Vernal pools and watercourses supporting riparian habitat are also afforded a level of protection.

Although it is focused on biological conservation, the Three-Centers Future does produce some impacts. As with the other futures, rural residential housing requires fire breaks which result in the transformation of the shrub communities into annual grasses and forbs. Although rural residential development in some shrub habitats is prohibited, some sage and chpparral is lost. This alteration increases habitat for some faunal species (i.e. those associated with grassland habitat), but decreases habitat for others (i.e., those associated with coastal sage scrub and chaparral).

Most of the biological impacts that do occur in this future result from the indirect consequences of development. The Three-Centers Future is an illustration of how the protection of vegetation alone is not sufficient for species conservation; the juxtaposition of new development in relationship to biological communities must also be assessed and factored into ecological management efforts. Many of the riparian species incur impacts as a result of species that are introduced with residential development. Cowbirds, domesticated animals, and other non-native species utilize various human induced land uses as jumping points for infiltration into native vegetation.

Impacts Common among the Futures

Each of the four futures considered in this study follows from a unique set of circumstances and each would, in turn, present unique challenges for the land managers who are responsible for the conservation of natural resources throughout the region. Yet, despite the differences for change, the futures share some consequences. Map 11.1 shows

areas of potential habitat loss for the threatened and endangered species modeled for analysis that are common to the Coastal, Northern, and Regional Low-Density Futures when allocated with 1,000,000 new residents. For reference, the species are the willowy monardella, the fairy shrimp, the California gnatcatcher, the western snowy plover, the southwestern willow flycatcher, and the least Bell's vireo. The Three-Centers Scenario, which is not included in this analysis, protects high quality potential habitat for these species, so only a small amount of potential habitat is lost through indirect impact. The map shows a speckled pattern of loss throughout the area. Losses in the northwestern part of the region are significant in that they occur near MCB Camp Pendleton, NWS Fallbrook, the Cleveland National Forest, and The Nature Conservancy's Santa Rosa Plateau Ecological Reserve. These areas maintain a significant amount of natural habitat and serve as a resource to maintain biodiversity. The losses could potentially contribute to the genetic isolation of these remaining large natural patches. The losses in the southwestern part of the study area are notable in that development pressures there have left only a few natural areas. The futures also share a loss of potential habitat for some threatened and endangered species between I-5 and I-15, north of MCAS Miramar.

Map 11.2 shows significant natural areas that are lost to development in all four of the scenarios when allocated with 1,000,000 new residents. The criteria used for this summary are:

1) From the landscape ecological pattern assessment, lost large-patch natural vegetation.

2) From the landscape ecological pattern assessment, lost corridors.

3) From the species richness assessment, decreased potential native species.

The ordering is hierarchical so that the later concerns are mapped over earlier concerns. For example, if a given parcel is part of a large patch (listed as item 1) and loses native species richness

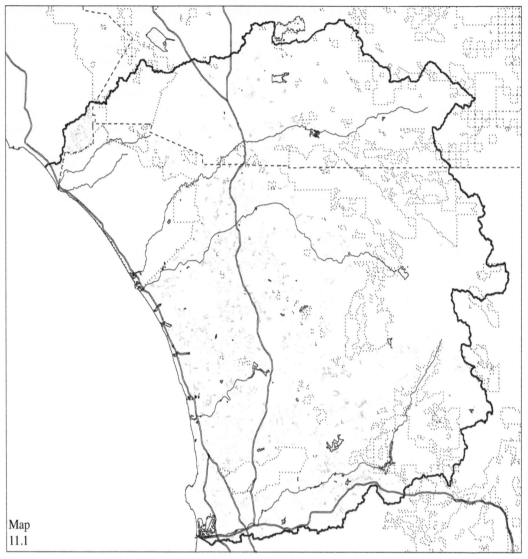

Map
11.1

Coastal, Northern, and Regional Low-Density Futures
1,000,000 Allocation
Common Loss of Threatened and Endangered Species Potential Habitat

Threatened and Endangered Species
23993 acres (1.4%)

Interstates

Rivers

Federal Lands

Counties

— See color insert following page 204 —

(listed as item 3), only the loss of species richness is mapped.

Beginning in the northern part of the study area, the speckled pattern around Temecula shows a loss of some potential native species richness. There is also a reduction in the large natural vegetation patch between MCB Camp Pendleton and adjacent Cleveland National Forest and The Nature Conservancy's Santa Rosa Plateau Ecological Reserve. Again, this loss could contribute to the biologic isolation of these large natural areas. The relative flat area around Lake Wohlford shows drops in contiguous natural vegetation, stream corridors, and native species richness. To the northeast of MCAS Miramar and, more immediately, to the north of the San Vicente Reservoir, a considerable amount of land is transformed through development. Losses of large natural patches and stream corridors encircle a zone that is pervaded with small pockets of reduced potential native species richness. These changes would contribute to the ecological isolation of MCAS Miramar.

Implications for the Management of Department of Defense Lands

As a set, the consequences of these four possible futures could have some bearing on all of the region's stakeholders, but, in practical terms, the impacts would not necessarily be equally shared. The geographical conditions throughout the study area are varied and so not all concerns are equally distributed. For example, the loss of California gnatcatcher habitat will contribute to a loss of regional biodiversity, but it will have little impact on the management of the forests around Palomar Mountain, since gnatcatchers do not nest or forage in montane woodlands.

Given these differences, it is useful to consider the implications of each of the futures from the specific perspectives of individual stakeholders. Here, the relative impact of the four futures on military lands can serve as an example of such an analysis. Similar analyses could be done for other agencies and organizations which are responsible for natural resources conservation, such as the U.S. Forest Service, the U.S. Bureau of Land Management, the California Coastal Commission,

the California Department of Fish and Game, the California Department of Parks and Recreation, the county and regional planning agencies, The Nature Conservancy, and many more.

Table 11.3 shows the percentage of regional potential habitat that would be managed by the Department of Defense (MCB Camp Pendleton, MCAS Miramar, and NWS Fallbrook) in each of the alternative futures. In some cases, particularly those associated with the wildlife conservation oriented Three-Centers Future, there is little or no change. But in most cases, the installations assume a larger percentage of potential habitat for native species. Some of these increases are numerically small, but even small changes can have significant impacts on land use and management strategies given the protective status of some species. Some increases are clearly significant. In the Regional Low-Density 1,000k Future, the percentage of on-base fairy shrimp potential habitat increases from 70.1% to 82.3%. Also in the Regional Low-Density 1,000k Future, the percentage of on-base California gnatcatcher habitat within the region increases from 36.4% to 51.5%. In all of the futures, DoD lands become responsible for a greater percentage of San Diego coast horned lizard potential habitat. While this species is not listed for protection by either the state or federal governments, its loss of potential habitat may be indicative of broader changes that can, in turn, lead to additional natural resources management concerns. The coast horned lizard's loss of habitat is largely a result of a loss of food: the native ants on which it feeds are out-competed by invasive Argentine ants that follow development. Other species that depend on native ants for food (and other larger species which depend on these species) may also become more dependent on DoD lands.

It should also be noted in Table 11.3 that in some futures, the amount of potential habitat on DoD lands decreases. Examples include: In the 1,000k Regional Low-Density Future, over twenty acres of least Bell's vireo potential habitat is lost. In the 1,000k Coastal Future over 20 acres of willowy monardella potential habitat is lost. In the Coastal, Northern, and Regional Low-Density Futures, between 29 and 76 acres of fairy shrimp potential habitat are lost. In the Coastal 1,000k Future and in the Northern and Regional Low-Density Scenarios,

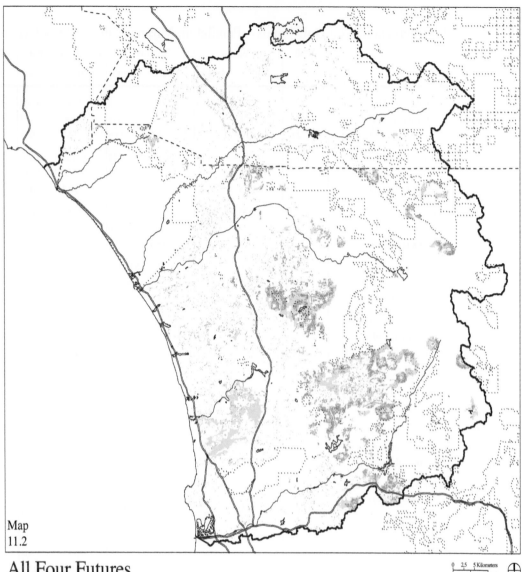

Map
11.2

All Four Futures
1,000,000 Allocation
Common Biological Impacts

0   2.5   5 Kilometers
0   2.5   5 Miles

Contiguous Natural Vegetation
64975 acres (3.8%)

Stream Corridor
1354 acres (0.1%)

Native Species Richness
42620 acres (2.5%)

Federal Lands

Counties

Interstates

Rivers

— See color insert following page 204 —

Table 11.3

Potential Habitat on DoD Lands—Pendleton, Miramar, and Fallbrook (in acres)

	2000	Coastal		Northern		Reg. Low-Density		Three-Centers	
		500k	1,000k	500k	1,000k	500k	1,000k	500k	1,000k
Total Willowy Monardella	8,401	7,704	7,026	7,815	7,128	7,591	6,908	8,401	8,401
Amount on DoD Lands	2,482	2,478	2,461	2,472	2,468	2,480	2,474	2,482	2,482
% on DoD Lands	**29.5**	**32.2**	**35.0**	**31.6**	**34.6**	**32.7**	**35.8**	**29.5**	**29.5**
Total California Sycamore	52,872	51,328	49,734	49,583	46,025	51,157	49,400	50,752	48,622
Amount on DoD Lands	5,332	5,326	5,316	5,330	5,325	5,324	5,316	5,332	5,330
% on DoD Lands	**10.1**	**10.4**	**10.7**	**10.7**	**11.6**	**10.4**	**10.8**	**10.5**	**11.0**
Total Fairy Shrimp	11,633	10,635	9,991	10,842	10,264	10,588	9,809	11,633	11,633
Amount on DoD Lands	8,150	8,121	8,163	8,135	8,116	8,137	8,074	8,150	8,150
% on DoD Lands	**70.1**	**76.4**	**81.7**	**75.0**	**79.1**	**76.9**	**82.3**	**70.1**	**70.1**
Total Argentine Ant	530,766	620,016	719,358	712,696	884,448	746,882	930,485	649,594	782,136
Amount on DoD Lands	28,542	28,692	28,952	28,626	28,738	28,676	29,995	28,608	28,918
% on DoD Lands	**5.4**	**4.6**	**4.0**	**4.0**	**3.2**	**3.8**	**3.2**	**4.4**	**3.7**
Total Coast Horned Lizard	721,305	651,479	577,957	580,554	462,700	554,663	410,209	626,933	522,011
Amount on DoD Lands	122,283	122,137	121,882	122,201	122,090	122,148	120,853	122,218	121,912
% on DoD Lands	**17.0**	**18.7**	**21.1**	**21.0**	**26.4**	**22.0**	**29.5**	**19.5**	**23.4**
Total R-crowned Sparrow	343,882	340,403	334,191	334,964	317,807	331,511	310,183	339,924	331,315
Amount on DoD Lands	37,113	37,114	37,105	37,111	37,109	37,113	37,047	37,112	37,100
% on DoD Lands	**10.8**	**10.9**	**11.1**	**11.1**	**11.7**	**11.2**	**11.9**	**10.9**	**11.2**
Total Calif. Gnatcatcher *	235,252	219,554	202,101	216,139	204,567	213,883	150,302	232,963	227,483
Amount on DoD Lands	77,936	77,874	77,802	77,895	77,886	77,905	77,407	77,935	77,913
% on DoD Lands	**33.1**	**35.5**	**38.5**	**36.0**	**38.1**	**36.4**	**51.5**	**33.5**	**34.3**
Total W Snowy Plover *	1,831	1,779	1,750	1,787	1,765	1,799	1,785	1,831	1,831
Amount on DoD Lands	830	830	827	828	828	830	828	830	830
% on DoD Lands	**45.3**	**46.7**	**47.3**	**46.3**	**46.9**	**46.1**	**46.4**	**45.3**	**45.3**
Total SW Willow Flycatcher	15,352	15,035	14,099	14,665	13,160	14,065	11,544	14,670	13,783
Amount on DoD Lands	2,787	2,785	2,787	2,787	2,787	2,787	2,764	2,786	2,769
% on DoD Lands	**18.2**	**18.5**	**19.8**	**19.0**	**21.2**	**19.8**	**23.9**	**19.0**	**20.1**
Total Least Bell's Vireo	28,061	27,279	25,400	26,756	24,030	25,415	21,253	26,868	25,363
Amount on DoD Lands	4,452	4,452	4,451	4,453	4,451	4,451	4,431	4,453	4,452
% on DoD Lands	**15.9**	**16.3**	**17.5**	**16.6**	**18.5**	**17.5**	**20.9**	**16.6**	**17.6**
Total Western Meadowlark	199,130	207,121	224,646	215,031	240,754	259,256	326,151	216,797	233,322
Amount on DoD Lands	28,797	28,789	28,616	28,773	28,725	28,754	29,057	28,784	28,807
% on DoD Lands	**14.5**	**13.9**	**12.7**	**13.4**	**11.9**	**11.1**	**8.9**	**13.3**	**12.3**
Total Great Horned Owl	767,404	785,875	794,076	831,218	875,516	815,749	870,717	835,751	917,913
Amount on DoD Lands	47,020	47,544	48,163	47,215	47,586	47,315	47,315	47,128	47,234
% on DoD Lands	**6.1**	**6.0**	**6.1**	**5.7**	**5.4**	**5.8**	**5.5**	**5.6**	**5.1**
Total Cougar	980,933	940,750	894,228	888,964	763,720	872,508	738,674	923,144	854,205
Amount on DoD Lands	93,511	92,485	92,459	93,500	92,484	93,497	91,956	93,503	93,442
% on DoD Lands	**9.5**	**9.8**	**10.3**	**10.5**	**12.1**	**10.7**	**12.4**	**10.1**	**10.9**

* Includes potential habitat of varying quality.

Table 11.4

Key Native Vegetation on Non-Department of Defense Lands (in acres)

	2000	Coastal		Northern		Reg. Low-Density		Three-Centers	
		500k	1,000k	500k	1,000k	500k	1,000k	500k	1,000k
Coastal Sage Scrub	206,672	188,491	168,484	181,103	160,377	179,771	120,928	191,626	180,678
Change from 2000		-18,181	-38,188	-25,569	-46,295	-26,901	-85,744	-15,046	-25,994
% Change from 2000		**-8.8**	**-18.5**	**-12.4**	**-22.4**	**-13.0**	**-41.5**	**-7.3**	**-12.6**
Chaparral	677,745	653,121	623,001	614,455	534.736	601,778	553,092	636,246	590,299
Change from 2000		-24,624	-54,744	-63,290	-143,009	-75,967	-124,653	-41,499	-87,446
% Change from 2000		**-3.6**	**-8.1**	**-9.3**	**-21.1**	**-11.2**	**-18.4**	**-6.1**	**-12.9**
C.S Scrub-Chaparral Mix	23,143	22,234	21,167	20,597	18,662	20,707	15,618	21,622	19,966
Change from 2000		-909	-1,976	-2,546	-4,481	-2,436	-7,525	-1,521	-3,177
% Change from 2000		**-3.9**	**-8.5**	**-11.0**	**-19.4**	**-10.5**	**-32.5**	**-6.6**	**-13.7**

2–3 acres of western snowy plover potential habitat is lost. And in all of the futures, gnatcatcher potential is lost. These losses for federally listed Endangered Species are especially notable since no change in on-base land use is anticipated in the scenarios. Instead, the losses result from development on adjacent or nearby private lands. These examples serve to illustrate how off-base land use change can impact on-base natural resources conservation management.

Related to the changes in the relative amounts of habitat managed by the Department of Defense are changes in vegetation. Table 11.4 shows the amount of coastal sage scrub and chaparral located on non-Department of Defense managed lands. These vegetation communities are notably important for native species. Coastal sage scrub is habitat for the threatened California gnatcatcher, a species that, as noted earlier, is often mentioned in discussions on the tensions between conservation and development initiatives. In the Regional Low-Density 1,000k Future over 41% of the non-DoD managed coastal sage scrub is lost. In the Coastal 1,000k Future, only 18% is lost; however, while this figure may seem relatively small, it should be noted that this loss includes plant communities near the coast that are preferred by the gnatcatcher.

Concluding Remarks

If this had been a normative planning study which set out to array a spectrum of options of how the future could be made to meet some pre-established set of values, then this space would be used to provide some suggestion that it was now the responsibility of the local stakeholders to choose which future they want to see materialize. This study however, is not such an investigation. Instead, it identified several development-related critical uncertainties that are largely beyond the control of local decision makers and speculated on how these external pressures might drive local conditions. As such, this study is perhaps better understood as a kind of geographic risk analysis.

To once more reiterate the intentions and uses of this study, by focusing on critical uncertainties rather than broad and collective goals, the possible futures which are considered in this report reflect concerns about what tomorrow *might* become, not preferences for what it *should* become. In each of the scenarios, there may be what some would call winners and losers. Similarly, it is also possible that a person reading this report would very much want to live in some of the described alternative futures or, equally possible, not live in any of them. While such assessments could mark the success or failure of a comprehensive regional planning initiative, they are incidental to the goals of this current work. Instead, the four alternative futures developed for analysis are best understood as different contextual models against which the risks and opportunities of local actions might be considered. For example, can possible actions to manage the lands of a military base succeed given different possible changes to the larger landscape? While concerns over possible regional growth pressures on military land management were the impetus of this project, the underlying concerns are common to other organizations, agencies, and institutions that are responsible for the conservation or preservation of the region's natural resources.

The critical uncertainties identified in this investigation remain difficult to predict, yet their potential consequences on the area could be significant. Scenarios of the future, such as have been investigated in this report, place these issues within a framework of understanding and can serve as a means to explore how local decisions might fare over the long term. It is the hope of the researchers that the approach taken in this study and its results can be used by all of the region's stakeholders as they look to make their own decisions.

NOTES:

[1] Glen Martin, "No New Dams Foreseen for Water Needs; Southern California Would Bear Brunt of Interior's Plan," *San Francisco Chronicle* (December 29, 1992), p. A23.

[2] "Baja California Studies 300MW Wind Possibilities," *Business News Americas* (October 26, 2002).

[3] "Mexico–Energy: Mexico to Sell Solar–Power Plant Rights in Mexicali," EFE News Services (U.S.), Inc. (February 5, 2003).

[4] Gerry Hadden, "Generating U.S. Electricity in Mexico," *All Things Considered*, National Public Radio (May 29, 2003). Available on-line at <http://discover.npr.org/features/feature.jhtml?wfld=1279431>.

[5] E.O. Wilson, "The Current State of Biodiversity" in E.O. Wilson and Frances M. Peter, eds., *Biodiversity* (Washington, D.C.: National Academy Press, 1986), pp. 3–18.

[6] Gary K. Meffe, C. Ronald Carroll, and contributors, *Principles of Conservation Biology*, 2nd edition (Sunderland, Massachusetts: Sinauer Associates, 1994); Michael E. Soulé, ed., *Viable Populations for Conservation* (Cambridge, England: Cambridge University Press, 1987).

[7] Jerry F. Franklin, "Preserving Biodiversity: Species, Ecosystems, or Landscape," *Ecological Applications* 3:2 (1993), pp. 202–205.

Bibliography

Ackoff, Russell. *Creating the Corporate Future: Plan or Be Planned For* (New York: John Wiley & Sons, 1981).

Adams, Charles F. and Howard B. Fleeter, Yul Kim, Mark Freeman, Imgon Cho. "Flight from Blight and Metropolitan Suburbanization Revisited," *Urban Affairs Review* 31:4 (March 1996), pp. 529–543.

Akcakaya, H. Resit and Jonathan L. Atwood. "A Habitat–based Metapopulation Model of the California Gnatcatcher," *Conservation Biology* 11:2 (April 1997), pp. 422–434.

Alberta Environment and Olson+Olson Planning & Design. *The Southern Rockies Landscape Planning Pilot Study: Summary Report* (Edmonton, Canada: Alberta Environment, Land and Forest Service, Integrated Resource Management Division, 2000).

Alexander, Roy and Andrew C. Milligan, eds. *Vegetation Mapping* (New York: John Wiley & Sons, 2000).

Alternatives for Washington, 11 volumes (Olympia, Washington: Office of Program Planning and Fiscal Management, 1974–1976)

American Ornithologists' Union. *Check–list of North American Birds*, 6th edition (Washington, D.C.: American Ornithologists' Union, 1983).

American Rivers, Press Release. "San Pedro River Named One of the Nation's Most Endangered Rivers" (Washington, D.C.: April 12, 1999). Available on-line at <http://www.amrivers.org/pressrelease/pressmersanpeddro1999.htm>.

An Original Bill to Provide for the Designation and Conservation of Certain Lands in the States of Arizona and Idaho, and For Other Purposes.

Public Law 100–696 (November 18, 1988), 16 United States Code § 460xx(a) (1994). Also known as Arizona–Idaho Conservation Act of 1988.

Andrews, Kenneth R. *The Concept of Corporate Strategy* (Homewood, Illinois: Dow–Jones Irwin, 1970).

Arnold, J.R. and R. Wiener. *The U.S. Army Corps of Engineers and Natural Resources Management on Army Installations 1941 – 1987* (Fort Belvoir, Virginia: Engineering and Housing Support Center and Office of History, U.S. Army Corps of Engineers, 1989).

Atwood, Jonathan L. "California Gnatcatchers and Coastal Sage Scrub: The Biological Basis for Endangered Species Listing" in J. E. Keeley, ed., *Interface between Ecology and Land Development in California* (Los Angeles, California: Southern California Academy of Sciences, 1993), pp. 149–169.

Atwood, Jonathan L. and Jeffrey S. Bolsinger. "Elevational Distribution of California Gnatcatchers in the United States," *Journal of Field Ornithology* 63:2 (Spring 1992), pp. 159–168.

Bahre, Conrad. "Human Impacts on the Grasslands of Southeastern Arizona" in *The Desert Grassland*, Mitchel P. McClaran and Thomas R. Van Devender, eds. (Tucson, Arizona: The University of Arizona Press, 1995), pp. 230–264.

Bahre, Conrad. *A Legacy of Change* (Tucson, Arizona: The University of Arizona Press, 1991).

Bailey, Harold P. *The Climate of Southern California* (Berkeley, California: University of

California Press, 1966).

Bailey, Robert G. *Description of the Ecoregions of the United States*, U.S. Forest Service Miscellaneous Publication No. 1391 (Ogden Utah: U.S. Department of Agriculture, 1995).

"Baja California Studies 300MW Wind Possibilities," *Business News Americas* (October 26, 2002).

Barel, Yves. *Prospective et Analyse de Systèmes* (Paris: La Documentation Française, 1971).

Barnes, Jr., Harry H. *Characteristics of Natural Channels*, U.S. Geological Survey Water Supply Paper 1849 (Washington, D.C: U.S. Government Printing Office, 1967). Available on-line <http://wwwrcamnl.wr.usgs.gov/sws/fieldmethods/Indirects/nvalues/>. The images from Barnes can also be found on-line at <http://manningsn.sdsu.edu/>.

Barney, G.O. "The Global 2000 Report to the President and the Threshold 21 Model: Influences of Dana Meadows and System Dynamics," *System Dynamics Review* 18:2 (Summer 2002), pp. 123–136.

Basiuk, Victor. "Technology, Western Europe's Alternative Futures, and American Policy," *Orbis—A Journal of World Affairs* 15:2 (Summer 1971), pp. 485–506.

Bauder, Ellen, D. Ann Kreager, and Scott C. McMillan. *Vernal Pools of Southern California: Recovery Plan* (Portland, Oregon: U.S. Department of the Interior Fish and Wildlife Service, 1998).

Beck, Ulrich. *Risk Society: Towards a New Modernity*, trans. Mark Ritter, (London: Sage Publications, 1992).

Beck, Ulrich and Johannes Willms, *Conversations with Ulrich Beck*, trans. Michael Pollak, (Cambridge: Polity, 2004).

Begon, Michael, John L. Harper, and Colin R. Townsend. *Ecology: Individuals, Populations, and Communities*, 2nd edition (Boston, Massachusetts: Blackwell Scientific Publications, 1990).

Beier, Paul. "Determining Minimum Habitat Areas and Habitat Corridors for Cougars," *Conservation Biology* 7:1 (1993), pp. 94 – 108.

Beier, Paul. "Dispersal of Juvenile Cougars in Fragmented Habitats," *Journal of Wildlife*

Management 59:2 (1995), pp. 228–237.

Beiman, Leo, Jerome Friedman, Charles J. Stone, and R. A. Olshen. *Classification and Regression Trees* (Boca Raton, Florida:CRC Press, 1984).

Beimborn, Edward, Rob Kennedy, and William Schaefer. *Inside the Blackbox: Making Transportation Models Work for Livable Communities* (New York: Citizens for a Better Environment and the Environmental Defense Fund, 1996).

Bell, Daniel. "Twelve Modes of Prediction—A Preliminary Sorting of Approaches in the Social Sciences," *Dædalus—Proceedings of the American Academy of Arts and Sciences* 93:3 (Summer 1964), pp. 845–888.

Bell, Wendell. *Foundations of Future Studies: Human Science for a New Era, Volume 1—History, Purposes, and Knowledge* (New Brunswick, New Jersey: Transaction Publishers, 1996).

Bengston, David N., George Xu, and David P. Fan. "Attitudes Toward Ecosystem Management in the United States, 1992–1988," *Society and Natural Resources* 14:6 (July 2001), pp. 471–487.

Benjamin, M.T., M. Sudol, D. Vorsatz, and A.M. Winer. "A Spatially and Temporally Resolved Biogenic Hydrocarbon Emissions Inventory for the California South Coast Air Basin," *Atmospheric Environment* 31 (1997), pp. 3087–3100.

Benjamin, M.T., M. Sudol, L. Bloch, and A.M. Winer. "Low–Emitting Urban Forests: A Taxonomic Methodology for Assigning Isoprene and Monoterpene Emission States. *Atmospheric Environment* 30 (1996), pp. 1437–1452.

Beres, Louis Rene and Harry R. Targ. *Constructing Alternative World Futures: Reordering the Planet* (Cambridge, Massachusetts: Schenkman Publishing Company, Inc., 1977).

Berger, Gaston. "The Prospective Attitude" in André Cournand and Maurice Lévy, eds., *Shaping the Future: Gaston Berger and the Concept of Prospective* (New York: Gordon and Breach Science Publishers, 1973), pp. 245 – 249; Originally published in *Prospective*, Issue No. 1 (1958), pp. 1–10.

Berger, Gaston. "Culture, Quality, and Freedom"

in André Cournand and Maurice Lévy, eds., *Shaping the Future: Gaston Berger and the Concept of Prosepctive*, (New York: Gordon and Breach Science Publishers, 1973), pp. 29 – 34; Originally published in *Prospective*, Issue No. 4 (1959), pp. 1–11.

Biggs, Reinnette, Ciara Raudsepp–Hearne, Carol Atkinson–Palombo, Erin Bohensky, Emily Boyd, Georgina Cundill, Helen Fox, Scott Ingram, Kasper Kok, Stephanie Spehar, Maria Tengo, Dagmar Timer, and Monika Zurek. "Linking Futures across Scales: A Dialog on Multiscale Scenarios," *Ecology and Society* 12:1 [Article 17] (2007) <<http://www.ecologyandsociety.org/vol12/iss1/art17>>.

Bilgin, Pinar. "Alternative Futures for the Middle East," *Futures* 33:5 (June 2001), pp. 423–436.

Biodiversity Research Consortium. *Objectives* (May 21, 1997). Available on-line at <http://bufo.geo.orst.edu/brc/objectives.html>.

Black, Charles and Paul H. Zedler. "An Overview of 15 years of Vernal Pool Restoration and Construction Activities in San Diego County, California" in Wiltham, C.W., E.T. Bauder, D. Belk, W.R. Ferren Jr., and R. Ornduff, eds., *Ecology, Conservation, and Management of Vernal Pool Ecosystems*, Proceedings from a 1996 Conference (Sacramento, California: California Native Plant Society, 1998), pp. 195–205.

Bluet, Jean–Claude and Josée Zémor. "Prospective Géographique Méthode et Directions de Recherche," *Metra* 9:1 (March 1970), pp. 111–127.

Bolger, Douglas T., Thomas A. Scott, and John T. Rotenberry. "Breeding Bird Abundance in an Urbanizing Landscape in Coastal Southern California," *Conservation Biology* 11:2 (April 1997), pp. 406–421.

Bonnett, Thomas W. and Robert L. Olson. *Scenarios of State Government in the Year 2010* (Washington, D.C.: Council of Governors' Policy Advisors, 1993).

Boulding, Kenneth E. "World Society: The Range of Possible Futures," in Elise and Kenneth E. Boulding, *The Future: Images and Processes* (Thousand Oaks, California: Sage Publications, 1995), pp. 39–56.

Braden, Gerald T., Robert L. McKernan, and Shawn M. Powell. "Effects of Nest Parasitism by the Brown–headed Cowbird on Nesting Success of the California Gnatcatcher," *Condor* 99:4 (November 1997), pp. 858–865.

Brothers, Timothy S. *Riparian Species Distributions in Relation to Stream Dynamics, San Gabriel River, California*, Ph.D. Dissertation (Los Angeles, California: University of California, 1985).

Brown, Bryan T. "Rates of Brood Parasitism by Brown–headed Cowbirds on Riparian Passerines in Arizona," *Journal of Field Ornithology* 65:2 (1994), pp. 160–168.

Brown, David E. *Biotic Communities: Southwestern United States and Northwestern Mexico* (Salt Lake City, Utah: University of Utah Press, 1994).

Brown, Lester. *World without Borders* (New York: Random House, 1972).

Bunn, Derek W. and Ahti A. Salo. "Forecasting with Scenarios," *European Journal of Operations Research* 68 (1993), pp. 291–303.

Burcham, L.T. "Historical Backgrounds of Range Land Use in California," *Journal of Range Management* 9 (1956), pp. 81–86.

Burcham, L.T. *California Range Land* (Sacramento, California: California Division of Forestry, 1957).

Burden, Frank R. *Environmental Monitoring Handbook* (New York, New York: McGraw-Hill, 2002).

Burt, William H. and Richard P. Grossenheider. *A Field Guide to the Mammals*, 3rd edition (New York: Houghton Mifflin Company, 1990).

Cahn, Matthew Alan. *Environmental Deceptions: The Tension between Liberalism and Environmental Policymaking in the United States* (Albany, New York: State University of New York Press, 1995).

Calder, I.R. "Hydrologic Effects of Land–Use Change" in D.R. Maidment, ed., *Handbook of Hydrology* (New York: McGraw–Hill, 1993).

California Air Resources Board. *Emission Inventory Procedural Manual, Vol. III: Methods for Assessing Area Source Emissions* (Sacramento, California: California Air Resources Board, Sacramento, 1997).

California Air Resources Board. *User's Guide for EMFAC2002* (Sacramento, California: California Air Resources Board, Sacramento, 2002). Software and documentation available online at <http://www.arb.ca.gov/msei/onroad/latest_version.htm>.

California Defense Retention and Conversion Act of 1999. California Statute § 425 (September 16, 1999). Also known as the Knight Bill.

Callaway, Ragan M. and Frank W. Davis. "Vegetation Dynamics, Fire, and the Physical Environment in Coastal Central California," *Ecology* 74:5 (1993), pp. 1567–1578.

Capistrano, Doris, Cristian K. Samper, Marcus J. Lee, and Ciara Raudsepp-Hearne, eds. *Ecosystems and Human Well-Being: Multiscale Assessments* [Millennium Ecosystem Assessment, volume 4] (Washington, D.C.: Island Press, 2005).

Carpenter, Stephen R. "Ecological Futures: Building an Ecology of the Long Now," *Ecology* 83:8 (August 2002), pp. 2069–2083.

Carpenter, Steve R., Prabhu L. Pingali, Elena M. Bennett, and Monika B. Zurek, eds. *Ecosystems and Human Well-Being: Scenarios* [Millennium Ecosystem Assessment, volume 2] (Washington, D.C.: Island Press, 2005).

Caughley, Graeme. *Conservation Biology in Theory and Practice* (Cambridge, Massachusetts: Blackwell Science, 1996).

Cerda, Xim, Javier Retana, and Sebastia Cros. "Thermal Disruption of Transitive Hierarchies in Mediterranean Ant Communities," *Journal of Animal Ecology* 66 (1997), pp. 363–374.

Chapman, Anne W. *The National Training Center Matures: 1985 – 1993* (Fort Monroe, Virginia: United States Army Training and Doctrine Command, 1997).

Chatman, Seymour. *Story and Discourse: Narrative Structure in Fiction and Film* (Ithaca, New York: Cornell University Press, 1980).

Chow, Ven Te., ed. *Handbook of Applied Hydrology: A Compendium of Water-Resources Technology* (New York: McGraw-Hill, 1964).

Cialdini, Robert B. *Influence: The Psychology of Persuasion* (New York: Quill–William Morrow and Company, 1993).

Clements, Frederick E. *Plant Succession: An Analysis of the Development of Vegetation*, Carnegie Institute of Washington Publication 242 (Washington, D.C.: Carnegie Institute of Washington, 1916).

Cole, F. Russell, Arthur C. Medeiros, Lloyd L. Loope, and William W. Zuehlke. "Effects of the Argentine Ant on Arthropod Fauna of Hawaiian High–elevation Shrubland," *Ecology* 73:4 (1992), pp. 1313–1322.

Cole, H.S.D., Christopher Freeman, Marie Jahoda, and K.L.R. Pavit, eds. *Models of Doom: A Critique of the Limits to Growth* (New York: Universe Books, 1973).

Collins, Paul W. "Rufous–crowned Sparrow" in A. Poole and F. Gill, eds., *The Birds of North America* 372 (Philadelphia, Pennsylvania: The Birds of North America, Inc. 1999).

Corell, Steven W. *Groundwater Flow Model Scenarios of Future Groundwater and Surface Water Conditions: Sierra Vista Subwatershed of the Upper San Pedro Basin–Southeastern Arizona*, Supplement to Modeling Report No. 10 (Phoenix, Arizona: Arizona Department of Water Resources, November 1996).

Commission for Environmental Cooperation. *Ribbon of Life: An Agenda for Preserving Transboundary Migratory Bird Habitat on the Upper San Pedro River* (Montreal, Canada: Communications and Public Outreach Department of the CEC Secretariat, 1999). Available on-line at <http://www.cec.org/files/PDF/sp–engl_EN.pdf>.

Congressional Research Service. *Ecosystem Management and Federal Agencies*, Report # 94–339–ENR (Washington, D.C.: Library of Congress, April 19, 1994).

Costanza, Robert, Alexey Voinov, Roelof Boumans, Thomas Maxwell, Ferdinando Vila, Lisa Wainger, and Helena Voinov. "Integrated Ecological Economic Modeling of the Patuxent River Watershed, Maryland," *Ecological Monographs* 72:2 (May 2002), pp. 203–231.

Costanza, Robert. "Four Visions of the Century Ahead: Will It be Star Trek, Ecotopia, Big Government, or Mad Max?" *The Futurist* 33:2 (February 1999), pp. 23–28.

Costanza, Robert. "Visions of Alternative (Unpredictable) Futures and Their Use in Policy

Analysis," *Conservation Ecology* (On-line Edition) 4:1, Article No. 5 (2000).

Cramer, J.S. *The Logit Model: An Introduction for Economists* (London: Edward Arnold, 1991).

Cronon, William. *Nature's Metropolis: Chicago and the Great West* (New York: Norton, 1991).

Daly, Herman E. and John B. Cobb, Jr. *For the Common Good: Redirecting the Economy toward Community the Environment, and a Sustainable Future*, 2nd edition (Boston, Massachusetts: Beacon Press, 1994).

Dator, Jim A. and Sharon J. Rodgers. *Alternative Futures for the State Courts of 2020* (Washington, D.C.: State Justice Institute; Chicago, Illinois: American Judiciary Society, 1991).

de Jouvenel, Bertrand. "Utopia for Practical Purposes" in Frank E. Manuel, ed., *Utopias and Utopian Thought* (Boston: Beacon Press, 1967), pp. 219–235.

de Lange, Aart R. "A Dynamic Input–Output Model for Investigating Alternative Futures— Applications to the South–African Economy," *Technological Forecasting and Social Change* 18:3 (1980), pp. 235–245.

de Neufville, Richard and Ralph L. Keeney. "Use of Decision Analysis in Airport Development for Mexico City," in Alvin W. Drake, Ralph L. Keeney, and Philip M. Morse, eds., *Analysis of Public Systems* (Cambridge, Massachusetts: Massachusetts Institute of Technology Press, 1972), pp. 497–519.

Deméré, Thomas A. "Faults and Earthquakes in San Diego County." Available on-line at <http://www.sdnhm.org/research/paleontology/sdfaults.html>.

Dobkin, D. "B487 Rufous–crowned Sparrow, *Aimophila ruficeps*," *California Wildlife Habitat Relationships System*, Version 7.0, spp. B487 (Sacramento, California: California Department of Fish and Game, 2001).

Dodson, R.D. "Advances in Hydrologic Computation" in David R. Maidment, ed., *Handbook of Hydrology* (New York: McGraw–Hill, 1993).

Downs, Anthony. "Alternative Forms of Future Urban Growth in the United States," *Journal of the American Planning Institute* 36:1 (January 1970), pp. 3–11.

Downs, Anthony. "Alternative Futures for the American Ghetto," *Dædalus—Proceedings of the American Academy of Arts and Sciences* 97:4 (Fall 1968), pp. 1331–1378; also re–printed in *The Appraisal Journal* 36:4 (October 1968), pp. 486–531.

Downs, Anthony. *Opening Up the Suburbs: An Urban Strategy for America* (New Haven, Connecticut: Yale University Press, 1973).

Dramstad, Wenche E., James D. Olson, and Richard T. T. Forman. *Landscape Ecology Principles in Landscape Architecture and Land–use Planning* (Washington, DC: Island Press, 1996).

Ducot, C. and G.J. Lubben. "A Typology for Scenarios," *Futures* 12:1 (February 1980), pp. 51–57.

Dunne, T. and L.B. Leopold. *Water in Environmental Planning* (San Francisco: W.H. Freeman, 1978).

Ecological Applications 6:3 (August 1996), pp. 692–974.

Edwards Jr., Thomas C., Collin G. Homer, Scott D. Bassett, Allan Falconer, R. Douglas Ramsey, and Doug W. Wight. *Utah Gap Analysis: An Environmental Information System*, Final Project Report 95–1 (Logan, Utah: Utah Cooperative Fish and Wildlife Research Unit, Utah State University, 1995).

Ehrlich, Paul R. and Edward O. Wilson. "Biodiversity Studies: Science and Policy," *Science* 253:5021 (August 16, 1991), pp. 758–762.

Elkington, John and Alex Trisoglio. "Developing Realistic Scenarios for the Environment: Lessons from Brent Spar," *Long Range Planning* 29:6 (December 1996), pp. 762–769.

Emery, F.E. and E.L. Trist. "The Causal Texture of Organizational Environments," *Human Relations* 18:1 (February 1965), pp. 21–32.

Emery, F.E. and E.L. Trist. *Towards a Social Ecology: Contextual Appreciations of the Future in the Present* (New York: Plenum Press, 1973).

Environ International Corporation. *User's Guide for CAMx* (Novato, California: Environ, 2003). Software and documentation are available online at <http://www.camx.com>.

Environmental Modeling Systems, Inc. *The Watershed Modeling System* (South Jordan, Utah: 2002). Also available on-line <http://www.environmental-center.com/software/ems–I/watershed.htm>.

Evered, Roger. "So What Is Strategy?" *Long Range Planning* 16:3 (June 1983), pp. 57–72.

Ewing, Reid. "Is Los Angeles-Style Sprawl Desirable?" *Journal of the American Planning Association* 63:1 (Winter 1997), pp. 107–126.

Fahey, Liam and Robert M. Randall. "What Is Scenario Learning?" in Liam Fahey and Robert M. Randall, eds., *Learning from the Future: Competitive Foresight Scenarios* (New York: John Wiley & Sons, 1998), pp. 3–21.

Farrand Jr., John. *National Audubon Society Field Guide to North American Birds: Western Region*, 2nd edition (New York: Alfred A. Knopf, Inc., 1994).

Fernandez–Armesto, Felipe. *Civilizations: Culture, Ambition, and the Transformation of Nature* (New York: Free Press, 2001).

Ferren Jr., Wayne R., David M. Hubbard, Anuja K. Parikh, and Nathan Gale. "Review of Ten Years of Vernal Pool Restoration and Creation in Santa Barbara, California," in Wiltham, C.W., E.T. Bauder, D. Belk, W.R. Ferren Jr., and R. Ornduff, eds., *Ecology, Conservation, and Management of Vernal Pool Ecosystems*, Proceedings from a 1996 Conference (Sacramento, California: California Native Plant Society, 1998), pp. 206–216.

Fisher, Robert N., Andrew V. Suarez, and Ted J. Case. "Spatial Patterns in the Abundance of the Coastal Horned Lizard," *Conservation Biology* 16:1 (2002), pp. 211–214.

Flaspohler, David J., Brian R. Bub, and Beth A. Kaplin. "Application of Conservation Biology Research to Management," *Conservation Biology* 14:6 (December 2000), pp. 1898–1902.

Flather, Curtis H., Linda A. Joyce, and Carol A. Bloomgarden. *Species Endangerment Patterns in the United States*, U.S. Forest Service General Technical Report # RM–241 (Fort Collins, Colorado: U.S. Department of Agriculture, Forest Service, Rocky Mountain Forest and Range Experiment Station, 1994).

Forest and Bird. "Argentine Ants—Lets Stop

the Invasion" (2003). Available on-line at <http://www.forestandbird.org.nz/Biosecurity/argentinants.asp>.

Forman, Richard T.T. *Land Mosaics: The Ecology of Landscapes and Regions* (Cambridge, U.K.: Cambridge University Press, 1995).

Forman, Richard T.T. and Michel Godron. *Landscape Ecology* (New York: John Wiley & Sons, 1986).

Forrester, Jay W. *World Dynamics* (Cambridge, Massachusetts: Wright-Allen Press, 1971).

Fralish, James S. and Scott B. Franklin. *Taxonomy and Ecology of Woody Plants in North American Forests* (New York: John Wiley & Sons, Inc., 2002).

Franklin, Jerry F. "Preserving Biodiversity: Species, Ecosystems, or Landscape," *Ecological Applications* 3:2 (1993), pp. 202–205.

Franzreb, K. E. *Ecology and Conservation of the Endangered Least Bell's Vireo* (Sacramento, California: U.S. Fish and Wildlife Service Endangered Species Office, 1989).

Freilich, Robert H. and Bruce G. Peshoff. "The Social Costs of Sprawl," *Urban Lawyer* 29:2 (Spring 1997), pp. 183–198.

Friedman, Jerome H. "Multivariate Adaptive Regression Splines," *The Annals of Statistics* 19:1 (March 1991), pp. 1–141.

Gates, J. Edward and Daniel R. Evans. "Cowbirds Breeding in the Central Appalachians: Spatial and Temporal Patterns and Habitat Selection," *Ecological Applications* 8:1 (February 1998), pp. 27–40.

George, T. Luke, Lowell C. McEwen, and Brett E. Peterson. "Effects of Grasshopper Control Programs on Rangeland Breeding Bird Populations," *Journal of Range Management* 48 (1995), pp. 336–342.

Gertler, Alan W., Julide Kahyaoglu, Darko Koracin, Menachem Luria, William Stockwell, and Erez Weinroth, "Development and Validation of a Predictive Model to Assess the Impact of Coastal Operations on Urban Scale Air Quality," Strategic Environmental Research and Development Program (SERDP) Project No. CP–1253 (Arlington, Virginia: SERDP, July 31, 2006).

Giddens, Anthony. *The Consequences of Modernity*,

(Palo Alto, California: Stanford University Press, 1990).

Gleason, Henry. "The Individualistic Concept of the Plant Association," *Bulletin of the Torrey Botanical Club* 53 (1926), pp. 1–20.

Goldberg, Stephen R. "Reproduction of the Coast Horned Lizard, *Phrynosoma coronatum*, in Southern California," *The Southwestern Naturalist* 28:4 (1983), pp. 478–479.

Gonzalez, Richard J., Jeff Drazen, Stacie Hathaway, Brent Bauer, and Marie Simovich. "Physiological Correlates of Water Chemistry Requirements in Fairy Shrimps (*Anostraca*) from Southern California," *Journal of Crustacean Biology* 16:2 (1996), pp. 315–322.

Goodman, Sherri W. "Foreword" in Michele Leslie, Gary K. Meffe, Jeffrey L. Hardesty, and Diane L. Adams, *Conserving Biodiversity on Military Lands: A Handbook for Natural Resources Managers* (Arlington, Virginia: The Nature Conservancy, 1996).

Grell, G.A., J. Dudhia, and D.R. Stauffer. *A Description of the Fifth-Generation Penn State/ NCAR Mesoscale Model (MM5)*, NCAR/TN–398+STR (Boulder, Colorado: National Center for Atmospheric Research, 1995), 122 pp.

Gritesevskyi, Andrii and Nebojsa Nakicenovic. "Modeling Uncertainty of Induced Technological Change," *Energy Policy* 28:13 (November 2000), pp. 907–921.

Guenther, A.B., P.R. Zimmermann, and P.C. Harley. "Isoprene and Monoterpene Emission Rate Variability: Model Evaluations and Sensitivity Analyses," *JGR Atmospheres* 98 (D7), pp. 12609–12617.

Hadden, Gerry. "Generating U.S. Electricity in Mexico," *All Things Considered*, National Public Radio (May 29, 2003). Available on-line at <http://discover.npr.org/features/feature. jhtml?wfld=1279431>.

Haire, S. L., C. E. Bock, B. S. Cade, and B. C. Bennett. "The Role of Landscape and Habitat Characteristics in Limiting Abundance of Grassland Nesting Songbirds in an Urban Open Space," *Landscape and Urban Planning* 48 (2000), pp. 65–82.

Halvorson, William L. "Changes in Landscape Values and Expectations: What Do We Want

and How Do We Measure It?" in R. Gerald Wright, ed., *National Parks and Protected Areas* (Cambridge, Massachusetts: Blackwell Science, Inc., 1996), pp. 15–30.

Hamblin, Ann, ed. *Visions of Future Landscapes, Proceedings of the 1999 Australian Academy of Science Fenner Conference on the Environment, 2–5 May 1999* (Canberra, Australia: Bureau of Rural Science, 2000). Available on-line at <http://www.brs.gov.au/events/fenner/vision. pdf>.

Hanlon, Jr., Edward. Testimony in U.S. House of Representatives, Committee on Government Reform, *Hearing on Challenges to National Security: Constraints on Military Training*, 107th Congress, 1st session (Washington, D.C.: U.S. Government Printing Office, May 9, 2001). Available on-line at <http://frwebgate.access. gpo.gov/cig–bin/getdoc.cgi?dbname=107_ house_hearings&docid=f:75041.pdf>.

Hansen, Kevin. *Cougar, the American Lion* (Flagstaff, Arizona: Northland Publishing, 1992).

Hare, F. Kenneth. "Future Environments: Can They Be Predicted?" *Transactions—Institute of British Geographers* 10:2 (1985), pp. 131–137.

Harley, P.C., V. Fridd-Stroud, J. Greenberg, A. Guenther, and P. Vasconcellos. *JGR Atmospheres* 103 (1998), pp. 25479–25486.

Hastings, Rodney and Raymond Turner. *The Changing Mile* (Tucson, Arizona: The University of Arizona Press, 1965).

Hathaway, Stacie A. and Marie A. Simovich. "Factors Affecting the Distribution and Co–occurrence of Two Southern Californian Anostracans (*Branchiopoda*), *Branchinecta sandiegonensis* and *Streptocephalus woottoni*," *Journal of Crustacean Biology* 16:4 (1996), p. 675.

Heady, Harold F. "Valley Grassland" in Michael G. Barbour and Jack Major, eds., *Terrestrial Vegetation of California* (New York: John Wiley & Sons, 1990), pp. 491–514.

Helmer, Olaf. *Analysis of the Future: The Delphi Method*, RAND Corporation Paper P–3558 (Santa Monica, California: RAND Corporation, 1967).

Hess, Bill. "Environmental Drive–By Study," *Sierra*

Vista Herald/Bisbee Daily Review (June 15, 1997), p. 1A.

Hess, Bill. "Nazism, Communism, Now Environmentalism," *Sierra Vista Herald/Bisbee Daily Review* (June 24, 1997), p. 1A.

Heugens, Pursey R.M.A.R. and Johannes van Oosterhout. "To Boldly Go Where No Man Has Gone Before: Integrating Cognitive and Physical Features in Scenario Studies," *Futures* 33:10 (December 2001), pp. 861–872.

Hickman, James C. *The Jepson Manual: Higher Plants of California* (Berkeley, California: University of California Press, 1993).

Hirschhorn, Larry. "Scenario Writing: A Developmental Approach," *Journal of the American Planning Association* 46:2 (April 1980), pp. 172–182.

Holland, R. F. *Preliminary Descriptions of the Terrestrial Natural Communities of California* (Sacramento, California: The Resources Agency, Department of Fish and Game, Natural Heritage Division, 1986).

Hölldobler, Bert and Edward O. Wilson. *The Ants* (Cambridge, Massachusetts: Harvard University Press, 1990).

Holway, David A. "Distribution of the Argentine Ant (*Linepithema humile*) in Northern California," *Conservation Biology* 9:6 (December 1995), pp. 1634–1637.

Holway, David A. "Effect of Argentine Ant Invasions on Ground-dwelling Arthropods in Northern California Riparian Woodlands," *Oecologia* 116 (1998), pp. 252–258.

Holway, David A. "Factors Governing Rate of Invasion: A Natural Experiment Using Argentine Ants," *Oecologia* 115 (1998), pp. 206–212.

Holway, David A. and Ted J. Case. "Mechanisms of Dispersed Central-place Foraging in Polydomous Colonies of the Argentine Ant," *Animal Behaviour* 59 (2000), pp. 433–441.

Hooge, Philip N., Mark T. Stanback, and Walter D. Koeing. "Nest-site Selection in the Cooperatively Breeding Acorn Woodpecker," *The Auk* 116:1 (1999), pp. 45–54.

Hopkins, Lewis D. and Marisa A. Zapata, Engaging the Future: Forecasts, Scenarios, Plans, and Projects (Cambridge, Massachusetts: Lincoln Institute of Land Policy, 2007).

Horgan, John. "It's Not Easy Being Green," *New York Times Book Review* 147:50,670 (January 12, 1997), p. 8.

Horie, Y., S. Sidawi, and R. Ellefsen. *Inventory of Leaf Biomass and Emission Factors for Vegetation in the South Coast Air Basin*, Technical Report, Contract No. 90163 (Diamond Bar, California: South Coast Air Quality Management District, 1991).

Houston, C. Stuart, Dwight G. Smith, and Christoph Rohner. "Great Horned Owl: *Bubo virginianus*" in A. Poole and F. Gill, eds., *The Birds of North America* 372 (Philadelphia, Pennsylvania: The Birds of North America, Inc. 1998).

Hulse, David, Joseph Eilers, Kathryn Freemark, Cheryl Hummon, and Denis White. "Planning Alternative Future Landscapes in Oregon: Evaluating the Effects of Water Quality and Biodiversity," *Landscape Journal* 19:1&2 (2000), pp. 1–19.

Human, Kathleen G. and Deborah M. Gordon. "Effects of Argentine Ants on Invertebrate Biodiversity in Northern California," *Conservation Biology* 11:5 (October 1997), pp. 1242–1248.

Hunter, Malcolm L. *Fundamentals of Conservation Biology*, 2nd edition. (Malden, Massachusetts, Blackwell Science, 2002).

Janis, Irving L. *Victims of Groupthink: A Psychological Study of Foreign–Policy Decisions and Fiascoes* (Boston, Massachusetts: Houghton Mifflin Company, 1972).

Jantsch, Erich. *Technological Forecasting in Perspective* (Paris: Organisation for Economic Co–Operation and Development, 1967).

Jayaweera, Mahesh and Takashi Asadea. "Impacts of Environmental Scenarios on Chlorophyll–*a* in the Management of Shallow, Eutrophic Lakes Following Biomanipulation: An Application of a Numerical Model," *Ecological Engineering* 5 (1995), pp. 445–468.

Johnson, Douglas H. and Lawrence D. Igl. "Area Requirements of Grassland Birds: A Regional Perspective," *The Auk* 118:1 (2001), pp. 24–34.

Johnson, Michael P. "Environmental Impacts of Urban Sprawl: A Survey of the Literature and Proposed Research Agenda," *Environment and Planning A* 33:4 (April 2001), pp. 717–735.

Julien, P.Y.. B. Dahafian, and F.L. Ogden. "Raster–
Based Hydrologic Modeling of Spatially-Varied
Surface Runoff," *Water Resources Bulletin*
31(1995), pp. 523–536.

Jungermann, Helmut. "Inferential Processes in
the Construction of Scenarios" *Journal of
Forecasting* 4:4 (1985), pp. 321–327.

Kahane, Adam. "Imagining South Africa's Future:
How Scenarios Helped Discover Common
Ground," in Liam Fahey and Robert M. Randall,
eds., *Learning from the Future: Competitive
Foresight Scenarios* (New York: John Wiley &
Sons, 1998), pp. 325–332.

Kahn, Herman and Anthony J. Wiener. *The Year
2000: A Framework for Speculation on the Next
Thirty Years* (New York: Macmillan Company,
1967).

Kahn, Herman and Irwin Mann. *Techniques of
Systems Analysis*, RAND Report # RM–1829
(Santa Monica, California: RAND Corporation,
December 3, 1956).

Kahn, Herman. *On Alternative World Futures:
Issues and Themes—Draft*, Hudson Institute
Report # HI–525–D/3 (Croton–on–Hudson, New
York: The Hudson Institute, April 14, 1966).

Kaivo–oja, Jari. "Alternative Scenarios of Social
Development: Is Analytical Sustainability
Policy Analysis Possible? How?," *Sustainable
Development* 7 (1999), pp. 140–150.

Katibeh, Edwin F. "A Brief History of Riparian
Forests in the Central Valley of California," in
Richard E. Warner and Kathleen M. Hendrix,
eds., *California Riparian Ecosystems: Ecology,
Conservation, and Productive Management*
(Berkeley, California: University of California
Press, 1984), pp. 23–29.

Keeley, Jon E. "Chaparral" in Michael G. Barbour
and W. Dwight Billings, eds., *North American
Terrestrial Vegetation*, (Cambridge, England:
Cambridge University Press, 2000), pp. 203–
253.

Keeney, Ralph L. "A Decision with Multiple
Objectives: The Mexico City Airport," *Bell
Journal of Economics and Management Science*
4:1 (Spring 1973), pp. 101–107.

Keeney, Ralph L. and Howard Raiffa. *Decisions
with Multiple Objectives: Preferences and Value
Tradeoffs* (New York: John Wiley & Sons,

1976).

Kendeight, S. Charles. "Birds of a Prairie
Community," *Condor* 43 (1941), pp. 165–174.

Kepner, William G., C.M Edmonds, and Christopher
J. Watts. *Remote Sensing and Geographic
Information Systems for Decision Analysis
in Public Resource Administration: A Case
Study of 25 Years of Landscape Change in
a Southwestern Watershed*, EPA Report #
EPA/600/R–02/039 (Las Vegas, Nevada: U.S.
Environmental Protection Agency, Office of
Research and Development, National Exposure
Research Laboratory, June 2002).

Kerr, Jeremy T. "Species Richness, Endemism,
and the Choice of Areas for Conservation,"
Conservation Biology 11:5 (October 1997), pp.
1094–1100.

Keynes, John Maynard. "The General Theory of
Employment," *Quarterly Journal of Economics*,
51 (1937), pp. 209–223.

Keynes, John Maynard. *The General Theory of
Employment, Interest and Money* (New York:
Harcourt Brace, 1936 [reprint 1964].

Kleiner, Art. *The Age of Heretics: Heroes, Outlaws,
and the Forerunners of Corporate Change* (New
York: Currency Doubleday, 1996).

Knick, Steven T. and John T. Rotenberry.
"Landscape Characteristics of Fragmented
Shrubsteppe Habitats and Breeding Passerine
Birds," *Conservation Biology* 9:5 (1995), pp.
1059–1071.

Kok, K., R. Biggs, and M. Zurek. "Methods for
Developing Multiscale Participatory Scenarios:
Insights from Southern Africa and Europe,"
Ecology and Society 12:1 [Article 8] (2007)
<http://www.ecologyandsociety.org/vol12/iss1/
art8>.

Kucera, T. "B553 California Gnatcatcher, *Polioptila
californica*," *California Wildlife Habitat
Relationships System*, Version 7.0, spp. B553
(Sacramento, California: California Department
of Fish and Game, 2001).

Küchler, A. Will. *Vegetation Mapping* (New York:
Ronald Publishing, 1967).

Lafferty, Kevin D. "Disturbance to Wintering
Western Snowy Plovers," *Biological
Conservation* 101 (2001), pp. 315–325.

Lambin, E.F. and J.J. Geist. *Land–Use and Land–*

Cover Change: Local Processes and Global Impacts (New York: Springer, 2006).

Landis, John, Michael Reilly, Robert Twiss, Howard Foster, and Patricia Frontiera. *Forecasting and Mitigating Future Urban Encroachment Adjacent to California Military Installations: A Spatial Approach* (Sacramento, California: California Technology, Trade, and Commerce Agency, June 20, 2001). Available online at <http://www.regis.berkeley.edu/cttca/finaldocs/cttca_report062101.pdf>. Landrieu–Zémor, Josée. "Une Méthode d'Analyse Prospective son Élaboration dan le cadre d'un Scénario Tendanciel Français," *Metra* 10:4 (December 1971), pp. 569–626.

Lanyon, Wesley E. "*Sturnella neglecta*" in A. Poole and F. Gill, eds., *The Birds of North America* 104 (Philadelphia, Pennsylvania: The Birds of North America, Inc. 1994).

Lanyon, Wesley E. "The Comparative Biology of the Meadowlarks (*Sturnella*) in Wisconsin," *Nuttal Ornithological Club* 1 (1957).

Leites, Nathan C. *A Study of Bolshevism* (Glencoe, Illinois: The Free Press, 1953).

Leites, Nathan C. *The Operational Code of the Politburo* (New York: McGraw–Hill, 1951).

Leites, Nathan C. and Constantin Melnik. *The House without Windows: France Selects a President* (Evanston, Illinois: Row, Peterson & Company, 1958).

Leslie, Michele, Gary K. Meffe, Jeffrey L. Hardesty, and Diane L. Adams, *Conserving Biodiversity on Military Lands: A Handbook for Natural Resources Managers* (Arlington, Virginia: The Nature Conservancy, 1996).

Liam Fahey and Robert M. Randall, eds. *Learning from the Future: Competitive Foresight Scenarios* (New York: John Wiley & Sons, 1998).

Liddell–Hart, Basil H. *Strategy* (New York: Praeger, 1965).

Linstone, Harlod A. and Murray Turoff, eds. *The Delphi Method: Techniques and Applications* (Reading, Massachusetts: Addison–Wesley Publishing Company, 1975).

Liotta, P.H. and Allan W. Shearer. "Zombie Concepts and Boomerang Effects: Uncertainty, Risk, and Security Intersection though the Lens of Environmental Change," in *Environmental Change and Human Security: Recognizing and Acting on Hazard Impacts*, P.H. Liotta, David A. Mouat, William G. Kepner, and Judith Lancaster (Eds.), (Dordecht, the Netherlands: Springer–Verlag, expected 2009).

Liverman, Diana M., Robert G. Varaday, Octavio Chávez, and Roberto Sánchez. "Environmental Issues along the United States–Mexico Border: Drivers of Change and Responses of Citizens and Institutions," *Annual Review of Energy and the Environment* 24 (1999), pp. 607–643. Available on-line at <http://las.arizona.edu/liverman/acree.pdf>.

McArthur, David S. "Geomorphology of San Diego County," in Philip R. Pryde, ed., *San Diego: An Introduction to the Region* (Dubuque, Iowa: Kendall Hunt Publishing Company, 1984), pp. 13–30.

McPherson, Guy R. "The Role of Fire in the Desert Grasslands" in Mitchel P. McClaran and Thomas R. Van Devender, eds., *The Desert Grassland* (Tucson, Arizona: The University of Arizona Press, 1995), pp. 130–151.

Madden, Elizabeth M., Andrew J. Hansen, and Robert K. Murphy. "Influence of Prescribed Fire History on Habitat and Abundance of Passerine Birds in Northern Mixed–grass Prairie," *The Canadian Field–Naturalist* 113 (1999), pp. 627–640.

Marien, Michael. "Herman Kahn's Things to Come," *The Futurist* 7:1 (February 1973), pp. 7–15.

Markley, Oliver W. *Alternative Futures: Contexts in Which Social Indicators Must Work*, EPRC Research Note 6747–11 (Menlo Park, California: Stanford Research Institute, Educational Policy Research Center, 1971).

Marshall, Robert M. and Scott H. Stoleson. "Threats" in Deborah M. Finch and Scott H. Stoleson, eds., *Status, Ecology, and Conservation of the Southwestern Willow Flycatcher*, General Technical Report RMRS–GTR–60, (Ogden, Utah: U.S. Department of Agriculture, Forest Service, Rocky Mountain Research Station, 2000), pp. 13–24.

Martin, Glen. "No New Dams Foreseen for Water Needs; Southern California Would Bear Brunt of Interior's Plan," *San Francisco Chronicle* (December 29, 1992), p. A23.

Mayer, Kenneth E. and William F. Laudenslayer, Jr. *A Guide to Wildlife Habitats of California* (Sacramento, California: Department of Forestry and Fire Protection, 1988).

Meadows, Donella H., Dennis L. Meadows. Jorgen Randers, and William W. Behrens III. *The Limits to Growth: A Report from the Club of Rome's Project on the Predicament of Mankind* (New York: Universe Books, 1972).

Meadows, Michael E. "Soil Erosion in the Swartland, Western Cape Province, South Africa: Implications of Past and Present Policy and Practice," *Environmental Science & Policy* 6:1 (February 2003), pp. 17 –28.

Meffe, Gary K. and C. Ronald Carroll. *Principles of Conservation Biology*, 2nd edition (Sunderland, Massachusetts: Sinauer Associates, 1994).

"Mexico–Energy: Mexico to Sell Solar-Power Plant Rights in Mexicali," EFE News Services (U.S.), Inc. (February 5, 2003).

Michael, Don. *On Learning to Plan—and Planning to Learn* (San Francisco: Jossey-Bass, 1973).

Mieszkowski, Peter and Edwin S. Mills. "The Causes of Metropolitan Suburbanization," *Journal of Economic Perspectives* 7:3 (Summer 1993), pp. 135–147.

Millennium Assessment, Ecosystems & Human Well–Being: Synthesis (Washington, D.C.: Island Press, 2005).

Mooney, Harold A. "Southern Coastal Scrub" in Michael G. Barbour and Jack Major, eds., *Terrestrial Vegetation of California* (New York: John Wiley & Sons, 1990), pp. 471–490.

Morey, S. "R029, Coast Horned Lizard, *Phrynosoma coronatum*," *California Wildlife Habitat Relationships System*, Version 7.0, spp. R029 (Sacramento, California: California Department of Fish and Game, 2001).

Morrell, Thomas E. and Richard H. Yahner. "Habitat Characteristics of Great Horned Owls in South-Central Pennsylvania," *Journal of Raptor Research* 28:3 (1994), pp. 164–170.

Morrison, Scott A. and Douglas T. Bolger. "Lack of an Urban Edge Effect on Reproduction in a Fragmentation-sensitive Sparrow," *Ecological Applications* 12:2 (April 2002), pp. 398–411.

Mouat, David, Mary E. Cablk, Jamie DeNormandie, Thomas C. Edwards, Jr., Robert Fisher, Manuel Gonzalez, Jill S. Heaton, Lori Hunter, Kimberly Karish, A. Ross Kiester, Robert Lilieholm, S. Mark Meyers, Natalie Robbins, Matt Stevenson, and Richard Toth. *Analysis and Assessment of Military and Non-Military Impacts on Biodiversity in the California Mojave Desert* (Reno, Nevada: Desert Research Institute, 2002).

National Association of Home Builders. *Smart Growth: Building Better Places to Live, Work and Play* (Washington, D.C.: National Association of Home Builders, 2000).

Noss, Reed F. and Allen Y. Cooperrider. *Saving Nature's Legacy: Protecting and Restoring Biodiversity* (Washington, D.C.: Island Press, 1994).

Nuclear Energy Agency. *Systematic Approaches to Scenario Development* (Paris: Organisation for Economic Co–operation and Development, 1992).

O'Neill, Karen. *Rivers by Design: State Power and the Origins of U.S. Flood Control* (Durham, North Carolina: Duke University Press, 2006).

Office of the Under Secretary of Defense (Acquisition and Technology). "To Deputy Under Secretary of Defense (Environmental Security) and Others," Subject: Implementation of Ecosystem Management in DoD (August 8, 1994).

Ogden, F.L. *CASC2D Reference Manual* (Storrs, Connecticut: University of Connecticut, Department of Civil and Environmental Engineering, 1997).

Ogden, F.L. and P.Y. Julien. "CASC2D: A Two-Dimensional, Physically–Based, Hortonian, Hydrologic Model" in V.J. Singh and D. Freverts, eds., *Mathematical Models of Small Watershed Hydrology and Applications*, (Littleton, Colorado: Water Resources Publications, 2002).

Omernik, J.M. "Map Supplement: Ecoregions of the Coterminous United States," *Annals of the Association of American Geographers* 77 (1987), pp. 118–125.

Orme, Amalie J. "The Mediterranean Environment of Greater California" in Antony R. Orme, ed.,

The Physical Geography of North America, (Oxford, England: Oxford University Press, 2002), pp. 402–424.

Ozbekhan, Hasan. "The Future of Paris: A Systems Study in Strategic Urban Planning," *Philosophical Transactions of the Royal Society London—A* 287 (1977), pp. 523–544.

Ozbekhan, Hasan. "The Triumph of Technology: 'Can' Implies 'Ought'" in Stanford Anderson, ed., *Planning For Diversity and Choice: Possible Futures and Their Relations to the Man-Controlled Environment* (Cambridge, Massachusetts: Massachusetts Institute of Technology Press, 1968), pp. 204–219.

Page, Gary W., Lynne E. Stenzel, and Christine A. Ribic. "Nest Site Selection and Clutch Predation in the Snowy Plover," *The Auk* 102 (April 1985), pp. 347–353.

Page, Gary W. and Lynne E. Stenzel. "The Breeding Status of the Snowy Plover in California," *Western Birds* 12:1 (1981), pp. 1–40.

Persky, Joseph and Wim Wiewel. *When Corporations Leave Town: The Costs and Benefits of Metropolitan Job Sprawl* (Detroit, Michigan: Wayne State University Press, 2000).

Peterson, Roger T. *Field Guide to Western Birds*, 3rd edition (New York: Houghton Mifflin Company, 1990).

Pianka, Eric R. and William S. Parker. "Ecology of Horned Lizards: A Review with Special Reference to *Phrynosoma platyrhinos*," *Copeia* 1975:1 (1975), pp. 141–162.

Pirages, Dennis C. and Paul R. Ehrlich. *Arc II: Social Response to Environmental Imperatives* (San Francisco: W.H. Freeman, 1973).

Polite, C. "B265 Great Horned Owl, *Bubo virginianus*" *California Wildlife Habitat Relationships System*, Version 7.0, spp. B265 (Sacramento, California: California Department of Fish and Game, 2001).

Powell, Abby. "Western Snowy Plovers and California Least Terns" (1999). Available on-line at <http:biology.usgs.gov/s+t/SNT/nofram/ca168.htm>.

Primack, Richard B. *Essentials of Conservation Biology*, 2nd edition (Sunderland, Massachusetts: Sinauer Associates, 1998).

Pryde, Philip R. "Climate, Soils, Vegetation, and Wildlife" in Philip R. Pryde. ed., *San Diego: An Introduction to the Region* (Dubuque, Iowa: Kendall Hunt Publishing Company, 1984), pp. 31–49.

Pryde, Philip R. "Introduction" in Philip R. Pryde, ed., *San Diego: An Introduction to the Region*, (Dubuque, Iowa: Kendall Hunt Publishing Company, 1984), pp. 1–12.

Purer, Edith A. "Ecological Study of Vernal Pools, San Diego County," *Ecology* 20:2 (June 1937), pp. 217–229.

Rasmussen, Larry L. *Earth Community Earth Ethics* (Maryknoll, New York: Orbis Books, 1996).

Rawls, W.J., D.L. Brakensiek, and N. Miller. "Green-Ampt Infiltration Parameters from Soils Data," *Transactions of the American Society of Agricultural Engineers* 26 (1983), pp. 62–70.

Rigney, M. "B154 Snowy Plover, *Charadrius alexandrinus*," *California Wildlife Habitat Relationships System*, Version 7.0, spp. B154 (Sacramento, California: California Department of Fish and Game, 2001).

Ringland, Gill. *Scenario Planning: Managing for the Future* (New York: John Wiley & Sons, 1998).

Robbins, W.W. "Alien Plants Growing without Cultivation in California," *California Agricultural Experiment Station Bulletin*, 637 (1940).

Rohner, Christoph and Charles J. Krebs. "Owl Predation on Snowshoe Hares: Consequences of Antipredator Behavior," *Oecologia* 108 (1996), pp. 303–310.

Rohner, Christoph. "Non-territorial "Floaters" in Great Horned Owls: Space Use during a Cyclic Peak of Snowshoe Hares," *Animal Behavior* 53 (1997), pp. 901–912.

Rosen, Walter G. "Letter regarding: 'Coining a Catchword,'" *New York Times Book Review* 147:50,712 (February 23, 1997), p. 2.

Rubenson, David, Marc Dean Millot, Gwen Farnsworth, and Jerry Aroesty. *More Than 25 Million Acres? DoD As a Federal, Natural, and Cultural Resource Manager*, RAND Corporation Report # MR–715–OSD (Santa Monica, California: RAND Corporation, 1996).

Rubinoff, Daniel. "Evaluating the California Gnatcatcher as an Umbrella Species for

Conservation of Southern California Coastal Sage Scrub," *Conservation Biology* 15:5 (October 2001), pp. 1374–1383.

Rushmer, Robert F. *Humanizing Health Care— Alternative Futures for Medicine* (Cambridge, Massachusetts: Massachusetts Institute of Technology Press, 1975).

Russell, Kenneth R. "Mountain Lion" in John L. Schmidt and Douglas L. Gilbert, eds., *Big Game of North America* (Harrisburg, Pennsylvania: Stockpole Books, 1978), pp. 207–225.

Ryan, John C. *Life Support: Conserving Biological Diversity*, Worldwatch Paper 108 (Washington, D.C.: Worldwatch Institute, April 1992).

Salata, L. *Status of the Least Bell's Vireo on Camp Pendleton, California* (Laguna Niguel, California: U.S. Fish and Wildlife Service, unpublished report).

San Diego Association of Governments and Regional Environmental Consultants. *Draft Comprehensive Species Management Plan for the Least Bell's Vireo, Vireo bellii pusillus* (San Diego, California: SANDAG, 1988).

San Diego Association of Governments. *2000 Land Use*. Available on-line at <http://www.sandag. org/resources/maps_and_gis/gis_downloads/ land.asp>.

San Diego Association of Governments. *San Diego Region Generalized Vegetation Map* (San Diego, California: San Diego Association of Governments, 1997).

Sanders, Nathan J., Kasey E. Barton, and Deborah M. Gordon. "Long–term Dynamics of the Distribution of the Invasive Argentine Ant, *Linepithema humile*, and Native Ant Taxa in Northern California," *Oecologia* 127 (January 2001), pp. 128–129.

Sauer, Carl O. "The Agency of Man on the Earth" in William L. Thomas, ed., *Man's Role in Changing the Face of the Earth* (Chicago, Illinois: University of Chicago Press, 1956), pp. 49–69.

Sawyer, John O. and Todd Keeler-Wolf. *A Manual of California Vegetation* (Sacramento, California: California Native Plant Society, 1995).

Scheid, Gerald A. "Habitat Characteristics of Willowy Monardella in San Diego County: Site

Selection for Transplants" in Thomas S. Elias, ed., *Conservation and Management of Rare and Endangered Plants* (Sacramento, California: California Native Plant Society, 1987), pp. 501–506.

Scholes R.J. and R. Biggs. *Ecosystem Services in Southern Africa: A Regional Assessment* (Pretoria, South Africa: Council for Scientific and Industrial Research, 2004).

Schoute, Job F. Th., Peter A. Finke, Frank R. Veeneklaas, and Henk P. Wolfert, eds. *Scenarios Studies for the Rural Environment* (Dordrecht, The Netherlands: Kluwer Academic Publishers, 1995).

Schwartz, Peter. *The Art of the Long View: Paths to Strategic Insight for Yourself and Your Company* (New York: Currency–Doubleday, [1991] 1996).

Scott, Annie. "Your Say: Environmental Scenario Planning," *Scenario & Strategy Planning* 3:6 (2002). Available on-line at <http://www. ssplanning.com/Articles/DisplayArticle. asp?PageID=97133882>.

Scott, J. Michael, Blair Csuti, James D. Jacobi, and Jack E. Estes. "Species Richness: A Geographic Approach to Protecting Future Biological Diversity," *BioScience* 37 (1987), pp. 782–788.

Sedgwick, James A. and Fritz L. Knopf. "A High Incidence of Brown–headed Cowbird Parasitism of Willow Flycatchers," *Condor* 90 (1988), pp. 253–256.

Shearer, Allan W. "Approaching Scenario–Based Studies: Three Perceptions about the Future and Considerations for Landscape Planning," *Environment and Planning B: Planning & Design* 32:1 (January 2005). pp. 67–87.

Shearer, Allan W., David A. Mouat, Scott D. Bassett, Michael W. Binford, Craig W. Johnson, and Justin A. Saarinen, "Examining Development– Related Uncertainties for Environmental Management: Strategic Planning Scenarios in Southern California," *Landscape and Urban Planning* 77 (2006), pp. 359–381.

Sheppard, Eric and Robert B. McMaster, eds., *Scale and Geographic Inquiry: Nature, Society, and Method* (Malden, Massachusetts: Blackwell, Publishing, 2004)

Shimwell, D.W. *The Description and Classification*

of Vegetation (Seattle, Washington: University of Washington Press, 1971).

Sibley, David A. *The Sibley Guide to Birds* (New York: Alfred A. Knopf, Inc., 2000).

Sidner, Ronnie. "A Bat Boom at Fort Huachuca," *Endangered Species Bulletin* 25:6 (November – December 2000), pp. 12–13. Available on-line at <http://endangered.fws.gov/esb/2000/11–12–13.pdf>.

Sogge, Mark K. "Breeding Season Ecology" in Deborah M. Finch and H. Stoleson, eds, *Status, Ecology, and Conservation of the Southwestern Willow Flycatcher*, General Technical Report RMRS–GTR–60, (Ogden, Utah: U.S. Department of Agriculture, Forest Service, Rocky Mountain Research Station, 2000), pp. 57–70.

Sogge, Mark K., Timothy J. Tibbitts, and Jim Petterson. "Status and Ecology of the Southwestern Willow Flycatcher in the Grand Canyon," *Western Birds* 28:3 (1997), pp. 142–157.

Soulé, Michael E. "Land Use Planning and Wildlife Maintenance," *Journal of the American Planning Association* 57:3 (1991), pp. 313–323.

Soulé, Michael E., ed. *Viable Populations for Conservation* (Cambridge, U.K.: Cambridge University Press, 1987).

Southerland, M.T. "Environmental Impacts of Dispersed Development from Federal Infrastructure Projects," *Environmental Monitoring and Assessment* 94 (2004), pp. 163–178.

SRI Consulting Business Intelligence, "Scenario Planning." Available on-line at <http://www.sric–bi.com/consulting/ScenarioPlan.shtml>.

State/NCAR Mesoscale Model (MM5), NCAR/TN–398+STR (Boulder, Colorado: National Center for Atmospheric Research, 1995), 122 pp.

State of California. *Historical Population Estimates and Components of Change, July 1, 1970 – 1990* (Sacramento, California: California Department of Finance, June 2002).

State of California. *Interim County Population Projections* (Sacramento, California: California Department of Finance, June 2001).

State of California. *Revised Historical City, County, and State Population Estimates,* 1991–2000, with 1990 and 2000 Census Counts (Sacramento, California: California Department of Finance, March 2002).

Stebbins, Robert C. *Western Reptiles and Amphibians*, 2nd Edition (New York: Houghton Mifflin Company, 1985).

Stein, Mikhala. *Exploring Alternative Futures: Scenario Planning as a Tool for Third Parties in Conflict Management*, unpublished Policy Analysis Exercise (Cambridge, Massachusetts: Harvard University, John F. Kennedy School of Government, 2000).

Steinitz, Carl, Chad Adams, Lauren Alexander, James De Normandie, Ruth Durant, Lois Eberhart, John Felkner, Kathleen Hickey, Andrew Mellinger, Risa Narita, Timothy Slattery, Clotilde Viellard, Yu–Feng Wang, and E. Mitchell Wright. *An Alternative Future for the Region of Camp Pendleton, California* (Cambridge, Massachusetts: Harvard Graduate School of Design, 1997).

Steinitz, Carl, Hector Arias, Scott Bassett, Michael Flaxman, Tomas Goode, Thomas Maddock, III, David Mouat, Richard Peiser, and Allan Shearer. *Alternative Futures for Changing Landscapes: The Upper San Pedro River Basin in Arizona and Sonora* (Washington, D.C.: Island Press, 2003).

Steinitz, Carl and Peter Rogers. *A Systems Analysis Model of Urban Change: An Experiment in Interdisciplinary Education*, MIT Report No. 20 (Cambridge, Massachusetts: Massachusetts Institute of Education Press, 1970).

Steinitz, Carl, Michael Binford, Paul Cote, Thomas Edwards, Jr., Stephen Ervin, Craig Johnson, Ross Kiester, David Mouat, Douglas Olson, Allan Shearer, Richard Toth, and Robin Wills. *Biodiversity and Landscape Planning: Alternative Futures for the Region of Camp Pendleton, California* (Cambridge, Massachusetts: Harvard University Graduate School of Design, 1996).

Steinitz, Carl. "A Framework for the Theory Applicable to the Education of Landscape Architects (and Other Environmental Design Professionals)," *Landscape Journal* 9:2 (Fall 1990), pp. 136–143.

Steinitz, Carl. "Design Is a Verb, Design Is a Noun,"

Landscape Journal 14:2 (Fall 1995), pp. 188–200.

Stevenson, Matthew R., Richard E. Toth, Thomas C. Edwards, Jr., Lori Hunter, Robert J. Lilieholm, Kimberly S. Karish, James DeNormandie, Manuel Gonzalez, and Mary Cablk, "What If...? Alternative Futures for the California Mojave Desert," (2002). Available on-line at <http://www01.giscafe.com/TechPapers/Papers/paper036/p192.htm>.

Stevenson, Tony. "Netweaving Alternative Futures—Information Technocracy or Communicative Community?" *Futures* 30:2–3 (March – April 1998), pp. 189–198.

Stilger, Robert L. "Alternatives for Washington" in Clement Bezold, ed., *Anticipatory Democracy: People in the Politics of the Future*, (New York: Random House, 1978), pp. 88–99.

Stilgoe, John R. *Common Landscape of North America, 1580 to 1845* (New Haven: Yale University Press, 1982).

Stilgoe, John R. *Borderland: Origins of the American Suburb, 1820 – 1939* (New Haven, Connecticut: Yale University Press, 1988).

Street, Penny. "Scenario Workshops: A Participatory Approach to Sustainable Urban Living?" *Futures* 29:2 (March 1997), pp. 139–158.

Suarez, Andrew V., David A. Holway, and Ted J. Case. "Patterns of Spread in Biological Invasions Dominated by Long-Distance Jump Dispersal: Insights from Argentine Ants," *Proceedings of the National Academy of Sciences* 98:3 (January 2001), pp. 1095–1100.

Suarez, Andrew V., Douglas T. Bolger, and Ted J. Case. "Effects of Fragmentation and Invasion on Native Ant Communities in Coastal Southern California," *Ecology* 79 (1998), pp. 2041–2056.

Suarez, Andrew V., Jon Q. Richmond, and Ted J. Case. "Prey Selection in Horned Lizards following the Invasion of Argentine Ants in Southern California," *Ecological Applications* 10:3 (2000), pp. 711–725.

Takacs, David. *The Idea of Biodiversity: Philosophies of Paradise* (Baltimore, MD: The Johns Hopkins University Press, 1996).

Taleb, Nassim Nicholas. *The Black Swan: The Impact of the Highly Improbable* (New York: Random House, 2007).

Tarbuck, Edward J. and Frederick K. Lutgens. *Earth Science* (New York: Macmillan College Publishing Co., 1994).

The Endangered Species Act of 1973 as amended, Public Law 93–205, 16 United States Code 1531–1544, 87 Statute 884 (December 28, 1973).

The National Environmental Policy Act of 1969 as amended, Public Law 91–190, 42 United States Code 4321–4347 (January 1, 1970); subsequently amended.

"The Nature Conservancy's Protection Initiative for the 1990s," *Nature Conservancy* 41:28 (1991), pp. 28–29.

The Sikes Act of 1960 as amended, Public Law 86–797, 16 United States Code 670(a)–670(o), 74 Statute 1052 (September 15, 1960); subsequently amended.

Turner, Jr., J. C. *Physiological Ecology of the Great Horned Owl, Bubo virginianus pacificus*, M.S. Thesis (Fullerton, California: California State College at Fullerton, 1969).

U.S. Bureau of Land Management. *The Upper San Pedro River Basin of the United States and Mexico: A Resource Directory and an Overview of Natural Resource Issues Confronting Decision–Makers and Natural Resources Managers*, BLM Report # BLM/AZ/PT–98/021 (Phoenix, Arizona: Bureau of Land Management, Arizona State Office, May 1998).

U.S. Census Bureau. Census 2000, Summary File 3, Area Name: California (2002). Available online at <www.census.gov>.

U.S. Census Bureau. 2002 Tiger/Line Files (2002). Available online at <http://www.census.gov/geo/www/tiger/tiger2002/tgr2002.html>.

U.S. Department of Defense. *Instruction 4715.3*, "Environmental Conservation Program" (May 3, 1996).

U.S. Environmental Protection Agency, *Alternative Futures for Environmental Policy Planning: 1975–2000*, EPA Report # EPA–540/9–75–027, prepared by Stanford Research Institute, Center for the Study of Social Policy (Washington, D.C.: Environmental Protection Agency, Office of Pesticide Programs, October 1975).

U.S. Environmental Protection Agency.

Working Papers in Alternative Futures and Environmental Quality (Washington, D.C.: Environmental Protection Agency, Environmental Studies Division, May 1973).

U.S. Environmental Protection Agency–Science Advisory Board. *Reducing Risk: Setting Priorities and Strategies for Environmental Protection*, Environmental Protection Agency Report # SAB–EC–90–021 (Washington, D.C.: U.S. Environmental Protection Agency, 1990).

U.S. Fish and Wildlife Service. *Least Bell's Vireo Draft Recovery Plan* (Portland, Oregon: Fish and Wildlife Service Reference Service, 1998).

U.S. Fish and Wildlife Service. *Southwestern Willow Flycatcher Draft Recovery Plan* (Albuquerque, New Mexico: Fish and Wildlife Service Reference Service, 2001).

U.S. Fish and Wildlife Service. *Southwestern Willow Flycatcher Draft Recovery Plan* (Albuquerque, New Mexico: Fish and Wildlife Service Reference Service, 2001).

U.S. Fish and Wildlife Service. *Standards for the Development of Habitat Suitability Index Models*, 103, ESM (Washington, D.C.: U.S. Fish and Wildlife Service, Division of Ecological Services, 1981).

U.S. Geological Survey. EROS Data Center, Sioux Falls, South Dakota, Landsat Enhanced Thematic Mapper + image, Path: 040, Row: 037, imaged on November 18, 2000 (obtained January 4, 2001).

U.S. Geological Survey. *7.5 minute digital elevation models*. Available on-line at <http://edcsns17. cr.usgs.gov/EarthExplorer>.

U.S. Geological Survey. *California Hydrologic Data Report–11046000 Santa Margarita River at Ysidora, CA*. Available on-line <http://ca.water.usgs.gov/archive/ waterdata/99/11046000.html>.

U.S. Geological Survey. *Peak Streamflow for California*. Available on-line <http://www.usgs. gov/ca/nwis/peak>, search for 11046000 (the site number for the Ysidora Gage on the Santa Margarita River).

U.S. Marine Corps, BO–ESA Consultation on Riparian and Estuarine Programmiatic Conservation Plan (Camp Pendleton, San Diego County, California: U.S. Marine Corps, October 30, 1995).

University of Arizona. "Argentine Ants: Urban Integrated Pest Management." Available on-line at <http://ag.arizona.edu/urbanipm/insects/ants/ argentineants.html>.

van der Heijden, Kees. *Scenarios: The Art of Strategic Conversation* (New York: John Wiley & Sons, 1996).

van Dijk, Terry. "Scenarios of Central European Land Fragmentation," *Land Use Policy* 20:2 (April 2003), pp. 149–158.

van Notten, Philip, Jan Rotmans, Marjolein B.A. van Asselt, and Dale S. Rothman. "An Updated Scenario Typology," *Futures* 35:5 (2003), pp. 423–443.

Varaday, Robert G., Margaret Ann Moote, and Robert Merideth. "Water Management Options for the Upper San Pedro Basin: Assessing the Social and Institutional Landscape," *Natural Resources Journal* 40:2 (Spring 2000), pp. 223–235.

Venn, Tyron J. Robbie L. McGavin, and Howard M. Rogers. "Managing Woodlands for Income Maximization in Western Queensland, Australia: Clearing for Grazing versus Timber Production," *Forest Ecology and Management* 185:3 (November 18, 2003), pp 291–306.

Vieux, Baxter E. *Distributed Hydrologic Modeling Using GIS*, Water Science and Technology Library, Volume 38 (Dordrecht, Netherlands: Kluwer Academic Publishers, 2001).

Wack, Pierre. "Scenarios: Shooting the Rapids," *Harvard Business Review* 63:6 (November – December 1985), pp. 139–150.

Wack, Pierre. "Scenarios: Uncharted Waters Ahead," *Harvard Business Review* 63:5 (September – October 1985), pp. 72–89.

Washburn, A. Michael and Thomas E. Jones. "Anchoring Futures in Preferences" in Nazli Choucri and Thomas W. Robinson, eds., *Forecasting in International Relations: Theory, Methods, Problems, Prospects* (San Francisco: W.H. Freeman and Company, 1978), pp. 95–115.

Way, M. J., M. E. Cammell, M. R. Paiva, and C. A. Collingwood. "Distribution and Dynamics of the Argentine Ant *Linepithema* (*Iridomyrmex*) *humile* (*Mayr*) in Relation to Vegetation, Soil Conditions, Topography, and Native Competitor

Ants in Portugal," *Insectes Sociaux* 44 (1997), pp. 425–428.

WEST Consultants. *Final Report: Santa Margarita River Hydrology, Hydraulics, and Sedimentation Study* (San Diego, California: WEST Consultants, Inc., 2000).

Western Exterminator Company. "Argentine Ant." Available on-line at <http://www.west–ext.com/argentine_ant.html>.

Wiens, J. A. and J. T. Rotenberry. "Habitat Associations and Community Structure in Shrubsteppe Environments," *Ecological Monographs* 51 (1981), pp. 21–41.

Wilson, E.O. "The Current State of Biodiversity" in E.O. Wilson and Frances M. Peter, eds., *Biodiversity* (Washington, D.C.: National Academy Press, 1986), pp. 3–18.

Woodwell, John C. "A Simulation Model to Illustrate Feedbacks among Resource Consumption, Production, and Factors of Production in Ecological–Economic Systems," *Ecological Modelling* 112:2–3 (October 15, 1998), pp. 227–247.

Young, James A. and B. Abbott Sparks. *Cattle in the Cold Desert* (Reno, Nevada: University of Nevada Press, 2002).

Zimbardo, Philip G. and Michael R. Leippe. *The Psychology of Attitude Change and Social Influence* (New York: McGraw–Hill, 1991).

Zink, Robert M., George F. Barrowclough, Jonathan L. Atwood, and Rachelle C. Blackwell–Rago. "Genetics, Taxonomy, and Conservation of the Threatened California Gnatcatcher," *Conservation Biology* 14:5 (October 2000), pp. 1394–1405.

Zurek, Monika B. and Thomas Henrichs. "Linking Scenarios across Geographical Scales in International Environmental Assessments," *Technological Forecasting and Social Change* 74 (2007), pp. 1282–1295.

Index